Quantum Physics: Concepts and Applications

Quantum Physics: Concepts and Applications

Edited by Jenson Lee

Larsen & Keller
www.larsen-keller.com

Quantum Physics: Concepts and Applications
Edited by Jenson Lee
ISBN: 979-8-88836-075-0 (Hardback)

Larsen & Keller
Published by Larsen and Keller Education,
5 Penn Plaza,
19th Floor,
New York, NY 10001, USA

Cataloging-in-Publication Data

Quantum physics : concepts and applications / edited by Jenson Lee.
p. cm.
Includes bibliographical references and index.
ISBN 979-8-88836-075-0
1. Quantum theory. 2. Physics. 3. Mechanics. 4. Thermodynamics. I. Lee, Jenson.
QC174.12 .R43 2023
530.12--dc23

Table of Contents

Preface

It is often said that books are a boon to mankind. They document every progress and pass on the knowledge from one generation to the other. They play a crucial role in our lives. Thus I was both excited and nervous while editing this book. I was pleased by the thought of being able to make a mark but I was also nervous to do it right because the future of students depends upon it. Hence, I took a few months to research further into the discipline, revise my knowledge and also explore some more aspects. Post this process, I begun with the editing of this book.

Quantum physics is a branch of science that focuses on quantum mechanics, which seeks to explain the behavior of matter and energy. Some of the major concepts within this field are superposition, wave-particle duality and uncertainty principle. Superposition refers to the phenomena of an object existing as a combination of multiple possible states simultaneously. The uncertainty principle is a mathematical concept which states that it is impossible to know the position and velocity of an object at the same time, regardless of the method used for measurement. The wave-particle duality describes the behavior of light and matter wherein it behaves as a wave or a particle depending on the method of measurement. The book aims to shed light on quantum physics and the recent researches in this field. From theories to research to practical applications, studies related to all contemporary topics of relevance to this field have been included herein. The book is a vital tool for all researching and studying this area of physics.

I thank my publisher with all my heart for considering me worthy of this unparalleled opportunity and for showing unwavering faith in my skills. I would also like to thank the editorial team who worked closely with me at every step and contributed immensely towards the successful completion of this book. Last but not the least, I wish to thank my friends and colleagues for their support.

Editor

1

Get Rid of Nonlocality from Quantum Physics

Andrei Khrennikov

International Center for Mathematical Modeling in Physics and Cognitive Sciences, Linnaeus University, SE-351 95 Växjö, Sweden; Andrei.Khrennikov@lnu.se

Abstract: This paper is aimed to dissociate nonlocality from quantum theory. We demonstrate that the tests on violation of the Bell type inequalities are simply statistical tests of local incompatibility of observables. In fact, these are tests on violation of *the Bohr complementarity principle.* Thus, the attempts to couple experimental violations of the Bell type inequalities with "quantum nonlocality" is really misleading. These violations are explained in the quantum theory as exhibitions of incompatibility of observables for a single quantum system, e.g., the spin projections for a single electron or the polarization projections for a single photon. Of course, one can go beyond quantum theory with the hidden variables models (as was suggested by Bell) and then discuss their possible nonlocal features. However, conventional quantum theory is local.

Keywords: Bell type inequality; nonlocality; action at a distance; incompatibility; Landau identity; complementarity principle; Bell operator; squared Bell operator; Hertz-Boltzmann Bild methodology; prequantum classical statistical field theory

---

1. Introduction

As is well known, the original EPR-argument [1] was fundamentally coupled with the Bohr complementarity principle [2–4] (see Section 6.1). Einstein, Podolsky, and Rosen reasoned against the completeness of quantum mechanics (QM) by showing that the "elements of reality" corresponding to two incompatible observables (e.g., position and momentum) can be assigned to the same physical system. However, their argument was purely theoretical (even merely philosophical) and it was impossible to check the EPR-statement experimentally. Bohr pointed to the latter in his reply to Einstein [5]; he considered the EPR-argument as metaphysical. Although nonlocality was mentioned in the EPR-paper, it was considered as just a possible alternative to incompleteness of QM.

Nonlocality was emphasized for the first time by by Bohm. It was elevated by Bell (who admired Bohmian mechanics) through the argument based on violation of Bell's inequality [6–8].

Our aim is to perform the *genuine quantum mechanical analysis of the derivation of the CHSH-inequality* considered as an inequality for correlations of quantum observables—*the quantum CHSH-inequality*. Thus, we do not try to go beyond QM. We are interested in the quantum mechanical interpretation of experimental violation of the CHSH-inequality. We show that, in fact, the degree of violation is straightforwardly coupled to the degree of incompatibility of observables, the norms of commutators. In particular, in the scenario with two spatially separated systems, these are tests of local incompatibility, i.e., *incompatibility of observables on a single subsystem of a compound system* (Theorem 1, Section 4.2).

We remark that the CHSH-inequality [9] is derived for classical correlations expressed in the framework of hidden variables (established by Bell [6–8]) and by using the calculus of classical probabilities (the Kolmogorov probability theory). The quantum CHSH-inequality is derived for quantum correlations by using the operator formalism.

We stress that the CHSH combination of correlations,

$$\langle \mathcal{B} \rangle = \frac{1}{2}[\langle A_1 B_1 \rangle + \langle A_1 B_2 \rangle + \langle A_2 B_1 \rangle - \langle A_2 B_2 \rangle], \quad (1)$$

has three different interpretations:

- Classical (hidden variables) correlations, $\langle \mathcal{B} \rangle_{\text{CL}}$.
- Experimental correlations, $\langle \mathcal{B} \rangle_{\text{EXP}}$.
- Quantum mechanical correlations, $\langle \mathcal{B} \rangle_{\text{QM}}$.

Following Bell, one can compare the classical theoretical quantity $\langle \mathcal{B} \rangle_{\text{CL}}$ with its experimental counterpart $\langle \mathcal{B} \rangle_{\text{EXP}}$. The majority of the quantum foundation and information community proceeds in this way. This way leads to operating with the notion of nonlocality and action at a distance.

Is there any reason to couple this mysticism with quantum theory?

It is more natural to start with the quantum theoretical analysis of quantity $\langle \mathcal{B} \rangle_{\text{QM}}$. It is easy to explain under what circumstances it can be bounded by 1 (or exceed 1). (Thus so-called classical bound has the purely quantum origin.) This quantum mechanical explanation is purely local. Thus, it is really incorrect in quantum theory to speak about its nonlocality or associate with it any kind of action at a distance.

It is well known that (by the complementarity principle) it is impossible to measure jointly two spin coordinates for electron. Therefore, $\langle \mathcal{B} \rangle_{\text{QM}}$ can exceed 1. If somebody does not believe in this prediction of QM, it would be natural to check violation of the principle of complementarity for a single electron (or photon), e.g., to check violation of the Heisenberg uncertainty relation (in the form of the Robertson inequality).

As emphasized above, here I proceed by using solely the formalism of QM, cf. with probabilistic analysis of the incompatibility interpretation of the Bell type inequalities in [10–24] and especially [17]—the probabilistic version of the present paper. See also the recent preprint of Griffiths [25], where incompatibility of quantum observables is emphasized; see the recent works of Boughn [26,27], where the nonlocality viewpoint on quantum theory is critically analyzed and the role of the ontological vs. information interpretations of the wave function in discussions on "quantum nonlocality" is emphasized.

Foundational issues such as the complementarity principle, incompatibility, nonlocality, realism, and hidden variables, are discussed in more detail in Section 6.

2. Measuring the Degree of Incompatibility via the CHSH-Test

We show that the degree of violation the quantum CHSH-inequality can be considered as a measure of incompatibility in two pairs of quantum observables, A_1, A_2 and B_1, B_2. This is the simple consequence of the *the Landau identity* [28,29] (see Equation (8)). In quantum theory incompatibility is mathematically expressed as noncommutativity. Thus, by testing incompatibility, we test the degree of noncommutativity, or in other words, the "magnitudes" of observables corresponding to commutators,

$$\hat{M}_A = i[\hat{A}_1, \hat{A}_2], \ \hat{M}_B = i[\hat{B}_1, \hat{B}_2]. \quad (2)$$

We use the hat-symbol to denote operators.

The incompatibility-magnitude can be expressed via the maximal value of averages of commutator-operators, i.e., by their norms, for example,

$$\sup_{\|\psi\|=1} |\langle \psi | \hat{M}_A | \psi \rangle| = \|\hat{M}_A\|. \quad (3)$$

By interpreting quantity $\langle \psi | \hat{M}_A | \psi \rangle$ as the theoretical counterpart of experimental average $\langle M_A \rangle_\psi$ of observable M_A, we can measure experimentally the incompatibility-magnitude, i.e., norm $\|\hat{M}_A\|$ from measurements of commutator-observable M_A. The main foundational problem is that measurement of such commutator-observables is challenging. Recently some progress was demonstrated on the basis of weak measurements, but generally we are not able to measure commutator-quantities.

We remark that (from the quantum mechanical viewpoint) the CHSH-test estimates the product of incompatibility-magnitudes for the A-observables and B-observables, i.e., the quantity $\|\hat{M}_A\|\|\hat{M}_B\|$. However, by considering the B-observables as axillary and selecting them in a proper way (for example, such that the B-commutator is a simple operator), we can use the CHSH-test to get the experimental value for the incompatibility-magnitude $\|\hat{M}_A\|$.

3. Incompatibility as Necessary Condition of Violation of Quantum CHSH-Inequality

3.1. General Case: Without Referring to the Tensor Product Structure

Consider the Bohm–Bell type experiments. Four observables A_1, A_2, B_1, B_2 taking values ± 1 are considered. It is assumed that observables in each pair $A_i, B_j, i, j = 1, 2$, can be measured jointly, i.e., A-observables are compatible with B-observables. However, the observables in pairs A_1, A_2 and B_1, B_2 are incompatible, i.e., they cannot be jointly measured. Thus, probability distributions $p_{A_iB_j}$ are well defined theoretically in QM and they can be verified experimentally; probability distributions $p_{A_1A_2}$ and $p_{B_1B_2}$ are not defined in QM and, hence, the question of their experimental verification does not arise.

We consider quantum observables represented by Hermitian operators. In QM, mathematical compatibility is represented by commutativity of operators, i.e., in the Bohm–Bell type experiments

$$[\hat{A}_i, \hat{B}_j] = 0, i, j = 1, 2, \tag{4}$$

and generally $[\hat{A}_1, \hat{A}_2] \neq 0,\ [\hat{B}_1, \hat{B}_2] \neq 0$. The quantum theoretical CHSH-correlation function has the form:

$$\langle \mathcal{B} \rangle = \frac{1}{2}[\langle \hat{A}_1 \hat{B}_1 \rangle + \langle \hat{A}_1 \hat{B}_2 \rangle + \langle \hat{A}_2 \hat{B}_1 \rangle - \langle \hat{A}_2 \hat{B}_2 \rangle]. \tag{5}$$

(Here and everywhere below the index QM pointing to the quantum formalism is omitted.)

It is compared with the experimental CHSH-correlation function.

In the quantum framework, the CHSH-correlation function can be expressed with the aid of the Bell-operator:

$$\hat{\mathcal{B}} = \frac{1}{2}[\hat{A}_1(\hat{B}_1 + \hat{B}_2) + \hat{A}_2(\hat{B}_1 - \hat{B}_2)] \tag{6}$$

as

$$\langle \mathcal{B} \rangle = \langle \psi | \hat{\mathcal{B}} | \psi \rangle. \tag{7}$$

By straightforward calculation, one can derive at the Landau identity:

$$\hat{\mathcal{B}}^2 = I - (1/4)[\hat{A}_1, \hat{A}_2][\hat{B}_1, \hat{B}_2]. \tag{8}$$

Thus, if *at least one of the commutators* equals to zero, i.e.,

$$[\hat{A}_1, \hat{A}_2] = 0, \tag{9}$$

or

$$[\hat{B}_1, \hat{B}_2] = 0, \tag{10}$$

then the following inequality holds:

$$|\langle \mathcal{B} \rangle| \leq 1. \tag{11}$$

To derive this inequality, we used solely the quantum formalism. The inequality is the consequence of compatibility for at least one pair of observables, A_1, A_2 or B_1, B_2. Thus, although formally

Equation (11) coincides with the standard CHSH-inequality, it has totally different meaning. It is better to call Equation (11) *the quantum CHSH inequality.*

Thus, *compatibility of the A-observables or the B-observables is sufficient for validity of the quantum CHSH-inequality* (for all quantum states) or in other words *conjunction of incompatibilities of the A-observables and the B-observables is the necessary condition for its violation* (for some quantum state).

3.2. Compound Systems

States of a compound quantum system $S = (S_A, S_B)$ are represented in tensor product $H_{AB} = H_A \otimes H_B$ of the state spaces H_A and H_B of subsystems S_A and S_B. Observables are given by operators

$$\hat{A}_i = \hat{\mathbf{A}}_i \otimes I, \; \hat{B}_i = I \otimes \hat{\mathbf{B}}_i, \tag{12}$$

where Hermitian operators $\hat{\mathbf{A}}_i$ and $\hat{\mathbf{B}}_i$ act in H_A and H_B, respectively. They represent observables $\mathbf{A}_i, \mathbf{B}_i$ on subsystems S_A, S_B of S. For spatially separated systems, we call them *local observables.*

This tensor representation automatically implies commutativity of operators $\hat{A}_i$ with operators $\hat{B}_j$, i.e., Equation (4) holds. We remark that the mathematical condition of incompatibility is reduced to condition $[\hat{\mathbf{A}}_1, \hat{\mathbf{A}}_2] \neq 0$ and $[\hat{\mathbf{B}}_1, \hat{\mathbf{B}}_2] \neq 0$. For spatially separated systems, it is natural to call incompatibility of the observables on S_A (on S_B) *local incompatibility*. Section 3.1 implies that *conjunction of local incompatibilities* is the necessary condition for violation of the quantum CHSH-inequality.

We remark that the mathematical formalism of this section is applicable to description of any kind of observables "respecting" the tensor product structure of observables. A physical system S need not be composed of two physical subsystems.

4. Incompatibility as Sufficient Condition of Violation of Quantum CHSH-Inequality

4.1. General Case: Without Referring to the Tensor Product Structure

Assume that A-observables as well as B-observables are incompatible, i.e., corresponding operators do not commute:

$$[\hat{A}_1, \hat{A}_2] \neq 0 \text{ and } [\hat{B}_1, \hat{B}_2] \neq 0, \tag{13}$$

i.e.,

$$\hat{M}_A \neq 0 \text{ and } \hat{M}_B \neq 0, \tag{14}$$

where $\hat{M}_A = i[\hat{A}_1, \hat{A}_2]$, $\hat{M}_B = i[\hat{B}_1, \hat{B}_2]$. It is important to note that $[\hat{M}_A, \hat{M}_B] = 0$. We can write the Landau identity in the form

$$\hat{\mathcal{B}}^2 = I + (1/4)\hat{M}_{AB}, \tag{15}$$

where $\hat{M}_{AB} = \hat{M}_A \hat{M}_B$. If $M_{AB} = 0$, then, despite the incompatibility condition in Equation (13), the QCHSH-inequality cannot be violated. We proceed under condition

$$\hat{M}_{AB} \neq 0. \tag{16}$$

In our framework, this condition is not so restrictive. We consider the quantum CHSH-inequality as a statistical test of incompatibility. It is natural to estimate the degree of incompatibility in one pair of observables, e.g., in the A-pair. In this approach, the B-pair plays the axillary role and we can freely play with its selection. To obtain the condition in Equation (16), it is sufficient to select B-operators in such a way that the operator $\hat{M}_B$ is invertable. We especially highlight the case of compound systems (see Section 3.2). Here incompatibility of the A-observables and the B-observables, see Equation (14), automatically implies the condition in Equation (16)

Under the condition in Equation (16), there exists some common eigenvector ψ_{AB} such that $M_A \psi_{AB} = \mu_A \psi_{AB}, M_B \psi_{AB} = \mu_B \psi_{AB}$ and both eigenvalues are nonzero.

Suppose that $\mu_A > 0$ and $\mu_B > 0$. Then, this ψ_{AB} is an eigenvector of operator $\hat{\mathcal{B}}^2$ with eigenvalue $(1+\mu) > 1, \mu = \mu_A\mu_B$. Hence, $\|\hat{\mathcal{B}}^2\| \geq (1+\mu) > 1$ and

$$1 < (1+\mu) \leq \|\hat{\mathcal{B}}^2\| = \|\hat{\mathcal{B}}\|^2.$$

Since $\hat{\mathcal{B}}$ is Hermitian, we have

$$\|\hat{\mathcal{B}}\| = \sup_{\|\psi\|=1} |\langle\psi|\hat{\mathcal{B}}|\psi\rangle|.$$

Finally, we get that

$$\sup_{\|\psi\|=1} |\langle\psi|\hat{\mathcal{B}}|\psi\rangle| > \sqrt{1+\mu} > 1.$$

Thus, there exist pure quantum states such that the QCHSH-inequality is violated.

Now, suppose that $\mu_A > 0$, but $\mu_B < 0$. To change the sign of μ_B, it is sufficient to interchange the B-observables.

Thus, *conjunction of incompatibilities of the A-observables and the B-observables constrained by Equation (16) is sufficient for violation of the quantum CHSH-inequality.*

4.2. Compound Systems

Here, $H = H_A \otimes H_B$ and $\hat{A}_j = \hat{\mathbf{A}}_j \otimes I, \hat{B}_j = I \otimes \hat{\mathbf{B}}_j$, where Hermitian operators $\hat{\mathbf{A}}_j$ and $\hat{\mathbf{B}}_j$ act in H_A and H_B, respectively.

4.2.1. Incompatibility as Necessary and Sufficient Condition of Violation of the Quantum CHSH-Inequality

Here, the joint incompatibility-condition in Equation (13) is equivalent to incompatibility of observables on subsystems:

$$\hat{\mathbf{M}}_A = i[\hat{\mathbf{A}}_1, \hat{\mathbf{A}}_2] \neq 0 \text{ and } \hat{\mathbf{M}}_B = i[\hat{\mathbf{B}}_1, \hat{\mathbf{B}}_2] \neq 0. \tag{17}$$

We have $\hat{M}_{AB} = \hat{M}_A\hat{M}_B = \hat{\mathbf{M}}_A \otimes \hat{\mathbf{M}}_B$. As mentioned above, constraint $\hat{M}_{AB} \neq 0$ is equivalent to Equation (17). Section 3.1 implies that *conjunction of local incompatibilities* is the sufficient condition for violation of the quantum CHSH-inequality. Thus, we obtain:

Theorem 1 (Local incompatibility criteria of QCHSH-violation)**.** *Conjunction of local incompatibilities is the necessary and sufficient condition for violation of the quantum CHSH-inequality.*

4.2.2. Eigenvectors of the Bell Operator and Its Square

Consider the eigenvector consideration of Section 4.1. The vector $\psi_{AB} = \psi_A \otimes \psi_B$, where $\psi_A \in H_A, \psi_B \in H_B$, and $\hat{\mathbf{M}}_A\psi_A = \mu_A\psi_A, \hat{\mathbf{M}}_B\psi_B = \mu_B\psi_B$. We assume that $\mu = \mu_A\mu_B > 0$. Then,

$$\langle\psi_A \otimes \psi_B|\hat{\mathcal{B}}^2|\psi_A \otimes \psi_B\rangle > 1.$$

Thus, for the squared CHSH-observable $\hat{\mathcal{B}}^2$, the one-boundary can be violated for separable states. Here, entanglement of A and B observables plays no role.

To be more illustrative, let us restrict consideration to finite dimensional Hilbert spaces. There can be found states Ψ and Φ such that

$$\max_{\|\psi\|=1} |\langle\psi|\hat{\mathcal{B}}^2|\psi\rangle| = \langle\Psi|\hat{\mathcal{B}}^2|\Psi\rangle,$$

$$\max_{\|\psi\|=1} |\langle\psi|\hat{\mathcal{B}}|\psi\rangle = |\langle\Phi|\hat{\mathcal{B}}|\Phi\rangle.$$

The tricky thing is that generally $\Psi \neq \Phi$. The equality for norms, $\|\hat{\mathcal{B}}\| = \sqrt{\|\hat{\mathcal{B}}^2\|}$, does not imply equality of the max-optimization states.

Of course, max-states for $\mathcal{B}$ and $\mathcal{B}^2$ are connected: the former can be represented as linear combinations of the latter (the feature of all operators with degenerate spectrum. (as shown in [30], max-states for $\mathcal{B}$ can be represented even as mixtures of max-states for $\mathcal{B}^2$.)

In the experiments to violate the quantum CHSH-inequality, tremendous efforts were put to prepare ensembles of entangled states. The main reason for this is that the direct measurement of the observable represented by operator $\hat{\mathcal{B}}^2$ is challenging. In Appendix B, we present the abstract analog of the Bell experiments treated as experiments to measure the degree of incompatibility. The tensor product structure is excluded and, in particular, an analog of entangled states related to measurement of an observable and its square is considered.

5. CHSH-Correlation Function as Measure of Incompatibility

We start with consideration of observables respecting the tensor product structure on the state space $H = H_A \otimes H_B$. Consider the eigenbases (e_{Ak}) and (e_{Bk}) of operators $\hat{\mathbf{M}}_A$ and $\hat{\mathbf{M}}_B$ (acting in H_A and H_B, respectively) and the corresponding eigenvalues μ_{Aj}, μ_{Bj}.

Let $\|\hat{\mathbf{M}}_A\| = \max_j |\mu_{Aj}| = |\mu_{Ai_a}|, \|\hat{\mathbf{M}}_B\| = \max_j |\mu_{Bj}| = |\mu_{Bi_b}|$ and let $\mu_{Ai_a}\mu_{Bi_b} > 0$. Then, $\|\hat{\mathcal{B}}^2\| = (1 + \mu_{Ai_a}\mu_{Bi_b})$. Thus,

$$b = \|\hat{\mathcal{B}}\| = \sqrt{1 + \frac{1}{4}\|[\hat{\mathbf{A}}_1, \hat{\mathbf{A}}_2]\| \, \|[\hat{\mathbf{B}}_1, \hat{\mathbf{B}}_2]\|}, \tag{18}$$

where $\langle \mathcal{B} \rangle_\psi$ is given by Equation (5); b is the maximal possible value of CHSH-correlations. If eigenvalues μ_{Ai_a} and μ_{Bi_b} have different signs, then we interchange the B-observables.

From Equation (18), we get that

$$\|[\hat{\mathbf{A}}_1, \hat{\mathbf{A}}_2]\| \|[\hat{\mathbf{B}}_1, \hat{\mathbf{B}}_2]\| = 4(b^2 - 1).$$

The norm of commutator can be considered as a measure of incompatibility. Thus, the CHSH-correlation function gives the possibility to check experimentally the product of degrees of incompatibility for the A and B observables.

One may consider this way of measuring of incompatibility as too tricky. However, typically, to measure commutator-observable and then its average is challenging (by "measuring commutator-observable", we mean measuring observable represented mathematically by commutator operator scaled by i). Therefore, even such a tricky approach to this problem as measurement of the CHSH-correlation function deserves attention.

Now, consider B-observables as axillary. In this way, we are able to determine the degree of incompatibility for the A-observables by using some pair of axillary observables B_1, B_2. We can select the latter in such a a way that their commutator is a "good observable", so that it can be easily measured for any state, thus its average and hence the norm can be determined. Then, we can measure incompatibility of observables A_1 and A_2 by using the formula:

$$\|[\hat{\mathbf{A}}_1, \hat{\mathbf{A}}_2]\| = 4(b^2 - 1)/\|[\hat{\mathbf{B}}_1, \hat{\mathbf{B}}_2]\|. \tag{19}$$

Why is the use of tensor product states so useful for measuring the degree of incompatibility? By spitting a system into two subsystems it is easy to check compatibility of A and B observables, thus the possibility to define the CHSH-correlation function which can be measured in experiment.

6. Foundational Questions

6.1. *Bohr's Complementarity Principle*

Often, it is claimed that Bohr's writings and, in particular, about the complementarity principle are very difficult for understanding. For example, in Schilpp's volume [31], p. 674 (see also Plotnitsky [32], p. 108), one can find the following statement: *"Thus, Einstein was confessed, after decades of his exchanges with Bohr, that he was 'unable to attain ... the sharp formulation ... [of] Bohr's principle of complementarity"'*. This principle has the complex structure and composed of a few components. One of the problems is that typically this principle is reduced to just one of its components, namely, the incompatibility-component. Incompatibility has the most striking consequences for quantum theory and experiment. However, as separated from the body of the complementarity principle, incompatibility is difficult for understanding.

As emphasized in [17], the complementarity principle is in fact the principle of contextuality of quantities used in the quantum formalism, in the sense of coupling them to corresponding experimental contexts. Bohr did not use the notion "experimental context". He considered experimental conditions [2]:

"Strictly speaking, the mathematical formalism of quantum mechanics and electrodynamics merely offers rules of calculation for the deduction of expectations pertaining to observations obtained under well-defined experimental conditions specified by classical physical concepts."

By using the notion of experimental context as the synonymous of Bohr's experimental conditions we can present the complementarity principle as composed of the following components [17] (we remark that one has to be very careful by operating with the notion of contextuality. Nowadays, this notion is widely used in foundational discussion on the Bell type inequalities. In such discussions, the meaning of the notion contextuality does not coincide with Bohr's contextuality, as taking into account the experimental con- text to explain the mechanism of generating the values of a quantum observable. From Bohr's viewpoint, any single quantum observable is contextual. One may say that consideration of Bohr's contextuality in parallel with Bell's contextuality can be misleading. However, we can consider Bell's contextuality simply as a very special case of Bohr's contextuality).

- (B1): There exists the fundamental quantum of action given by the Planck constant h:
- (B2): The presence of h prevents approaching internal features of quantum systems.
- (B3): Therefore, it is meaningless (from the viewpoint of physics) to build scientific theories about such features.
- (B4): An output of any observable is composed of contributions from a system under measurement and the measurement device.
- (B5): Therefore, the complete experimental arrangement (context) has to be taken into account.
- (B6): There is no reason to expect that all experimental contexts can be combined. Therefore, there is no reason to expect that all observables can be measured jointly. Hence, there exist incompatible observables.

(B6) can be called the incompatibility principle; this is a consequence of (B4) and (B5). Typically, the complementarity principle is identified with (B6). However, such a viewpoint does not match Bohr's understanding of the complementarity principle, as the combination (B1)–(B6).

This is the good place to remark that (B6) is very natural. The existence of incompatible experimental contexts is not surprising. Compatibility of all experimental contexts would be really surprising.

6.2. *"Quantum Nonlocality"*

We briefly discuss the notion of (non)locality.

6.2.1. Relativistic Invariance

Everywhere in physics, besides the Bell inequality debates [6–9,30,33–36], locality is identified with the relativistic invariance of theory. Therefore, the statements on nonlocality of quantum theory can make the impression (and they do!) that there is something wrong with relativistic invariance. However, there is nothing wrong with relativistic invariance. Of course, QM (in particular, the Schrödinger equation) is not relativistically invariant and attempts to construct relativistically invariant QM (based on the Dirac equation) were not successful. However, *QM is an approximation of quantum field theory which is relativistically invariant* (see Bogolubov and Shirkov [37] and Haag [38] (especially Chapter 3, "Algebras of Local Observables and Fields")). To complete the picture, we remark that there is a non-relativistic quantum field theory (see for example, book [39]).

6.2.2. Hidden Variables and Action at a Distance

One can say that nonlocality is a consequence of "action at the distance" [6–8] (see, e.g., Shimony [40,41] and Jaeger [42,43] for the detailed presentation). This interpretation is based on the invention of hidden variables. However, the analysis presented in this paper shows clearly that, to proceed in this framework, *one has to start with rejection of the basic principle of QM, the complementarity principle.* It is not clear why violations of this principle should be sought for compound systems. Thus, before inventing hidden variables, it would be natural to find violations of say the Heisenberg uncertainty principle (in the form of Robertson inequality).

Moreover, the modern attempt to go beyond QM with hidden variables is too straightforward. Already in the 19th century, in the process of transition from Newtonian mechanics to classical field theory, physicists were confronted with the same problem as in the process of transition from classical physics to QM. It was resolved in the framework of Bild (image) methodology developed by Hertz and Boltzmann [44–46] (see Section 6.4 and papers [17,47,48]).

6.2.3. Nonlocality = Violation of the Bell Type Inequalities.

The common comment to my talks is that per definition "quantum nonlocality" is a violation of the Bell type inequalities. However, this viewpoint is really misleading. If one recognizes that such violation is just a signature of incompatibility, then it is strange to speak about nonlocality, instead of complementarity.

6.3. Obscuring Incompatibility by Tensor Product Structure of Observables

As pointed out, we concentrate our analysis on the CHSH-inequality [9]. In contrast to the previous studies (see, e.g., [30,33–36]), we do not emphasize the role of the tensor product structure for the state space and observables. We proceed in the general framework and the tensor product model is just a special case of this framework. The common emphasis of the tensor product structure obscures the crucial role played by incompatibility of observables. Mathematically the crucial role of incompatibility-noncommutativity for violation of the CHSH-inequality was clarified already by Landau [28,29] (see also [30,33–36]). However, the mathematical calculations presented in these works did not lead to reinterpretation of violation of the CHSH-inequality.

I would like to emphasize the crucial role played by works of Landau [28,29]. In fact, Landau's articles carried the same message as the present paper: the CHSH inequality is an experimental test of the principle of complementarity. (He even used the terminology "complementary observables", instead of "incompatible observables".) Unfortunately, his excellent mathematical work was not completed by an extended interpretational discussion. Surprisingly, nowadays, his works are practically forgotten (of course, qualified people are aware about papers [28,29]. However, generally, the members of the quantum foundational community practically never refer to these papers. During 20 years of debates on the Bell inequality in Växjö, I have never heard about them. I got to know about Landau's works from E. Dzhafarov, an expert in mathematical psychology. It happened say eight years ago and I also ignored the complementarity message of Landau. I was content to enjoy his mathematics). I see two reasons for this:

1. Landau used the abstract framework of $C^{\star}$-algebras and, for many "real physicists", this was not so attractive.
2. He emphasized the novel way to derive the Tsirelson bound and typically this paper is considered as devoted to this derivation, i.e., its crucial component, coupling of Bell's argument to Bohr's principle of complementarity was completely ignored.

In the present paper, I select the intermediate strategy for representation. On the one hand, I do not just follow Landau using the language of $C^{\star}$-algebras. On the other hand, I also do not want to follow the common path based on the tensor product representation. I proceed in the complex Hilbert space formalism, but generally without referring to the tensor product structure of operators. Mathematics is really simple. It is based on the interrelation of spetcral properties of the Bell operator $\mathcal{B}$ and its square $\mathcal{B}^2$ (In fact, I have the impression that the essence of CHSH-test is this spectral interplay between the spectral properties of a Hermitian operator and its square. I try to present this vision in the abstract form in Appendix B).

6.4. Herz–Boltzmann Bild-Methodology of Science

It is surprising that not only Bell, but even Einstein, Bohr, and Heisenberg did not know about the works of Hertz and Boltzmann [44–46] on so-called "Bild" (image) methodology for physical theories. According to Hertz and Boltzmann, when speaking about a scientific theory, one has to specify its type: descriptive theory or observational theory. The crucial point is that a descriptive theory need not be straightforwardly coupled with theory of observations. By extending the Hertz–Boltzmann methodology to the quantum domain, we recognize that QM is an observational theory. Theories with hidden variables are of the descriptive type. The same observational theory can be based on a variety of descriptive theories. Bell's type descriptive theories have very rigid coupling to QM, the observational theory. One can construct a variety of corresponding descriptive theories which are not constrained by the Bell type inequalities.

In this paper, we do not to discuss the Bild-methodology of Herz and Boltzmann [44–46] in much detail (see my recent article [48]). We only make the remark on the notion of realism. From the Bild-viewpoint, realism in physics as well as any other area of science is reduced solely to experimental facts. In QM, this is exactly Bohr's point of view. Thus, the only realistic component of QM are outcomes of measurements (Bohr's phenomena). Any physical theory (descriptive as well as observational) is only about human images of natural phenomena. At the same time, these images are created on the basis of human's interaction with nature.

Neither Einstein nor Bohr was not aware of the works of Hertz and Bolzmann. (In any event, they never cited these works.) Both Einstein and Bohr identified descriptive and observational theories. In fact, the EPR-paper [1] can be considered as the message that QM is not a descriptive theory. However, at the same time, Einstein–Podolsky–Rosen dreamed of a descriptive theory with the straightforward coupling to observations. According to Hertz and Boltzmann, the latter is generally impossible. In his reply [5], Bohr tried to explain that QM is an observational theory and such things as the EPR elements of reality do not belong to its domain. However, nobody was aware about Hertz–Boltzmann distinguishing of descriptive and observational theories. Therefore their discussion can be compared to conversation of the blind with the deaf.

Finally, we refer to an example of descriptive theory coupled to QM (treated as an observational theory). This is *prequantum classical statistical field theory* (PCSFT), which was developed by the author of this paper and coauthors [49] (see Appendix C).

6.5. *On Incompatibility Interpretation of the Bell Type Inequalities*

In this paper, we analyze the CHSH-inequality and conclude that this is a test of the complementarity principle. It seems that this analysis can be extended to other Bell type inequalities. The crucial mathematical step in this analysis is derivation of the analogs of the Landau identity (see Hardy [33] for generalization of the CHSH inequality to an N-measurement scheme and Cereceda's paper [34], where the very general case (including Mermin's inequalities) was studied in very detail).

In Appendix A, we show (independently of results based on the Landau identity that incompatibility for at least one pair of observables is the necessary condition for violation of any type of the Bell type inequalities.

7. Conclusions

The point of Bell's theorem is that a local hidden variables theory cannot reproduce the results of quantum theory. The implication is that only a nonlocal hidden variable theory (like Bohmian mechanics) can reproduce the correlations found in quantum theory (and in the real world). Here, clearly, "nonlocality" refers to hidden variables theories, not to quantum theory. The very common misconception is to (incorrectly) associate the term with quantum theory. (In particular, from this viewpoint, the comments of Aspect [50] and Wiseman [51] on the crucial experiments [52–54] are really misleading; cf. with the comment of Kupczynski [24] and the author of this paper [55]).) We hope that the argument presented in this paper has convincingly demonstrated that this association is wrong. The deeper message of this paper is that the Bell inequality can be reinterpreted as *a condition on the quantum compatibility of local observables.* If local commutators vanish, the correlations are bounded just as they are when hidden variables are assumed to be local. The Bell type inequalities have one interpretation for hidden variables theories (the classical case), and another lesser-known and very interesting one for quantum theory.

Consequently the outputs of experiments testing violation of the Bell type inequalities also can be interpreted in two different ways. The conventional interpretation is that these were classical vs. quantum physics tests. My interpretation is that such experiments were, in fact, the tests of local incompatibility of quantum observables. I claim that the latter interpretation does not diminish the foundational value of these breakthrough experiments [52–54,56,57]. Complementarity is the basic feature of quantum observables. Tests on this feature are of the great foundational importance. At the same time the conventional interpretation of these tests, local realism vs. quantum theory is, in fact, not so exciting. What is the meaning to test nowadays quantum against classical? The validity of quantum theory was confirmed by the huge body of experiments and technological applications.

We also point out that the Bell type experiments played the crucial role in development of quantum technology: creation of efficient sources of entangled systems and photo-detectors as well as transmission of quantum systems to long distances with minimal disturbance.

It is clear that to get rid of nonlocality from quantum theory is not a simple task. The present note is just a step towards the common acceptance of the local interpretation of QM.

This paper should not be considered as directed against attempts to go beyond QM, by introducing "hidden variables". However, in such attempts, one has to take into account the basic principles of QM an especially the complementarity principle (see the recent article of Khrennikov and Alodjants [18]). One also has to take into account the lessons of 19th century physics in the period of transition from Newtonian mechanics to field theory (Section 6.4).

Acknowledgments: I would like to thank Willem De Myunck, David Griffiths, Ehtibar Dzhafarov, and Marian Kupczynski for stimulating discussions and comments. This work was supported by the research project of the Faculty of Technology, Linnaeus University, "Modeling of complex hierarchic structures".

Appendix A. Incompatibility as Necessary Condition for Violation of Any Bell Type Inequality

Consider a family of quantum observables $D_1, ..., D_n$ represented by Hermitian operators $\hat{D}_1, ..., \hat{D}_n$. We restrict considerations to observables with discrete values; thus, operators have the purely discrete spectra. Denote by $\hat{E}_i(x)$ the orthogonal projector corresponding to the eigenvalue x of $\hat{D}_i$.

Suppose that the observables are pairwise compatible, i.e., (D_i, D_j) can be measured jointly for any quantum state ρ and jpd is well defined

$$p_{D_iD_j}(x, y; \rho) \equiv P(D_i = x, D_j = y; \rho). \tag{A1}$$

In QM, compatibility is mathematically represented via commutativity of operators, i.e., $[\hat{D}_i, \hat{D}_j] = 0$, and, hence, $[\hat{E}_i(x), \hat{E}_j(y)] = 0$. The quantum formalism gives the following formula for jpd (von Neumann [58]):

$$p_{D_iD_j}(x, y; \rho) = \mathrm{Tr}\ \text{æ}\mathrm{E_i(x)E_j(y)} = \mathrm{Tr}\ \text{æ}\mathrm{E_j(y)E_i(x)}. \tag{A2}$$

Now, we point to the really surprising feature of quantum measurement theory. If observables are pairwise compatible, i.e., each pair can be jointly measured with corresponding jpds $p_{ij}(x, y; \rho)$ given by Equation (A2), then they are also triple-wise compatible, quadruple-wise compatible and so on... Any family of observables, $D_{i_1}, ..., D_{i_m}$ can be jointly measured and the joint probability distribution is given by the formula:

$$p_{D_{i_1}...D_{i_m}}(x_1, ..., x_m; \rho) = \mathrm{Tr}\ \text{æ}\mathrm{E_{i_1}(x_1)...E_{i_m}(x_m)}. \tag{A3}$$

On the left-hand side of this formula, one can take any permutation of indexes. This implies:
$\mathbf{2} \Longrightarrow \mathbf{m}$: *pairwise compatibility* $\Longrightarrow$ *multiple compatibility.*
This is really astonishing. It is surprising that its specialty (from the general viewpoint of measurement theory) was not discussed in foundational literature.

We turn to Bell's inequalities. Now, we are endowed with $\mathbf{2} \Longrightarrow \mathbf{m}$ property of quantum observables. Consider the most general Bell-type framework. There are K groups of quantum observables:

$$D^k = (D_1^k, ..., D_{N_k}^k), k = 1, ..., K.$$

Mathematically, they are represented by Hermitian operators:

$$\hat{D}^k = (\hat{D}_1^k, ..., \hat{D}_{N_k}^k).$$

Suppose that, for different k, observables are compatible, i.e., in the mathematical framework:

$$[\hat{D}_i^n, \hat{D}_j^m] = 0, n \neq m.$$

Thus, jpds $p_{i_1...i_K} \equiv p_{D_{i_1}^1...D_{i_K}^K}$ are well defined and, hence, covariations as well

$$\langle D_{i_1}^1 \cdots D_{i_K}^K \rangle = \mathrm{Tr}\ \text{æ}\hat{\mathrm{D}}_{\mathrm{i_1}}^1 \cdots \hat{\mathrm{D}}_{\mathrm{i_K}}^{\mathrm{K}} = \sum \mathrm{x_1 \cdots x_K\ p_{i_1...i_K}(x_1, ..., x_K)}.$$

Consider some Bell-type inequality (e.g., the CHSH inequality or the Mermin inequality),

$$\sum_{i_1...i_K} t_{i_1...i_K} \langle D_{i_1}^1...D_{i_K}^K \rangle + \text{correlations of lower orders} \leq c, \tag{A4}$$

where $t_{i_1...i_K}$ are some real constants. This inequality may be violated only if at least one pair of observables, say (D_i^n, D_j^n), is incompatible, i.e., in the mathematical formalism

$$[\hat{D}_i^n, \hat{D}_j^n] \neq 0. \tag{A5}$$

Otherwise, jpd exists and the inequality in Equation (A4) cannot be violated.

Theorem A1. *Incompatibility is a necessary condition for violation of any Bell-type inequality.*

In the standard nonlocality discussions, it is assumed that there are K systems $S^k, k = 1, 2, ..., K$, and observables D^k are local observables for S^k. Endowed with this scheme, we analyze the possibility to violate the Bell-type inequality in Equation (A4). The necessary condition is that Equation (A5) holds true. This condition is local.

Appendix B. "Entanglement" in the Absence of the Tensor Product Structure

In Section 4.2.2, we consider compound systems and discuss the well known fact that eigenvectors of operator $\hat{\mathcal{B}}^2$ giving the max-value of its quadratic form can be selected as separable (non-entangled) states; they need not be eigenvectors of the Bell operator; its eigenvectors are linear combinations of the aforementioned separable states.

We want to show that the tensor product structure of states and operators is not crucial for the above consideration.

Consider any Hermitian operator $\hat{C}$ and its square $\hat{C}^2$. Let u be an eigenvector of the latter, i.e., $\hat{C}^2 u = \lambda u, \lambda > 0$, and let u is not an eigenvector of the former. Set $v = \hat{C}u/\sqrt{\lambda}$, i.e.,

$$\hat{C}u = \sqrt{\lambda}v, \ \hat{C}v = \sqrt{\lambda}u. \tag{A6}$$

Set

$$\psi_{\pm} = u \pm v. \tag{A7}$$

Then, $\hat{C}\psi_{\pm} = \pm\sqrt{\lambda}\psi_{\pm}$. Thus, $\psi_{\pm}$ are eigenvectors of $\hat{C}$.

If the quadratic form of $\hat{C}^2$ approaches its max-value on eigenstate $u/\|u\|$, then the quadratic form of $\hat{C}$ approaches its max-value on eigenstate $\phi_{\pm} = \psi_{\pm}/\|\psi_{\pm}\|$.

States $\psi_{\pm}$ can be considered as *generalization of entangled states*, which is to say entangled with respect to operator $\hat{C}$.

This consideration can be coupled to measurement theory. Consider some quantum observable D represented by Hermitian operator $\hat{D}$. (For simplicity, suppose that $\hat{D} \geq 0$.) Suppose that this observable is simple theoretically, by complex experimentally (Spectrum and eigenvectors of operator $\hat{D}$ can be easily found, but measurement of observable D is really challenging).

Consider also the observable C represented by Hermitian operator $\hat{C} \equiv \sqrt{\hat{D}}$. Suppose that this observable is complex theoretically, but rather simple experimentally (The structure of spectrum and eigenvectors of operator $\hat{C}$ is complicated, but measurement design for C is straightforward).

We are interested in the following problem: *Find experimentally the upper bound for the average* $\langle D\rangle_\psi$ *of observable D with respect to all possible states.* We stress that we are interested in the experimental verification of a theoretical prediction of QM.

We can easily find state u corresponding to max-eigenvalue λ of operator $\hat{D}$. Then, one of the max-states of operator $\hat{C}$ can be found with the aid of "C-entanglement":

$$\phi_{+} = (u + \lambda^{-1/2}\hat{C}u)/\|u + \lambda^{-1/2}Cu\|. \tag{A8}$$

Finally, we prepare an ensemble of systems in quantum state ϕ_{+} and perform C-measurement for these systems.

In the Bell-type scenario (for observables respecting the tensor product structure), $\hat{C} = \hat{\mathcal{B}}$ is the Bell operator, $\hat{D} = \hat{\mathcal{B}}^2$. In fact, the degree of incompatibility is encoded in the observable corresponding to operator $\hat{D}$. However, its straightforward measurement would involve measurement of observables corresponding to commutators. The latter is challenging. At the same time, eigenstates of D have the simple tensor product structure (separable states). They can easily be found. Then, eigenstates of the Bell operator can be generated as superpositions of Equation (A8).

Appendix C. Prequantum Classical Statistical Field Theory

The basic variables of PCSFT [49] are classical random fields defined on physical space. A random field can be considered as a function of two variables $\phi = \phi(x;\omega)$: x is the spatial variable (with three real coordinates); ω is a random parameter. We remark that random fields can be considered as random vectors valued in the complex Hilbert space $H = L_2(\mathbf{R}^3)$ of square integrable complex valued functions.

The key point of this theory is that covariance operator B of random field ϕ is identified (up to normalization by trace) with the density operator of QM:

$$B \to \rho = B/\mathrm{TrB}. \tag{A9}$$

The covariance operator is an element of the descriptive theory (PCSFT) and the density operator is the element of the observational theory (QM) (For a complex valued random field, its covariance operator B is a Hermitian positive operator with the finite trace. Thus, B has all features of a density operator, besides normalization Træ = 1).

We remark that here the trace of field's covariance operators equals to average of field's energy:

$$\mathrm{TrB} = \mathrm{E}||\text{Œ}(!)||^2, \tag{A10}$$

where E is mathematical expectation and

$$||\phi(\omega)||^2 = \int_{\mathbf{R}^3} |\phi(x;\omega)|^2 \, dx$$

is square of the L_2-norm of the field (for the concrete value of the random parameter ω). Thus, normalization (determining "descriptive $\to$ observational" correspondence) is with respect to field's energy.

Physical variables of PCSFT are quadratic forms of fields. Each quadratic form on $H = L_2$ is determined by a Hermitian operator, $\hat{A} : H \to H$. Hence, PCSFT variables have the form,

$$f_A(\phi)(\omega) = \langle \phi(\omega)|A|\phi(\omega)\rangle,$$

where $\phi(\omega) \equiv \phi(x;\omega) \in L_2$ for each ω. Quadratic forms are elements of the descriptive theory (PCSFT) and Hermitian operators are elements of the observational theory.

Averages calculated in PCSFT coincide with averages calculated in QM. However, the range of values of a quadratic form does not coincide with the range of values of the corresponding quantum observable, Hermitian operator (cf. with descriptive theories of the Bell type).

References

1. Einstein, A.; Podolsky, B.; Rosen, N. Can quantum-mechanical description of physical reality be considered complete? *Phys. Rev.* **1935**, *47*, 777–780. [CrossRef]
2. Bohr, N. *The Philosophical Writings of Niels Bohr*; Ox Bow Press: Woodbridge, UK, 1987.
3. Plotnitsky, A. *Epistemology and Probability: Bohr, Heisenberg, Schrödinger and the Nature of Quantum-Theoretical Thinking*; Springer: Berlin, Germany; New York, NY, USA, 2009.
4. Plotnitsky, A. *Niels Bohr and Complementarity: An Introduction*; Springer: Berlin, Germany; New York, NY, USA, 2012.
5. Bohr, N. Can quantum-mechanical description of physical reality be considered complete? *Phys. Rev.* **1935**, *48*, 696–702. [CrossRef]
6. Bell, J.S. On the Einstein Podolsky Rosen paradox. *Physics* **1964**, *1*, 195–200. [CrossRef]

7. Bell, J.S. *Speakable and Unspeakable in Quantum Mechanics*, 2nd ed.; Cambridge University Press: Cambridge, UK, 2004.
8. Bell, J.S. On the problem of hidden variables in quantum theory. *Rev. Mod. Phys.* **1966**, *38*, 450. [CrossRef]
9. Clauser, J.F.; Horne, M.A.; Shimony, A.; Holt, R.A. Proposed experiment to test local hidden-variable theories. *Phys. Rev. Lett.* **1969**, *23*, 880. [CrossRef]
10. Accardi, L. The probabilistic roots of the quantum mechanicalparadoxes. In *The Wave–Particle Dualism. A Tribute to Louis de Broglie on his 90th Birthday*; Diner S., Fargue D., Lochak G., Selleri F., Eds.; D. Reidel Publication Company: Dordrecht, The Netherlands, 1984; pp. 47–55.
11. Khrennikov, A. *Interpretations of Probability*, 2nd ed. (completed); De Gruyter: Berlin, Germany, 2009.
12. Khrennikov, A. Växjö interpretation-2003: Realism of contexts. In *Quantum Theory: Reconsideration of Foundations*; Khrennikov A., Ed.; Växjö Univ. Press: Växjö, 2004, pp. 323–338.
13. Khrennikov, A. The principle of supplementarity: A contextual probabilistic viewpoint to complementarity, the interference of probabilities, and the incompatibility of variables in quantum mechanics. *Found. Phys.* **2005**, *35*, 1655–1693. [CrossRef]
14. Khrennikov, A. Schrödinger dynamics as the Hilbert space projection of a realistic contextual probabilistic dynamics. *Europhys. Lett.* **2005**, *69*, 678–684. [CrossRef]
15. Khrennikov, A. Bell-Boole inequality: Nonlocality or probabilistic incompatibility of random variables? *Entropy* **2008**, *10*, 19–32. [CrossRef]
16. Khrennikov, A. *Contextual Approach to Quantum Formalism*; Springer: Berlin, Germany; New York, NY, USA, 2009.
17. Khrennikov, A. Bohr against Bell: Complementarity versus nonlocality. *Open Phys.* **2017**, *15*, 734–738. [CrossRef]
18. Khrennikov, A.; Alodjants, A. Classical (local and contextual) probability model for Bohm–Bell type experiments: No-Signaling as independence of random variables. *Entropy* **2019**, *21*, 157. [CrossRef]
19. De Muynck, W. *Foundations of Quantum Mechanics, an Empiricist Approach*; Springer: Dordrecht, The Netherlands, 2006.
20. Accardi, L. Some loopholes to save quantum nonlocality. In *Foundations of Probability and Physics-3*; AIP: Melville, NY, USA, 2005; pp. 1–20, doi:10.1063/1.1874552.
21. Nieuwenhuizen, T.M. Is the contextuality loophole fatal for the derivation of Bell inequalities? *Found. Phys.* **2011**, *41*, 580–591. [CrossRef]
22. Dzhafarov, E.N.; Kujala, J.V. Context-content systems of random variables: The contextuality-by default theory. *J. Math. Psych.* **2016**, *74*, 11–33. [CrossRef]
23. Kupczynski, M. Can Einstein with Bohr debate on quantum mechanics be closed? *Phil. Trans. Royal Soc. A* **2017**, *375*, 2016039. [CrossRef] [PubMed]
24. Kupczynski, M. Closing the door on quantum nonlocality. *Entropy* **2018**, *20*, 877. [CrossRef]
25. Griffiths, R.B. Quantum nonlocality: Myth and reality. *arXiv* **2019**, arXiv:1901.07050.
26. Boughn, S. Making sense of Bell's theorem and quantum nonlocality. *Found. Phys.* **2017**, *47*, 640–657. [CrossRef]
27. Boughn, S. There is no action at a distance in quantum mechanics, spooky or otherwise. *arXiv* **2018**, arXiv:1806.07925.
28. Landau, L.J. Experimental tests of general quantum theories. *Lett. Math. Phys.* **1987**, *14*, 33–40. [CrossRef]
29. Landau, L.J. On the violation of Bell's inequality in quantum theory. *Phys. Lett. A* **1987**, *120*, 54–56. [CrossRef]
30. Braunstein, S.L.; Mann, A.; Revzen, M. Maximal violation of Bell inequalities for mixed states. *Phys. Rev. Lett.* **1992**, *68*, 3259–3261. [CrossRef]
31. Schilpp, P.A. (Ed.) Remarks to the Essays Appearing in this Collective Volume. In *Albert Einstein: Philosopher-Scientist*; Tudor: New York, NY, USA, 1949; pp. 663–688.
32. Plotnitsky, A. *The Principles of Quantum Theory*; Springer: Berlin, Germany; New York, NY, USA, 2016.
33. Hardy, L. *N*-measurement Bell inequalities, N-atom entangled states, and the nonlocality of one photon. *Phys. Lett. A* **1991**, *160*, 1. [CrossRef]
34. Cereceda, J.L. Maximally entangled states and the Bell inequality. *Phys. Lett. A* **1996**, *212*, 123–129. [CrossRef]
35. Mermin, N.D. Extreme quantum entanglement in a superposition of macroscopically distinct states. *Phys. Rev. Lett.* **1990**, *65*, 1838–1840. [CrossRef] [PubMed]

36. Wolf, M.M.; Perez-Garcia, D.; Fernandez, C. Measurements incompatible in quantum theory cannot be measured jointly in any other local theory. *Phys. Rev. Lett.* **2009**, *103*, 230402. [CrossRef] [PubMed]
37. Bogoliubov, N.N.; Shirkov, N.N. *Introduction to Theory of Quantized Fields*; Interscience Publishers: Moscow, Russia, 1959.
38. Haag, R. *Local Quantum Physics. Fields, Particles, Algebras*; Springer: Berlin/Heidelberg, Germany, 1996.
39. Greiner, W.; Reinhardt, J. *Field Quantization*; Part II; Springer: Berlin/Heidelberg, Germany, 1996.
40. Shimony, A. Hidden-variables models of quantum mechanics (Noncontextual and contextual). In *Compendium of Quantum Physics*; Springer: Berlin/Heidelberg, Germany, 2009; pp. 287–291.
41. Shimony, A. Experimental test of local hidden variable theories. In *Foundations of Quantum Mechanics*; Academic: New York, NY, USA, 1971.
42. Jaeger, G. *Quantum Information: An Overview*; Springer: Berlin, Germany; New York, NY, USA, 2007.
43. Jaeger, G. *Quantum Objects: Non-Local Correlation, Causality and Objective Indefiniteness in the Quantum World*; Springer: Berlin, Germany; New York, NY, USA, 2013.
44. Hertz, H. *The Principles of Mechanics: Presented in a New Form*; Macmillan: London, UK, 1899.
45. Boltzmann, L. Uber die Frage nach der objektiven Existenz der Vorgnge in der unbelebten Natur. In *Populre Schriften*; Barth, J.A., Ed.; Vieweg+Teubner Verlag: Leipzig, Germany, 1905.
46. Boltzmann, L. On the development of the methods of theoretical physics in recent times. In *Theoretical Physics and Philosophical Problems*; McGuinness, B., Ed.; Springer: Dordrecht, The Netherlands, 1974; Volume 5, Vienna Circle Collection.
47. D'Agostino, S. Continuity and completeness in physical theory: Schrodinger's return to the wave interpretation of quantum mechanics in the 1950's. In *E. Schrodinger: Philosophy and the Birth of Quantum Mechanics*; Bitbol, M., Darrigol, O., Eds.; Editions Frontieres: Gif-sur-Yvette, France, 1992; pp. 339–360.
48. Khrennikov, A. Hertz's viewpoint on quantum theory. *Act. Nerv. Super.* **2019**, *61*, 24–30. [CrossRef]
49. Khrennikov, A. *Beyond Quantum*; Pan Stanford Publication: Singapore, 2014.
50. Aspect, A. Closing the door on Einstein and Bohr's quantum debate. *Physics* **2015, 8**, 123. [CrossRef]
51. Wiseman, H. Quantum physics: Death by experiment for localrealism. *Nature* **2015**, *526*, 649–650. [CrossRef] [PubMed]
52. Hensen, B.; Bernien, H.; Dreau, A.E.; Reiserer, A.; Kalb, N.; Blok, M.S.; Ruitenberg, J.; Vermeulen, R.F.; Schouten, R.N.; Abellan, C.; et al. Experimental loophole-free violation of a Belli nequality using entangled electron spins separated by 1.3 km. *Nature* **2015**, *526*, 682. [CrossRef]
53. Giustina, M.; Versteegh, M.A.; Wengerowsky, S.; H.; steiner, J.; Hochrainer, A.; Phelan, K.; Steinlechner, F.; Kofler, J.; Larsson, J.Å; Abellan, C.; et al. A significant-loophole-free test of Bell's theo-rem with entangled photons. *Phys. Rev. Lett.* **2015**, *115*, 250401. [CrossRef]
54. Shalm, L.K.; Meyer-Scott, E.; Christensen, B.G.; Bierhorst, P.; Wayne, M.A.; Stevens, M.J.; Gerrits, T.; Glancy, S.; Hamel, D.R.; Allman, M.S.; et al. A strong loophole-free test of local realism. *Phys. Rev. Lett.* **2015**, *115*, 250402. [CrossRef]
55. Khrennikov, A. After Bell. In *Fortschritte der Physik (Progress in Physics), Topical Issue—International Conference Frontiers of Quantum and Mesoscopic Thermodynamics, Prague, Czech Republic, 27 July–1 August 2015*; Wiley: Hoboken, NJ, USA, 2017; Volume 65.
56. Aspect, A.; Dalibard, J.; Roger, G. Experimental test of Bell's Inequalities using time-varying analyzers. *Phys. Rev. Lett.* **1982**, *49*, 1804–1807. [CrossRef]
57. Weihs, G.; Jennewein, T.; Simon, C.; Weinfurther, H.; Zeilinger, A. Violation of Bell's Inequality under Strict Einstein Locality Conditions. *Phys. Rev. Lett.* **1998**, *81*, 5039–5043. [CrossRef]
58. Von Neumann, J. *Mathematical Foundations of Quantum Mechanics*; Princeton University Press: Princeton, NJ, USA, 1955.

2

Foundations of the Quaternion Quantum Mechanics

Marek Danielewski [1,*] and Lucjan Sapa [2]

[1] Faculty of Materials Science & Ceramics, AGH UST, Mickiewicza 30, 30-059 Kraków, Poland
[2] Faculty of Applied Mathematics, AGH UST, Mickiewicza 30, 30-059 Kraków, Poland; sapa@agh.edu.pl
* Correspondence: daniel@agh.edu.pl

Abstract: We show that quaternion quantum mechanics has well-founded mathematical roots and can be derived from the model of the elastic continuum by French mathematician Augustin Cauchy, i.e., it can be regarded as representing the physical reality of elastic continuum. Starting from the Cauchy theory (classical balance equations for isotropic Cauchy-elastic material) and using the Hamilton quaternion algebra, we present a rigorous derivation of the quaternion form of the non- and relativistic wave equations. The family of the wave equations and the Poisson equation are a straightforward consequence of the quaternion representation of the Cauchy model of the elastic continuum. This is the most general kind of quantum mechanics possessing the same kind of calculus of assertions as conventional quantum mechanics. The problem of the Schrödinger equation, where imaginary '*i*' should emerge, is solved. This interpretation is a serious attempt to describe the ontology of quantum mechanics, and demonstrates that, besides Bohmian mechanics, the complete ontological interpretations of quantum theory exists. The model can be generalized and falsified. To ensure this theory to be true, we specified problems, allowing exposing its falsity.

Keywords: relativistic quaternion quantum mechanics; Cauchy-elastic solid; Schrödinger and Poisson equations; quaternions; Klein–Gordon equation

---

1. Introduction

Quantum mechanics is considered to be an irreducibly statistical theory, as a result unable to predict the behavior of individual processes. On the other hand, it has been increasingly used, with stunning success, to gain control over individual objects on an atomic scale. This situation motivates the research into the foundations, leading to a variety of approaches towards an adequate theoretical justification of individual phenomena. It is not agreed whether such an interpretation requires a modification of the standard quantum formalism or whether it can be achieved within that formalism [1]. The first possibility of an ontological, i.e., realist, quantum mechanics was introduced with the original de Broglie–Bohm theory.

Quaternion quantum mechanics, QQM, presented by us is ontic in the sense that it answers the central question of interpretation of quantum mechanics. It is directly related to being (the Cauchy elastic continuum) as well as to the basic categories of being and their relations. The main concepts of quaternion quantum mechanics (QQM) for both the general and mathematical audience are shown. The ideas coming from the quantum theory remain almost completely unfamiliar to most mathematicians who found it difficult to absorb physical ideas, mainly because of the absence of clear definitions and statements of the concepts involved. This paper attempts to overcome some of these gaps in communication. The subject is incredibly huge, and selections are unavoidable.

Quantum mechanics, where we are. The quantum mechanics foundation remains a subject of discussion ever since it was presented in the 1920s. An instantaneous process of the wave-function collapse doesn't fit together with the speed of light limit in Special Relativity. This is the "spooky action" that irritated Einstein so much about quantum mechanics. Present explanations base on assumption that collapse incompatibility has no observable consequences and it's philosophically permissible. However, the problem stays when one asks what happens with the mass and energy of a particle when its wave-function collapses. Notice that the instantaneous jump is not expected in General Relativity and the unclear "string theory" doesn't help.

The Copenhagen interpretation represents the main-stream view; yet recent years witness revived interest in the alternative deterministic interpretation, pioneered by Madelung, de Broglie, and Bohm. David Bohm and Basil Hiley developed an interpretation of quantum mechanics which gives a clear and intuitive interpretation of its meaning with no need of assuming a fundamental role for the human observer [2]. Their ontological interpretation means the causal interpretation of quantum mechanics and is a refinement and completion of de Broglie's 1926 pilot-wave model of non-relativistic quantum theory rediscovered by Bohm in 1952 [3]. This deterministic interpretation is still considered as basically equivalent to the Copenhagen orthodox understanding. It is noteworthy that high-level scientists and even S. Weinberg do not consider it important to mention the unclear issues of the Copenhagen interpretation, the topics that motivated Bohm. The importance of the Bohm approach, i.e., the fact that it consistently solves the measurement problem and allows the classical description of macroscopic objects, is frequently ignored.

John S Bell [4] has given a tremendous input to our understanding of reality by making clear that nonlocal features characterize natural processes. Even so, he also was expressing dissatisfaction with the conceptual and logical status of the theory and has indicated only two possible ways to solve problem [5]:

> *"Either the wavefunction, as given by the Schrödinger equation, is not everything, or it is not right."*

Bell's immense prestige has pushed many physicists to reexamine old problems. We are now observing a comeback of the imprecise and unprofessional thinking which have characterized the debate in the second quarter of the 20th century. Particularly noteworthy is the domination of an incorrect, accepted by many scientists interpretation concerning the real meaning of Bell's theorem [6]. What really matters is the fact that the derivation of Bell's inequality in no way whatsoever needs an assumption of realism. In spite of this fundamental fact, which everybody can verify by going carefully through the proof, a large part of the scientific community shares the wrong opinion that realism is among the basic assumption needed for the derivation of Bell's result. Bell himself has stressed this aspect and commented [7] that it is extremely difficult to eradicate this prejudice:

> *"My own first paper [4] on this subject starts with a summary of the EPR argument from locality to deterministic hidden variables. But the commentators have almost universally reported that it begins with deterministic hidden variables."*

Bell's inequality has nothing to do with realism, it straightforward identifies that what quantum phenomena impose to us is to accept the inescapable fact that natural processes involving entangled states of composite and far-away systems turn out to be unavoidably non-local.

Where we are today? There is a widely known remark by Richard Feynman in 1964:

> *"It is safe to say that no one understands quantum mechanics"* [8]

and Murray Gell-Mann statement in his lecture at the 1976 Nobel Conference that

> *"Niels Bohr brainwashed the whole generation of theorists into thinking that the job (of finding an interpretation of quantum mechanics) was done 50 years ago"* [9]

are both actual. The deterministic interpretation that was considered as basically equivalent to the Copenhagen orthodox understanding is not in use. It was reasoned that since this theory is empirically indistinguishable from the standard theory, it should be considered an example of "bad science". S Weinberg in a letter to S Goldstein [10,11] explicitly expressed such a way of thinking:

"At the regular weekly luncheon meeting today of our Theory Group, I asked my colleagues what they think of Bohm's version of quantum mechanics. The answers were pretty uniform and much what I would have said myself. First, as we understand it, Bohm's quantum mechanics uses the same formalism as ordinary quantum mechanics, including a wavefunction that satisfies the Schrödinger equation, but adds an extra element, the particle trajectory. The predictions of the theory are the same as for ordinary quantum mechanics, so, there seems little point in the extra complication, except to satisfy some a priori ideas about what a physical theory should be like ... In any case ... we are all too busy with our own work to spend time on something that doesn't seem likely to help us make progress with our real problems."

Here, it is important to mention that the predictive equivalence of the two theories is no more obvious [12]. It is safe to say that none of the existing interpretations and modifications of quantum mechanics truly solve the problem of how to derive it from fundamental laws, is it ontic, etc.

Quaternion quantum mechanics today. The first suggestion of quaternion quantum mechanics appears in a footnote of Birkhoff and J. von Neumann 1936 paper [13]. They suggest, in particular, that the physically significant statements in quantum mechanics actually constitute a sort of projective geometry, while the physically significant statements concerning a given system in classical dynamics constitute the Boolean algebra. This indicates that quantum mechanics has greater logical consistency than classical mechanics; a conclusion validated by the impossibility in general of measuring different quantities independently.

Yang has pointed out [14] that it is always possible to represent the pure states of a system of "general quantum mechanics" by rays in a vector space in a one-to-one manner, and for this, it is necessary and sufficient to employ suitable orthogonal vector subspaces of some Hilbert spaces, H, over the following fields of numbers:

$\mathbb{R}$, the real numbers,
$\mathbb{C}$, the complex numbers,
$\mathbb{Q}$, the quaternions.

This result suggests that it is not necessary to go beyond the three possibilities $\mathbb{R}, \mathbb{C}$ and $\mathbb{Q}$ for the representation of general quantum mechanics (the Hurwitz Theorem states that the real numbers $\mathbb{R}$, complexes $\mathbb{C}$, quaternions $\mathbb{Q}$ and octonions $\mathbb{O}$ are the only normed division algebras over the real numbers). A quantum mechanics coefficients assuming values that are quaternionic was proposed by Finkelstein et al. [15]. It was shown that a quaternion calculus exists that they called general quantum mechanics (as distinguished from complex quantum mechanics) and it is always possible to represent pure states of a system of general quantum mechanics by rays in a vector space over the quaternions, but not so over the real and complex numbers. These authors use Stone's theorem to explain the imaginary number "*i*" in the Schrödinger equation however, the central problem of finding feasible dynamics for quaternionic quantum theory has remained unsolved. More recently, the global effects in quaternionic quantum field theory [16] were applied to analyze the experimental status of quaternionic quantum mechanics [17].

The algebra of complex numbers, quaternions and octonions play also an important role in the physical interpretation of the standard model for electroweak interactions and quantum chromodynamics [18]. It is known that certain nonlinear Schrödinger (NLS) equations, in one or more space dimensions, possess space-localized solutions $\psi = \psi(t, x)$, e.g., solitons in the one-dimensional case. Bodurov has shown that the same result is valid for a large class of complex nonlinear wave equations and NLS equations [19,20]. Białynicki-Birula and Mycielski have found that NLS equation admits closed-form space localized solutions (gaussons) [21]. They have shown also that "*... in every electromagnetic field, sufficiently small gaussons move like classical particles*". Weng adopted the complex quaternion and octonion to formulate the field equations for electromagnetic and gravitational fields. The results reveal that the quaternion space is appropriate to describe the gravitational features [22]. Consistent with QQM are also the Three Wave Hypothesis by Horodecki that is based on de Broglie's particle-wave duality and the assumption of covariant æther [23,24]. Recently Gantner demonstrated the equivalence of complex and quaternionic quantum mechanics [25].

The Klein–Gordon and Schrödinger equations are important tools for describing quantum mechanics, respectively relativistic and non-relativistic. Their stationary versions allow us to find the values of quantized energy as the eigenvalues of self-adjoint operators on the Hilbert spaces. Adler studied downgraded quaternion-imaginary Lagrangian and showed that a quaternionic quantum field theory can be formulated when the numbers of bosonic and fermionic degrees of freedom are equal [26]. More recently he studied the quaternionic projective group representations and so-called trace dynamics in Hilbert spaces [27,28]. His idea of the trace dynamics relies on using a variational principle based on a Lagrangian constructed as a trace of noncommuting operator variables, making systematic use of cyclic permutation under the trace operation. In our paper we construct a Lagrangian with the use of the Cauchy–Riemann operator, acting on quaternionic valued functions. Nottale's contribution was the derivation of the physical and mathematical tools of quantum mechanics by using the bi-quaternion concept. His approach proposes an answer to the question of the origin of complex and bi-quaternionic numbers, and more generally of Clifford algebra in quantum mechanics [29]. Recently, a quaternionic commutator bracket was proposed by Arbab et al. [30].

Our quaternion Klein–Gordon and Schrödinger equations have much more physical information than their complex equivalents, i.e., make it a much richer theory. In this work we will derive the Schrödinger equation with the use of the variational calculus, minimalizing the suitable integral functional generated by our constructed Lagrangian. Further qualitative analysis of these quaternionic equations will be the subject of our future papers.

Summing up, the quaternion quantum mechanics has many new features which make it a much richer theory. It is caused generally by the noncommutativity of quaternion-valued wave functions. It has not yet been more fully developed mainly due to the problem of how to write the Schrödinger equation. Remains valid the remark by Lord Kelvin (alias William Thomson, who formulated the laws of thermodynamics) [31]:

> *"Quaternions came from Hamilton … and, though beautifully ingenious, have been an unmixed evil to those who have touched them in any way, including Clerk Maxwell."*

The Planck–Kleinert crystal. It was shown that quantum gravity effects modify Heisenberg's uncertainty principle to a generalized uncertainty principle (GUP) [32]. The GUP-induced corrections to the Schrödinger equation, when applied to a non-relativistic particle in a one-dimensional box, led to the quantization of length and the result was interpreted as a signal of fundamental discreteness of space itself. Similarly, corrections to the Klein–Gordon and Dirac equations gave rise to length, area and volume quantizations, again indicative of the fundamentally grainy nature of space. Such an approach modifies all quantum mechanical Hamiltonians [33]. When applied to an elementary particle, it implies that space which confines it must be quantized. This suggests that space itself is discrete, and that all measurable lengths are quantized in units of a fundamental length (which can be the Planck length).

The original arguments to implement the classical mechanics equations in the field of wave mechanics were given by Kleinert [34]. Kleinert combined Planck scale approaches to Quantum Gravity (such as String Theory and Doubly Special Relativity, black hole physics) that predict a minimum measurable length. The building blocks of the Kleinert continuum (solid) are Planck particles, m_P, that obey the laws of mass, momentum and energy conservation. Each particle exerts a short-range force at the Planck length. The Kleinert concept linked with the Cauchy model of the elastic continuum has been later analyzed with the arbitrary assumption of the complex potential field [35]. Recently, the Cauchy theory was rigorously combined with the Helmholtz decomposition of the vector field of deformations together with quaternion algebra [36] and such representation of the Cauchy equation of motion produced the Klein–Gordon wave equation [37].

In this paper, we present the fundamentals of the quaternion quantum theory, with the clear and precise specification of what the theory is basically about. The fundamental new results, explicitly the deformations represented in quaternion algebra and the family of waves in the elastic continuum are derived.

The essentials of quaternion algebra are given in Section 2.1, and the Cauchy model of the elastic continuum in Section 2.2. Both can be omitted by the experienced reader. In Section 3 we present the rigorous derivation of the quaternion representation of the Cauchy deformation field that allows considering multiple forms of waves and standing waves in ideal elastic solid. The final result is the vast possibility of waveforms in the elastic continuum.

2. Methods

2.1. Essentials of the Quaternion Algebra

The algebra of quaternions, $\mathbb{Q}$, owns all laws of algebra with unique properties [38]. The essential here are (1) the multiplication of quaternions that is noncommutative and (2) it allows quantifying twists and compression. In the next sections, we attempt to convince readers that quaternions are a physical reality. Not only helpful and convenient, the quaternions also allow entering and understanding the processes in continua, together with the wave mechanics.

Hamilton tried for ten years to create the $\mathbb{R}^3$ analog of the complex numbers. His unquestionable motivation was the physics of solids and liquids. Finally in 1843, while on a walk with his wife Helen, he realized that three distinct imaginary units are necessary. He carved a new idea on the Broom Bridge in Dublin, which today is immortalized by a commemorative plaque [39]. The beauty of quaternions was immediately recognized, James Clerk Maxwell in 1869 wrote [40]:

> *"The invention of the calculus of quaternions is a step towards the knowledge of quantities related to space which can only be compared for its importance, with the invention of triple coordinates by Descartes. The ideas of this calculus are fitted to be of the greatest in all parts of science."*

In spite of this, all Maxwell attempts to reformulate electromagnetism using quaternions were unsuccessful [31].

The simplified, trivial and unfortunately very common opinion tells that quaternions were invented as an extension to the complex numbers. Nowadays they are in mass use in the computation of rotations ... in every computer graphics film studio. In the same way, the calculus of imaginary numbers by many is considered as the justified useful tool, by no means the physical reality. Not many scientists agree that reformulating basic principles in terms of quaternion algebra allows a deeper understanding of physics.

Our review of basic definitions and formulas of the quaternion numbers and functions is shortened and limited to those used in the paper [38]. In the original Hamilton notation, a quaternion is regarded as the sum of a real (scalar) and imaginary (vector) part: $\sigma = \sigma_0 1 + \hat{\phi} = [\sigma_0, \hat{\phi}] \in \mathbb{Q}$. An arbitrary quaternion $\sigma \in \mathbb{Q}$ can be written in terms of its basic components

$$\sigma = (\sigma_0, \phi_1, \phi_2, \phi_3) \in \mathbb{Q}, \tag{1}$$

where the four σ_0, ϕ_i coefficients are real. In (1) we introduced notation that is right and convenient in the case of an ideal elastic continuum where only the compression, σ_0, and twist (torsion) emerge, explicitly: $\sigma_1 = \phi_1, \sigma_2 = \phi_2$ and $\sigma_3 = \phi_3$.

Rigorously, in the mathematical way, the quaternion algebra $\mathbb{Q}$ can be defined as follows. Let $\mathbb{R}^4$ be the four-dimensional Euclidean vector space with the orthonormal basis $\{e_0, e_1, e_2, e_3\}$, where $e_0 = (1,0,0,0)$, $e_1 = (0,1,0,0)$, $e_2 = (0,0,1,0)$, $e_3 = (0,0,0,1)$ and with the three-dimensional vector subspace $P = span\{e_1, e_2, e_3\}$. We introduce the multiplication law by the formula

$$a \cdot b = \left(a_0 b_0 - \hat{a} \circ \hat{b}\right) e_0 + \hat{a} \times \hat{b} + a_0 \hat{b} + \hat{a} b_0, \tag{2}$$

where $a = \sum_{i=0}^{3} a_i e_i$, $b = \sum_{i=0}^{3} b_i e_i \in \mathbb{R}^4$, $\hat{a} = \sum_{i=1}^{3} a_i e_i$, $\hat{b} = \sum_{i=1}^{3} b_i e_i \in P$ and $\circ$, $\times$ mean the scalar and vector products in P, respectively:

$$\hat{a} \circ \hat{b} = \sum_{i=1}^{3} a_i b_i,$$

$$\hat{a} \times \hat{b} = \det \begin{bmatrix} e_1 & e_2 & e_3 \\ a_1 & a_2 & a_3 \\ b_1 & b_2 & b_3 \end{bmatrix}.$$

Note that these products are analogous as in $\mathbb{R}^3$.

The multiplication given by (2) is noncommutative. The vector space $\mathbb{R}^4$ with the multiplication (2) is a noncommutative algebra with unity usually denoted by $\mathbb{Q}$ and it is named the quaternion algebra. In practice, the following algebraical notation is useful: $e_0 = 1$, $e_1 = i$, $e_2 = j$, $e_3 = k$. By the definition (2), the quaternion imaginary units obey the following relations:

$$i^2 = j^2 = k^2 = -1,\ ij = -ji = k,\ jk = -kj = i,\ ki = -ik = j. \tag{3}$$

Thus for an arbitrary quaternion $\sigma \in \mathbb{Q}$, we can write shortly

$$\sigma = \sigma_0 1 + \phi_1 i + \phi_2 j + \phi_3 k = \sigma_0 1 + \hat{\phi} \tag{4}$$

A conjugate quaternion is defined as

$$\sigma^* = \sigma_0 1 - \phi_1 i - \phi_2 j - \phi_3 k = \sigma_0 1 - \hat{\phi}. \tag{5}$$

It follows immediately from (2)–(5) that $\sigma \cdot \sigma^* = \sigma^* \cdot \sigma = \sigma_0^2 + \Sigma_{l=1}^{3} \phi_l^2$, and therefore the Euclidean norm

$$||\sigma|| = \sqrt{\sigma \cdot \sigma^*}. \tag{6}$$

Hence $\mathbb{Q}$ is a normed algebra.

Let $\Omega \subset \mathbb{R}^3$ be a bounded set. The so-called $\mathbb{Q}$-valued functions may be written as

$$\sigma(x) = \sigma_0(x)1 + \phi_1(x)i + \phi_2(x)j + \phi_3(x)k,\ x = (x_1, x_2, x_3) \in \Omega, \tag{7}$$

where the functions $\sigma_0(x)$, $\phi_l(x)$, $l = 1,2,3$ are real-valued. Properties such as continuity, differentiability, integrability and so on, which are ascribed to σ have to be possessed by all the components $\sigma_0(x)$, $\phi_1(x)$, $\phi_2(x)$, $\phi_3(x)$. In this manner, the Banach, Hilbert and Sobolev spaces of $\mathbb{Q}$-valued functions can be defined [38]. In the Hilbert space over $\mathbb{Q}$,

$$L^2(\Omega) = \left\{\sigma : \Omega \to \mathbb{Q} \Big| \int_\Omega \sigma_0^2 dx < \infty,\ \int_\Omega \phi_l^2 dx < \infty,\ l = 1,2,3\right\} \tag{8}$$

we introduce the inner product as follows

$$\langle \sigma_1, \sigma_2 \rangle = \int_\Omega \sigma_1^* \cdot \sigma_2 dx,\ \sigma_1, \sigma_2 \in L^2(\Omega). \tag{9}$$

In a similar way, we define the Sobolev spaces

$$H^k(\Omega) = \left\{\sigma : \Omega \to \mathbb{Q} \middle| \sigma, \sigma^{(1)}, \ldots, \sigma^{(k)} \in L^2(\Omega)\right\},\ k \in \mathbb{N}. \tag{10}$$

The definition of self-adjoint operators acting on these spaces is analogous as in the real and complex cases. Moreover, the theories of analytic functions, distributions, Fourier series, Lebesgue measure,

Gelfand triples, Laplace transform and many others on the vector space of $\mathbb{Q}$-valued functions over $\mathbb{Q}$ can be defined in a standard way as in the real and complex cases with analogous properties.

Similarly, the functions $\sigma(t,x)$ depending on time t may be considered.

We will use the Cauchy–Riemann operator D acting on the quaternion-valued functions

$$D\sigma = \left(-\text{div}\hat{\phi}\right)1 + \text{grad}\sigma_0 + \text{rot}\hat{\phi},\ \sigma = \sigma_0 1 + \hat{\phi}, \tag{11}$$

where grad $\sigma_0 = \frac{\partial\sigma_0}{\partial x_1}i + \frac{\partial\sigma_0}{\partial x_2}j + \frac{\partial\sigma_0}{\partial x_3}k$, $\text{div}\hat{\phi} = \frac{\partial\phi_1}{\partial x_1} + \frac{\partial\phi_2}{\partial x_2} + \frac{\partial\phi_3}{\partial x_3}$ and

$$\text{rot}\hat{\phi} = \det\begin{bmatrix} i & j & k \\ \frac{\partial}{\partial x_1} & \frac{\partial}{\partial x_2} & \frac{\partial}{\partial x_3} \\ \phi_1 & \phi_2 & \phi_3 \end{bmatrix}.$$

Under the constraint $\text{div}\hat{\phi} = 0$, D corresponds physically to the nabla operator ∇ in $\mathbb{R}^3$:

$$D\sigma = \text{grad}\sigma_0 + \text{rot}\hat{\phi},\ \sigma = \sigma_0 1 + \hat{\phi}. \tag{12}$$

The exponent function has its trigonometrical representation

$$e^{\sigma} = e^{\sigma_0}\left(\cos|\hat{\phi}| + \frac{\hat{\phi}}{|\hat{\phi}|}\sin|\hat{\phi}|\right), \sigma = \sigma_0 1 + \hat{\phi}, \tag{13}$$

where σ is a $\mathbb{Q}$-valued function.

Remark 1. *Note that $DD\sigma = -\Delta\sigma$, thus Equation (12) links quaternion quantum mechanics to reality in $\mathbb{R}^3$.*

Remark 2. *In further parts we will identify a vector $\sigma_0 1$ with a real number σ_0 so we will write for simplicity $\sigma = \sigma_0 + \hat{\phi}$. Therefore $\mathbb{Q}$ can be treated as the generalized simple sum $\mathbb{R} \oplus P$.*

2.2. Cauchy Classical Theory of Elasticity

Cauchy finished the theory of the ideal elastic continuum in 1822 [41,42], right away Poisson [43] studied the elementary waves. In 1885 Neumann [44] gave the proof of the uniqueness of solutions of some boundary-initial value problems and before the end of the XIX-th century, a rigorous completeness proof was given by Duhem [45]. Cauchy theory is the first real attempt to construct a theory of elasticity using the continuum approach, where the macroscopic phenomena are described in terms of field variables [46,47]. Important here is that:

1. mathematical analysis of the various phenomena;
2. the Cauchy and to the same degree the majority of physical problems cannot be reduced to vectorial models (the vector product does not permit the formulation of algebra with unity, for example, the division operation is not defined [48]);
3. the Hamilton algebra of quaternions, $\mathbb{Q}$, and Hamilton concept of the four-dimensional space allow combining Cauchy theory with the Helmholtz decomposition theorem.

In the following, we consider FCC structure, where the Poisson number $\nu = 0.25$, l_P denotes the dimension of the FCC elementary cell that consists of four interacting Planck particles showing the mass m_P.

1. The continuum is treated as a closed system occupying the constant volume $\Omega \subset \mathbb{R}^3$.
2. The continuum density, ρ_P, is high and we consider the small deformation limit only, $l_P = const.$, thus the density changes are negligible and $\rho_P = 4m_P/l_P^3 = const.$
3. The small deformation limit implies invariant transverse wave velocity: $c = \sqrt{0.4Y/\rho_P} = const.$, where Y is the Young modulus [46], Equation (16).

4. In agreement with the Helmholtz decomposition theorem [46], every lattice deformation $\mathbf{u}$ can be expressed as a sum of compression and twist, $\mathbf{u} = \mathbf{u}_0 + \mathbf{u}_\phi$.
5. We will consider here the long evolution times, $t \gg t_P$, where t_P is Planck time and the moving (translating) waves, v, where v is the velocity of the wave mass center as defined by Bodurov [19,20].

In this section, we do not consider the external fields. In such a continuum, the equation of motion relates to local acceleration due to the displacement with the field variables, the compression (div$\mathbf{u}$) and twist (rot$\mathbf{u}$)

$$\frac{\partial^2 \mathbf{u}}{\partial t^2} = 3c^2 \mathrm{graddiv}\mathbf{u} - c^2 \mathrm{rotrot}\mathbf{u}. \tag{14}$$

From Equation (14), the energy per mass unit in the deformation field follows [46,49]

$$e = \frac{\rho_E}{\rho_P} = \frac{1}{2}\dot{\mathbf{u}} \circ \dot{\mathbf{u}} + \frac{3}{2}c^2(\mathrm{div}\mathbf{u})^2 + \frac{1}{2}c^2 \mathrm{rot}\mathbf{u} \circ \mathrm{rot}\mathbf{u}, \tag{15}$$

where $\dot{\mathbf{u}} = \partial \mathbf{u}/\partial t$.

Equation (14) and relation (15) obey the Euler–Lagrange relation $\frac{\partial e}{\partial \mathbf{u}} - \frac{\mathrm{d}}{\mathrm{d}t}\left(\frac{\partial e}{\partial \dot{\mathbf{u}}}\right) = 0$. It means that one can derive the vector equation of motion from the scalar relation of energy conservation and vice versa. The scalar relation (15) and the vector Equation (14) rule the deformation in the ideal elastic continua. By the Helmholtz decomposition theorem, every deformation can be expressed by the compression, $\mathbf{u}_0$, and twist, $\mathbf{u}_\phi$, and if $\mathbf{u}$ belongs to the C^3 class of functions then $\mathbf{u} = \mathbf{u}_0 + \mathbf{u}_\phi$, where rot$\mathbf{u}_0 = 0$ and div$\mathbf{u}_\phi = 0$. Upon acting on Equation (14) by the divergence and rotation operators, we decompose it and get well known transverse and the longitudinal wave equations in the usual form $\mathbf{u}_{tt} = k\Delta\mathbf{u}$:

$$\begin{aligned} \mathrm{div}\left(\frac{\partial^2 \mathbf{u}}{\partial t^2} = 3c^2 \mathrm{graddiv}\mathbf{u} - c^2 \mathrm{rotrot}\mathbf{u}\right) &\Rightarrow \frac{\partial^2}{\partial t^2}\mathrm{div}\mathbf{u}_0 = 3c^2 \Delta \mathrm{div}\mathbf{u}_0, \\ \mathrm{rot}\left(\frac{\partial^2 \mathbf{u}}{\partial t^2} = 3c^2 \mathrm{graddiv}\mathbf{u} - c^2 \mathrm{rotrot}\mathbf{u}\right) &\overset{\Delta\mathbf{u} = \nabla(\nabla\cdot\mathbf{u}) - \nabla\times(\nabla\times\mathbf{u})}{\Rightarrow} \frac{\partial^2}{\partial t^2}\mathrm{rot}\mathbf{u}_\phi = c^2 \Delta \mathrm{rot}\mathbf{u}_\phi. \end{aligned} \tag{16}$$

The Cauchy equation of motion combined with the Helmholtz decomposition theorem results in four uncoupled second-order scalar differential equations, "quattro cluster", and implies the transverse and longitudinal waves in the Cauchy elastic solid. Note that these equations remain coupled with the relation of the energy density, Equation (15). In the next sections, using the quaternion representation of the Cauchy elastic continuum, the plentiful forms of the wave equations are derived.

3. Results

3.1. The Cauchy Deformation Field in the Quaternion Representation

The Cauchy classical theory of elasticity is an elegant starting point to show the physical reality and the significance and beauty of quaternions. The Hamilton algebra $\mathbb{Q}$ allows recoupling the compression and twist that were separated in (16). Upon denoting $\sigma_0 = \mathrm{div}\mathbf{u}_0 = (\sigma_0, 0, 0, 0)$ and $\hat{\phi} = \mathrm{rot}\mathbf{u}_\phi = (0, \phi_1, \phi_2, \phi_3)$ as well $\hat{\phi} = \phi_1 i + \phi_2 j + \phi_3 k$ we get

$$\begin{aligned} \frac{\partial^2 \sigma_0}{\partial t^2} &= 3c^2 \Delta \sigma_0, \\ \frac{\partial^2 \hat{\phi}}{\partial t^2} &= c^2 \Delta \hat{\phi}. \end{aligned} \tag{17}$$

Relation (15) takes the form

$$e = 1/2\dot{\mathbf{u}} \circ \dot{\mathbf{u}} + 3/2c^2\sigma_0^2 + 1/2c^2\hat{\phi} \circ \hat{\phi}. \tag{18}$$

The decomposition $\mathbf{u} = \mathbf{u}_0 + \mathbf{u}_\phi$ in Equation (16) results in four equations in Equation (17) and implies the existence of the deformation field $\sigma = \sigma_0 + \hat{\phi}$ that represents the twist and compression fields as a superposition of real (scalar compression σ_0) and imaginary (twist vector $\hat{\phi}$) field parts at each point

$$\sigma = \sigma_0 + \hat{\phi} \in \mathbb{Q} \text{ and } \sigma^* = \sigma_0 - \hat{\phi} \in \mathbb{Q}, \tag{19}$$

where the following constraint holds

$$\mathrm{div}\hat{\phi} = \mathrm{divrot}\mathbf{u}_\phi = 0. \tag{20}$$

Adding equations in Equation (17) and from constraint (20), we get the quaternion form of the motion equation

$$\frac{1}{c^2}\frac{\partial^2\sigma}{\partial t^2} - \Delta\sigma - 2\Delta\sigma_0 = 0, \text{ where } \sigma = \sigma_0 + \hat{\phi}. \tag{21}$$

Equation (21) must obey constraint (20) and require boundary conditions for a solution in the quaternion form. Since $\dot{\mathbf{u}} \circ \dot{\mathbf{u}} = \hat{\dot{u}} \circ \hat{\dot{u}} = -\hat{\dot{u}} \cdot \hat{\dot{u}} = \hat{\dot{u}} \cdot \hat{\dot{u}}^*$, where $\hat{\dot{u}} = \dot{u}_1 i + \dot{u}_2 j + \dot{u}_3 k$ and $\dot{\mathbf{u}} = (\dot{u}_1, \dot{u}_2, \dot{u}_3)$, the overall energy of the deformation field, the Formula (18) becomes in the quaternion form

$$e = \frac{1}{2}\hat{\dot{u}} \cdot \hat{\dot{u}}^* + \frac{1}{2}c^2\sigma \cdot \sigma^* + c^2\sigma_0^2. \tag{22}$$

The energy is conserved, so relation (22) leads to the nonlocal boundary condition for Equation (21) [37].

Remark 3. *The Cauchy theory of elastic continuum combined with the Helmholtz decomposition theorem and quaternion algebra results in second-order differential Equation (21) and constraint (20). It infers the transverse, longitudinal and multiple complex forms of waves. Equation (21) and the relation (22) satisfy the Eule–Lagrange differential equation, i.e., satisfy the fundamental equation of the calculus of variations.*

3.2. Quaternion Quantum Mechanics

In this section, we combine the Hamilton algebra of quaternions [50], the Cauchy classical mechanics in $\mathbb{R}^3$ [41,42] with Planck–Kleinert crystal concept [34,35]. The fundamental new results, explicitly the ontology of Quaternion Quantum Mechanics, the appearance of imaginary numbers in Schrödinger equation and the family of waves in elastic continua will follow.

The crystal hypothesis can be found in Maxwell paper published in 1856 "*A Dynamical Theory of the Electromagnetic Field*" [51] where Maxwell explicitly remarked on the ether:

> "*On our theory, it* (energy) *resides in the electromagnetic field, in the space surrounding the electrified and magnetic bodies, as well as in those bodies themselves, may be described according to a very probable hypothesis, as the motion and the strain of one and the same medium* (elastic ether)."

Less known, if not entirely forgotten, is the section on gravity where Maxwell wrote:

> " ... *assumption, therefore, that gravitation arises from the action of the surrounding medium leads to the conclusion that every part of this medium possesses, when undisturbed, an enormous intrinsic energy As I am unable to understand in what way a medium can possess such properties, I cannot go any further in this direction in searching for the cause of gravitation.*"

Maxwell was unsuccessful in formulating his theory in quaternion form. His idea of solid ether showing "enormous intrinsic energy" was unimaginable in the XIX-th century.

In simple words, we regard quantum space as an analog to the Cauchy elastic solid. Already, upon adding Equation (17), we obtained quaternion form of the motion equation, Equation (21). In this section, we show that upon splitting Equation (21) into the system of the wave and Poisson type equations, the multiple non-linear forms of the wave equation follow. That is, the quaternion motion equation produces the family of the non-linear stationary and quasi-stationary waves. The properties of Planck–Kleinert crystal (Maxwell elastic ether) are presented in Table 1.

Table 1. The physical constants of the Planck–Kleinert crystal (fcc ideal isotropic crystal).

Label Used in This Work	Planck Constants	Symbol for Unit	Value	SI Unit	Reference
Transverse wave velocity	Light velocity in vacuum	c	2.99792458×10^{8}	$m \cdot s^{-1}$	[52]
Lattice parameter	Planck length	l_P	$1.616229(38) \times 10^{-35}$	m	[52]
Poisson ratio		ν	0.25	-	[52]
Mass of the Planck particle	Planck mass	m_P	$2.176470(51) \times 10^{-8}$	kg	[52]
Planck–Kleinert crystal density		ρ_P	2.062072×10^{97}	$kg \cdot m^{-3}$	[52]
Duration of the internal process	Planck time	t_P	$5.39116(13) \times 10^{-44}$	s^{-1}	[52]
Young modulus	Intrinsic energy density	Y	$4.6332447 \times 10^{114}$	$kg \cdot m^{-1} s^{-2}$	$c = \sqrt{0.4Y/\rho_P}$

Remark 4. *The Quaternion Quantum Mechanics follows from rigorous quaternion representation of the Cauchy linear theory of elasticity, Equations (20) and (21).*

3.3. Relativistic Stationary Waves in the Cauchy Continuum: the Quaternion Klein–Gordon Equation

To do so, we consider the wave showing the overall energy $E_n = \text{const}$. Subsequently, Equation (21) can be written as a multisystem

$$\begin{cases} \frac{1}{c^2}\frac{\partial^2 \sigma}{\partial t^2} - (n+1)\Delta\sigma + k_n^2 \sigma \cdot \sigma^* = 0, \\ (n-2)\Delta\sigma_0 + n\Delta\hat{\phi} - k_n^2 \sigma \cdot \sigma^* = 0, \end{cases} \quad \text{for } n = 0, 1, 2, \ldots \tag{23}$$

where $k_n = 1/\lambda_n$ and $\lambda_n = f(E_n)$ denotes the wavelength. Note that by adding Equation in (23), the momentum balance is expressed again by a single partial differential Equation (21). The system (23) is a hyperbolic-elliptic quaternion representation of a the wave Equation (21) and have solutions of the form

$$\sigma = \sigma_0 + \hat{\phi} = \sigma_0 + \phi_1 i + \phi_2 j + \phi_3 k \in \mathbb{Q}. \tag{24}$$

The second equation in (23) is the Poisson type equation, and it describes/defines the compression potential as a function of energy density. The analysis of the system (23) in a case when $n = 0$ shows that wave equation is the quaternion Klein–Gordon type equation [53] and Poisson equation

$$\begin{cases} \left(\frac{1}{c^2}\frac{\partial^2}{\partial t^2} - \Delta\right)\sigma + k_n^2 \sigma^* \cdot \sigma = 0, \\ 2\Delta\sigma_0 = -k_n^2 \sigma^* \cdot \sigma. \end{cases} \tag{25}$$

Consider a quasi-stationary wave in the Planck–Kleinert crystal, Table 1, i.e., the particle showing the constant overall energy (equivalent mass), $m = Ec^{-2} = const$, E denotes overall wave energy and $c = l_P/t_P$ [37]. Thus, Equation (25) can be written

$$\begin{cases} \left(\frac{\partial^2}{\partial t^2} - c^2\Delta\right)\sigma + \frac{8\pi m}{m_P t_P^2}\sigma^* \cdot \sigma = 0, \\ 2c^2\Delta\sigma_0 = -\frac{8\pi m}{m_P t_P^2}\sigma \cdot \sigma^*, \end{cases} \tag{26}$$

and the first equation above in the compacted form as

$$\left(\frac{1}{c^2}\frac{\partial^2}{\partial t^2} - \Delta\right)\sigma + \frac{8\pi m}{\hbar t_P}\sigma^* \cdot \sigma = 0 \tag{27}$$

or in the covariant notation

$$\partial_\mu \partial^\mu \sigma + m\frac{8\pi m}{\hbar t_P}\sigma^* \cdot \sigma = 0, \tag{28}$$

where $\partial_\mu \partial^\mu = \frac{1}{c^2}\frac{\partial^2}{\partial t^2} - \Delta$ and $\hbar = m_P c^2 t_P = 1.0545727 \times 10^{-34}\left[\mathrm{kg} \cdot \mathrm{m}^2 \mathrm{s}^{-1}\right]$.

The Klein–Gordon equation fulfills the laws of special relativity, but contains two fundamental problems [54]. The first one is that it allows negative energies as a solution. As can be seen, the energy computed using Formulae (22) as well as Equation (28) is per definition always positive due to the constraint (20). Existence and properties of solutions to wave equations are studied for example in [55] and in the references therein.

The second equation in (26) is the Poisson equation and describes the irrotational, e.g., compression, potential in the deformation field. By introducing physical constants of the Planck–Kleinert crystal we get

$$c^2 \Delta \sigma_0 = -4\pi m \frac{1}{m_P t_P^2} \sigma \cdot \sigma^*. \tag{29}$$

The above relation can be expressed as a function of the local mass density [37]: $\rho = \rho_E / c^2$, where $\rho_E = mc^2 \sigma \cdot \sigma^* / l_P^3$

$$c^2 \Delta \sigma_0 = -4\pi \rho \frac{l_P^3}{m_P t_P^2} = -4\pi \rho G, \tag{30}$$

where using data in Table 1, the gravitational constant equals: $G = l_P^3 / \left(t_P^2 m_P\right) = 6.674082 \times 10^{-11}$ $\left[\mathrm{m}^3 \cdot \mathrm{kg}^{-1} \mathrm{s}^{-2}\right]$.

Remark 5. *The low deformation limit considered by us allows for the simplified assumption of the constant overall energy density as well as the constant transverse wave velocity. Consequently, the gravity in the simplified form of the Poisson equation follows. By considering the dependence of the energy density on deformation one can get the extended quaternion form of relation (29) or the geometrical theory of gravitation when the invariant velocity of a transverse wave is assumed.*

3.4. Non-Relativistic Waves in the Cauchy Continuum: the Quaternion Schrödinger Equation

The quasi-stationary wave means here that the wave can be treated as a particle in an arbitrary volume Ω. Such a wave has multiple properties, important here are:

1. the overall energy, $E = E^0 + Q$, where E^0 and Q are the ground and excess energies, and the overall energy density, ρ_E;
2. the equivalent mass interrelated to the wave overall energy;
3. the wave mass center and its translation velocity v [19].

In the following, the label kinetic will denote the wave translation in continuum: e.g., the wave velocity, v. The category dynamic will denote the local movements within the elastic continuum itself, i.e., the local kinetic energy density, k, caused by the lattice local velocity, $\hat{\dot{u}}$. From relation (22), upon substituting $\widetilde{\sigma}_0 = \sqrt{3}\sigma_0$, the overall wave energy can be expressed by the formula

$$\begin{aligned} E = E^0 + Q = \int_\Omega \rho_E \mathrm{d}x &= c^2 \int_\Omega \rho_P \left(\frac{1}{2c^2} \hat{\dot{u}} \cdot \hat{\dot{u}}^* + \frac{1}{2} \widetilde{\sigma} \cdot \widetilde{\sigma}^*\right) \mathrm{d}x \\ &= c^2 \int_\Omega \rho_P k \mathrm{d}x + c^2 \int_\Omega \rho_P s \mathrm{d}x \\ &= K + S, \end{aligned} \tag{31}$$

where $\widetilde{\sigma} = \widetilde{\sigma}_0 + \hat{\sigma} = \sqrt{3}\sigma_0 + \hat{\sigma}$.

The overall mass of the particle, m, and the particle mass density, ρ, follow from (31),

$$m = \int_\Omega \rho \mathrm{d}x = \int_\Omega \rho_P \left(\frac{1}{2c^2} \hat{\dot{u}} \cdot \hat{\dot{u}}^* + \frac{1}{2} \widetilde{\sigma} \cdot \widetilde{\sigma}^*\right) \mathrm{d}x. \tag{32}$$

We conclude that the "the overall particle mass" follows from the relations (31) and (32): $m = E/c^2$. The *mass* is ontic in that sense that the particle mass is identified with the overall dynamic energy of the wave in elastic quasi-continuum of the Planck particles, see also [56]. Note that in general when $Q > 0$, m differs from the mass at the ground state. In simple words, when a Planck particle in the ideal crystal due to an arbitrary cyclic process has the overall energy E^P, then the ratio $m = E^P/c^2$ we call the mass of the "particle localized on" this single Planck particle.

The waves in Ω may differ in shape and other essentials, the overall particle energy, relation (31), reveals the wave nature. The considered here quasi-stationary wave has to satisfy the relation (31) and at every position, the energy density is a sum of the dynamic, k, and strain, s, energy terms: $\rho_E = c^2\rho_P(k+s) = 0.4Y(k+s)$.

We start with an elementary situation when the velocity of the wave mass center v is low and constant, $v \ll c$. By using the extremum principle, namely the action concept, one can quantify the elementary properties of such waves. At every position in Ω:

1. the existence assumption of the quasi-stationary wave implies an equal duration of the periodic cycles in the whole volume occupied by the wave, $\Delta t = const$. Consequently implies, that the s- and k-actions are equal

$$\int_t^{t+\Delta t} s(\tau,x)\mathrm{d}\tau = \int_t^{t+\Delta t} k(\tau,x)\mathrm{d}\tau = \gamma(x)\Delta t; \tag{33}$$

2. the sum of the overall strain, S, and the kinetic energy, K, in relation (31) equals the overall wave energy $E = E^0 + Q$, and is time-invariant;
3. spans of the strain and the kinetic energy terms are equal,

$$[0, \max\{k(t,x)\}] = [0, \max\{s(t,x)\}] = \left[0, \rho_E(t,x)/\left(c^2\rho_P\right)\right] = [0, \rho_E(t,x)/(0.4Y)].$$

The relation (33) is valid for the whole Ω so

$$\int_t^{t+\Delta t}\int_\Omega s(\tau,x)\mathrm{d}x\mathrm{d}\tau = \int_t^{t+\Delta t}\int_\Omega k(\tau,x)\mathrm{d}x\mathrm{d}\tau = \gamma_\Omega\Delta t$$

and also for an arbitrary number of cycles: $t = n\Delta t$. Thus, from assumptions 2 and 3 above and the relation (31), it follows that both actions in Ω can be approximated by the discrete formula

$$\gamma_\Omega n\Delta t = \int_0^{n\Delta t}\int_\Omega s\mathrm{d}x\mathrm{d}\tau = \int_0^{n\Delta t}\int_\Omega k\mathrm{d}x\mathrm{d}\tau. \tag{34}$$

Taking into account that we consider time evolution in a case when $t \gg \Delta t$, the continuous expressions for both actions follow

$$\gamma_\Omega t = \int_0^t\int_\Omega s\mathrm{d}x\mathrm{d}\tau = \int_0^t\int_\Omega k\mathrm{d}x\mathrm{d}\tau. \tag{35}$$

Taking time derivative of the relations in (35) we get:

$$\int_\Omega s\mathrm{d}x = \int_\Omega k\mathrm{d}x \text{ for } t \gg \Delta t. \tag{36}$$

Both terms, s and k, in (36) oscillate and depend on the time and position. It will be useful to normalize the displacement term s in (31) with respect to the overall particle mass, relation (32). Note that because we restrict our analysis to the low velocities $v \ll c$, the translation energy has a minor impact on the overall wave mass and $m \cong m_0$. From the formulas (31), (32) and (36), the normalized strain energy density, s, equals

$$\int_\Omega \frac{\rho_P}{m}\widetilde{\sigma}\cdot\widetilde{\sigma}^* \mathrm{d}x = \int_\Omega \psi\cdot\psi^* \mathrm{d}x = 1, \text{ where } \psi = \sqrt{\frac{\rho_P}{m}}\widetilde{\sigma}. \tag{37}$$

Consequently from (31), (36) and (37)

$$0 = \int_\Omega \left(\rho_P c^2 \widetilde{\sigma}\cdot\widetilde{\sigma}^* - E\psi\cdot\psi^*\right)\mathrm{d}x \tag{38}$$

and also

$$0 = \int_\Omega \left[\rho_P \hat{\dot{u}}\cdot\hat{\dot{u}}^* - E\psi\cdot\psi^*\right]\mathrm{d}x. \tag{39}$$

Note that, the relations (31), (32) and (38) imply the relation between the overall energy of the wave, and equivalently the overall wave mass $E = mc^2$, and the probability density because the relation (37) is satisfied. Obviously, the both terms, $\psi = \sqrt{\frac{\rho_P}{m}}\widetilde{\sigma}$ and $\psi\cdot\psi^*$, vary in time.

Remark 6. *The relation (38) links up frequency of the wave with its overall energy:*

(1) *the excess, Q, and ground, E^0, energies are entangled in (38) and can't be separated;*
(2) *the overall wave energy can be increased by translation term, accordingly also all displacements and velocities are affected;*
(3) *the wave periodicity implies that by solving the relation (39), one should expect only the discrete values if excess energy Q;*
(4) *when the wave overall energy equals its ground energy, Q = 0, then relation (38) results in*

$$0 = \int_\Omega \left(\rho_P c^2 \widetilde{\sigma}^0\cdot\left(\widetilde{\sigma}^0\right)^* - E^0\psi\cdot\psi^*\right)\mathrm{d}x, \tag{40}$$

where $\widetilde{\sigma}^0$ denotes the displacement in the wave at the ground energy E^0.

3.5. Waves in the Time-Invariant Potential Field

Let us consider now the evolution of the wave as in the relation (31) in the time-invariant potential field, e.g., in the field generated by other particles. The overall energy is now a sum of the ground and excess energy

$$E = E^0 + Q = \int_\Omega \left(\frac{1}{2}\rho_P \hat{\dot{u}}\cdot\hat{\dot{u}}^* + \frac{1}{2}\rho_P c^2 \widetilde{\sigma}\cdot\widetilde{\sigma}^* + V(x)\psi\cdot\psi^*\right)\mathrm{d}x, \tag{41}$$

where $\sqrt{\widetilde{\sigma}\cdot\widetilde{\sigma}^*} = \|\widetilde{\sigma}^0 + \widetilde{\sigma}_Q\| = \|\widetilde{\sigma}\|$ and $\widetilde{\sigma} = \widetilde{\sigma}^0 + \widetilde{\sigma}_Q$.

We consider low excess energy, $v \ll c$. Consequently: (1) the impact of Q on the overall particle mass is marginal, $m \cong m_0$, and (2) the displacement $\widetilde{\sigma}$ in (41) can be normalized using the formula (37). Thus relation (41) becomes

$$\begin{aligned} E = E^0 + Q &= \int_\Omega \left(\tfrac{1}{2}\rho_P \hat{\dot{u}}\cdot\hat{\dot{u}}^* + \tfrac{1}{2}mc^2\psi\cdot\psi^* + V(x)\psi\cdot\psi^*\right)\mathrm{d}x \\ &= \tfrac{1}{2}mc^2 + \int_\Omega \left(\tfrac{1}{2}\rho_P \hat{\dot{u}}\cdot\hat{\dot{u}}^* + V(x)\psi\cdot\psi^*\right)\mathrm{d}x. \end{aligned} \tag{42}$$

Both, the E^0 and m are constant, thus it is enough to minimize the relation

$$Q = \int_\Omega \left(\frac{1}{2}\rho_P \hat{\dot{u}}\cdot\hat{\dot{u}}^* + V(x)\psi\cdot\psi^*\right)\mathrm{d}x. \tag{43}$$

The above relation contains the unknown kinetic velocity due to the potential $V(x)$. The Cauchy–Riemann operator of the deformation, $D\widetilde{\sigma}$, can be understood, by means of the relations (11) and (12), as an analogy of the gradient in $\mathbb{R}^3$. In classical dynamics, the potential gradient results in acceleration. For the quaternion representation of the deformation field, it is reasonable to guess that the local momentum in the crystal (i.e., the Planck particle momentum), $\hat{p} = m_p\hat{u}$, is related to the Cauchy–Riemann operator of the quaternion displacement, $D\widetilde{\sigma}$. Namely, the local lattice velocity is proportional to (1) the force that is the normalized Cauchy–Riemann derivative of the local displacement, $l_P D\widetilde{\sigma}$, and (2) the transverse wave velocity c. Accordingly

$$\hat{p} = -m_P c l_P D\widetilde{\sigma} = -\hbar D\widetilde{\sigma}, \tag{44}$$

where we introduced constants $\hbar = m_P c^2 t_P$ and $t_P = l_P/c$ i.e., is the time that transverse wave travels at the lattice distance: $l_P = ct_P$.

The momentum balance requires

$$\hat{u} = \frac{\hat{p}}{m} = -\frac{\hbar}{m} D\widetilde{\sigma}. \tag{45}$$

By introducing (45), the relation (43) becomes

$$Q = \int_\Omega \left(\rho_P \frac{\hbar^2}{2m^2}(D\widetilde{\sigma}) \cdot (D\widetilde{\sigma})^* + V(x)\psi \cdot \psi^*\right)\mathrm{d}x. \tag{46}$$

Normalization using (37) results in the functional

$$Q[\psi] = \int_\Omega \left(\frac{\hbar^2}{2m}(D\psi) \cdot (D\psi)^* + V(x)\psi \cdot \psi^*\right)\mathrm{d}x. \tag{47}$$

There are numerous methods for solving the above problem, e.g., the path integrals, the Hamilton Jacobi equation, etc. We minimize the functional $Q[\psi]$, that is the integral above, with respect to a quaternion function such that ψ satisfies the normalization introduced in the relation (37). We look for a differential equation that has to be satisfied by the ψ function to extremize (here minimize) the energies allowed by (47). Subsequently, we will show that the extremum problem leads to the quaternion analog of the time-independent Schrödinger equation.

Given the functional (47), we seek for the conditional extreme and use the Lagrange coefficients method combined with a procedure presented in Appendix A. Then, there exists a multiplayer $\lambda \neq 0$ such that ψ minimalizes the functional

$$\widetilde{Q}[\psi] = \int_\Omega \left[\frac{\hbar^2}{2m}(D\psi) \cdot (D\psi)^* + V(x)\psi \cdot \psi^* + \lambda\left(\frac{1}{|\Omega|} - \psi \cdot \psi^*\right)\right]\mathrm{d}x. \tag{48}$$

It follows from Appendix A that ψ satisfies the differential equation

$$-\frac{\hbar^2}{2m}\Delta\psi + V(x)\psi = \lambda\psi, \tag{49}$$

where a constant factor on the right-hand side can be considered as extra energy of the particle in the presence of the field $V = V(x)$.

For $E = \lambda$, Equation (49) is clearly the time-independent Schrödinger equation satisfied by the particle in the ground state of the energy E,

$$-\frac{\hbar^2}{2m}\Delta\psi + V(x)\psi = E\psi. \tag{50}$$

It has to be satisfied together with the condition

$$\mathrm{div}\hat{\psi} = 0 \text{ where } \psi = \psi_0 + \hat{\psi}. \quad (51)$$

Upon using the NIST data [52] of the Planck's natural units m_P, l_P, t_p and the light velocity c, the constant $\hbar$ introduced in relation (44) equals the Planck constant, $\hbar = 6.626069311 \times 10^{-34}$.

The particle (wave) mass center. The meaning assigned to "space-localized" is used in the sense given by the Bodurov definition [57]:

A singularity-free multi-component function $\sigma = (\sigma_0, \phi_1, \phi_2, \phi_3) \in \mathbb{Q}$ of the space $x = (x_1, x_2, x_3)$ and time t variables will be called space-localized if $\|\sigma(t,x)\| \to 0$ sufficiently fast when $\|x\| \to \infty$, so that its Hermitean norm, see Equation (9)

$$\langle \sigma, \sigma^* \rangle = \int_\Omega \left(\sigma_0^2 + \sum_{l=1}^{3} \phi_l \cdot \phi_l^* \right) \mathrm{d}x = \int_\Omega \sigma \cdot \sigma^* \mathrm{d}x < \infty, \quad (52)$$

remains finite for all time.

3.6. *Quaternion Time Dependent Schrödinger Equation*

By analogy to the complex time-dependent Schrödinger equation $i\hbar\frac{\partial \Psi}{\partial t} = -\frac{\hbar^2}{2m}\Delta\Psi + V(x)\Psi$, we propose the quaternion form

$$\frac{1}{3}(i+j+k)\hbar\frac{\partial \Psi}{\partial t} = -\frac{\hbar^2}{2m}\Delta\Psi + V(x)\Psi. \quad (53)$$

We will show now that by the substitution $\Psi(t,x) = e^{-(i+j+k)\frac{E}{\hbar}t}\psi(x)$, the Equation (53) leads to the time-independent Schrödinger Equation (50). Note that by the trigonometric Formula (13), we have

$$\Psi(t,x) = \left[\cos\left(\sqrt{3}\frac{E}{\hbar}t\right) - \frac{1}{\sqrt{3}}(i+j+k)\sin\left(\sqrt{3}\frac{E}{\hbar}t\right)\right]\psi(x), \quad (54)$$

$$\begin{aligned} \frac{\partial \Psi}{\partial t}(t,x) &= \left[-\sqrt{3}\tfrac{E}{\hbar}\sin\left(\sqrt{3}\tfrac{E}{\hbar}t\right) - (i+j+k)\tfrac{E}{\hbar}\cos\left(\sqrt{3}\tfrac{E}{\hbar}t\right)\right]\psi(x) \\ &= -(i+j+k)\tfrac{E}{\hbar}\left[\cos\left(\sqrt{3}\tfrac{E}{\hbar}t\right) - \tfrac{1}{\sqrt{3}}(i+j+k)\sin\left(\sqrt{3}\tfrac{E}{\hbar}t\right)\right]\psi(x) \\ &= -(i+j+k)\tfrac{E}{\hbar}\cdot e^{-(i+j+k)\frac{E}{\hbar}t}\psi(x). \end{aligned} \quad (55)$$

Obviously

$$\Delta\Psi(t,x) = e^{-(i+j+k)\frac{E}{\hbar}t}\Delta\psi(x). \quad (56)$$

Hence it is immediately seen that Equation (53) implies Equation (50). Consider the special case $\Psi_1 = \Psi_2 = \Psi_3$ and put $\widetilde{\Psi} := \Psi_1 = \Psi_2 = \Psi_3$. It follows from elementary calculations that $\Psi := \Psi_0 + \frac{i+j+k}{\sqrt{3}}\widetilde{\Psi}$ solves the quaternion Schrödinger Equation (53) if and only if $\Psi := \Psi_0 + i\widetilde{\Psi}$ solves the complex Schrödinger equation

$$\frac{1}{\sqrt{3}}i\hbar\frac{\partial \Psi}{\partial t} = -\frac{\hbar^2}{2m}\Delta\Psi + V(x)\Psi. \quad (57)$$

Remark 7. *The quaternion form of the time-dependent Schrödinger Equation (53) was given. In the very special diagonal case it was demonstrated that the quaternion and complex time dependent Schrödinger equations, (53) and (57), are equivalent in same sense. Moreover, it was shown that by suitable natural substitution, the time-dependent Schrödinger Equation (53) implies the quaternion stationary Schrödinger Equation (50).*

4. Summary

We presented the foundation of Quaternion Quantum Mechanics based on the Cauchy model of the elastic continuum. Cauchy model of an ideal elastic solid with the Helmholtz decomposition theorem and the quaternion algebra $\mathbb{Q}$ generates the transverse, longitudinal and multiple forms of waves. The quaternionic analog of vector formulation of the Cauchy model elucidates the coupling between the irrotational and solenoidal displacements in the deformation field (compression and torsion) and allows for a physical interpretation of the wave mechanics. The wave, i.e., the collective movement of the constituents forming the elastic Navier–Cauchy continuum, is considered equivalent to the particle. By combining the quaternion representation of the Cauchy model with the Planck–Kleinert crystal concept we presented the self-consistent formulation of the wave phenomena where the quantum space is regarded as an analog to the elastic solid. The presented Quaternion Quantum Mechanics follows from rigorous quaternion representation of the Cauchy linear theory of elasticity:

1. The quaternion form of the time-dependent Schrödinger Equation (53) was obtained and the special case when it solves complex Schrödinger equation was demonstrated. Thus, the origin of complex numbers in QM was explained.
2. The Klein–Gordon and Poisson equations were derived from assumptions that are independent of the postulates of quantum mechanics and prove the origin of the wave function. The problem of the indefinite probability of the density, present in the classical Klein–Gordon equation, is ruled out in its quaternion form.
3. All presented family of the quaternion wave functions is ontic, directly represent a state of elastic continuum showing properties of the Planck–Kleinert crystal.

The method used by us allows the self-consistent interpretation of the wave phenomena and yields the non-relativistic gravity field. It is obvious that it can be generalized and extended upon neglecting the assumptions of the small deformation limit that implies the constant density and the constant transport properties within the deformation field. The model can be falsified, see K. Popper [58]. Namely, the following hypotheses require formal verification:

1. Experimental verification to reveal superdeterminism, e.g., using the method proposed by Hossenfelder [59].
2. The experimental verification of the particle mass center, Equation (52).
3. The avoiding of the assumption of the small deformation limit implies the varying transverse wave velocity. The dependence of the energy density on deformation will result in:
    - the extended quaternion form of relation (29) and
    - the geometrical theory of gravitation when an invariant velocity of the transverse wave is avoided.
4. A reexamination of Schrödinger's charge density hypothesis. Namely, the Noether theorem can be used to decrease the order of the quaternion form of the Schrödinger equation [60].
5. The time-dependent Schrödinger requires the rigorous proof of the quaternion form of the Hamilton–Jacobi equation.

What is more, the results demonstrate that quaternions are much more comfortable than vectors, have huge advantages in the calculation of twist (and rotations) and can be regarded as the most concise representation of physical reality. Not only at the Planck scale, not only helpful and convenient, but quaternions also allow us to understand the processes in continua.

Author Contributions: Conceptualization, M.D.; methodology, M.D. and L.S.; formal analysis, M.D. and L.S.; mathematical investigation, L.S.; data curation, M.D.; writing—original draft preparation, M.D. and L.S.; writing—review and editing, M.D. and L.S.; funding acquisition, M.D. and L.S. All authors have read and agreed to the published version of the manuscript.

Acknowledgments: A remark by Hagen Kleinert that the complex potential was a key postulate in the 2007 paper has inspired this work. We are indebted to Bogusław Bożek for stimulating comments.

Appendix A

Let $\Omega \subset \mathbb{R}^3$ be an open bounded set with the smooth boundary $\partial\Omega$. Moreover, let $\alpha, \beta : \overline{\Omega} \to \mathbb{R}$ be given sufficiently regular functions, in a special case they can be constant. Define the real functional

$$F[\psi] = \int_\Omega [\alpha(x)\psi \cdot \psi^* + \beta(x)(D\psi) \cdot (D\psi)^*]\mathrm{d}x \tag{A1}$$

acting on the set $S = \left\{\psi : \overline{\Omega} \to H \text{ of the } C^2 \text{ class}\right\} \cap \{\psi = g \text{ on } \partial\Omega\}$, where g is a given function. The functional F can be written in the form

$$\begin{aligned} F[\psi] &= \int_\Omega \left[\alpha(x)\|\psi\|^2 + \beta(x)\left(\|\nabla\psi_0\|^2 + \|\mathrm{rot}\hat{\psi}\|^2\right) + 2\beta(x)\mathrm{rot}\hat{\psi} \circ \nabla\psi_0 + 2\beta(x)\left(\mathrm{div}\hat{\psi}\right)^2\right]\mathrm{d}x \\ &= \int_\Omega \Big\{\alpha(x)\left(\psi_0^2 + \psi_1^2 + \psi_2^2 + \psi_3^2\right) + \beta(x)\left[\left(\frac{\partial\psi_0}{\partial x_1}\right)^2 + \left(\frac{\partial\psi_0}{\partial x_2}\right)^2 + \left(\frac{\partial\psi_0}{\partial x_3}\right)^2\right] \\ &+\beta(x)\left[\left(\frac{\partial\psi_2}{\partial x_3} - \frac{\partial\psi_3}{\partial x_2}\right)^2 + \left(\frac{\partial\psi_3}{\partial x_1} - \frac{\partial\psi_1}{\partial x_3}\right)^2 + \left(\frac{\partial\psi_1}{\partial x_2} - \frac{\partial\psi_2}{\partial x_1}\right)^2\right] \\ &+2\beta(x)\left[\left(\frac{\partial\psi_2}{\partial x_3} - \frac{\partial\psi_3}{\partial x_2}\right)\frac{\partial\psi_0}{\partial x_1} + \left(\frac{\partial\psi_3}{\partial x_1} - \frac{\partial\psi_1}{\partial x_3}\right)\frac{\partial\psi_0}{\partial x_2} + \left(\frac{\partial\psi_1}{\partial x_2} - \frac{\partial\psi_2}{\partial x_1}\right)\frac{\partial\psi_0}{\partial x_3}\right] \\ &+\beta(x)\left(\frac{\partial\psi_1}{\partial x_1} + \frac{\partial\psi_2}{\partial x_2} + \frac{\partial\psi_3}{\partial x_3}\right)^2\Big\}\mathrm{d}x. \end{aligned}$$

Suppose that ψ minimizes F on S. We will show that ψ solves some differential equation. Let $\varphi : \overline{\Omega} \to H$ be any smooth function and $\varphi = 0$ on $\partial\Omega$, i.e., $\varphi \in C_0^\infty\left(\overline{\Omega}, H\right)$. Define a real function

$$f : \mathbb{R} \to \mathbb{R},\ f(\tau) = F[\psi + \tau\varphi] \tag{A2}$$

Obviously $f'(0) = 0$, because $\psi + \tau\varphi \in S$ and f has a minimum in $\tau = 0$.

We make the calculations:

$$\begin{aligned} f(\tau) = \ & \int_\Omega \Big\{\alpha(x)\left[\left(\psi_0 + \tau\varphi_0\right)^2 + \left(\psi_1 + \tau\varphi_1\right)^2 + \left(\psi_2 + \tau\varphi_2\right)^2 + \left(\psi_3 + \tau\varphi_3\right)^2\right] \\ & +\beta(x)\left[\left(\frac{\partial\psi_0}{\partial x_1} + \tau\frac{\partial\varphi_0}{\partial x_1}\right)^2 + \left(\frac{\partial\psi_0}{\partial x_2} + \tau\frac{\partial\varphi_0}{\partial x_2}\right)^2 + \left(\frac{\partial\psi_0}{\partial x_3} + \tau\frac{\partial\varphi_0}{\partial x_3}\right)^2\right] \\ & +\beta(x)\left[\left(\frac{\partial\psi_2}{\partial x_3} + \tau\frac{\partial\varphi_2}{\partial x_3} - \frac{\partial\psi_3}{\partial x_2} - \tau\frac{\partial\varphi_3}{\partial x_2}\right)^2\right. \\ & \left.+\left(\frac{\partial\psi_3}{\partial x_1} + \tau\frac{\partial\varphi_3}{\partial x_1} - \frac{\partial\psi_1}{\partial x_3} - \tau\frac{\partial\varphi_1}{\partial x_3}\right)^2 + \left(\frac{\partial\psi_1}{\partial x_2} + \tau\frac{\partial\varphi_1}{\partial x_2} - \frac{\partial\psi_2}{\partial x_1} - \tau\frac{\partial\varphi_2}{\partial x_1}\right)^2\right] \\ & +2\beta(x)\left[\left(\frac{\partial\psi_2}{\partial x_3} + \tau\frac{\partial\varphi_2}{\partial x_3} - \frac{\partial\psi_3}{\partial x_2} - \tau\frac{\partial\varphi_3}{\partial x_2}\right)\left(\frac{\partial\psi_0}{\partial x_1} + \tau\frac{\partial\varphi_0}{\partial x_1}\right)\right. \\ & +\left(\frac{\partial\psi_3}{\partial x_1} + \tau\frac{\partial\varphi_3}{\partial x_1} - \frac{\partial\psi_1}{\partial x_3} - \tau\frac{\partial\varphi_1}{\partial x_3}\right)\left(\frac{\partial\psi_0}{\partial x_2} + \tau\frac{\partial\varphi_0}{\partial x_2}\right) \\ & \left.+\left(\frac{\partial\psi_1}{\partial x_2} + \tau\frac{\partial\varphi_1}{\partial x_2} - \frac{\partial\psi_2}{\partial x_1} - \tau\frac{\partial\varphi_2}{\partial x_1}\right)\left(\frac{\partial\psi_0}{\partial x_3} + \tau\frac{\partial\varphi_0}{\partial x_3}\right)\right] \\ & +\beta(x)\left[\frac{\partial\psi_1}{\partial x_1} + \tau\frac{\partial\varphi_1}{\partial x_1} + \frac{\partial\psi_2}{\partial x_2} + \tau\frac{\partial\varphi_2}{\partial x_2} + \frac{\partial\psi_3}{\partial x_3} + \tau\frac{\partial\varphi_3}{\partial x_3}\right]^2\Big\}\mathrm{d}x, \end{aligned}$$

$$
\begin{aligned}
0 = f'(0) = \; & 2\int_\Omega \Big\{\alpha(x)\big(\psi_0\varphi_0+\psi_1\varphi_1+\psi_2\varphi_2+\psi_3\varphi_3\big)+\beta(x)\Big(\frac{\partial\psi_0}{\partial x_1}\frac{\partial\varphi_0}{\partial x_1}+\frac{\partial\psi_0}{\partial x_2}\frac{\partial\varphi_0}{\partial x_2}+\frac{\partial\psi_0}{\partial x_3}\frac{\partial\varphi_0}{\partial x_3}\Big)\\
&+\beta(x)\Big[\Big(\frac{\partial\psi_2}{\partial x_3}-\frac{\partial\psi_3}{\partial x_2}\Big)\Big(\frac{\partial\varphi_2}{\partial x_3}-\frac{\partial\varphi_3}{\partial x_2}\Big)+\Big(\frac{\partial\psi_3}{\partial x_1}-\frac{\partial\psi_1}{\partial x_3}\Big)\Big(\frac{\partial\varphi_3}{\partial x_1}-\frac{\partial\varphi_1}{\partial x_3}\Big)\\
&\quad+\Big(\frac{\partial\psi_1}{\partial x_2}-\frac{\partial\psi_2}{\partial x_1}\Big)\Big(\frac{\partial\varphi_1}{\partial x_2}-\frac{\partial\varphi_2}{\partial x_1}\Big)\Big]\\
&+\beta(x)\Big[\Big(\frac{\partial\varphi_2}{\partial x_3}-\frac{\partial\varphi_3}{\partial x_2}\Big)\frac{\partial\psi_0}{\partial x_1}+\Big(\frac{\partial\psi_2}{\partial x_3}-\frac{\partial\psi_3}{\partial x_2}\Big)\frac{\partial\varphi_0}{\partial x_1}+\Big(\frac{\partial\varphi_3}{\partial x_1}-\frac{\partial\varphi_1}{\partial x_3}\Big)\frac{\partial\psi_0}{\partial x_2}\\
&\quad+\Big(\frac{\partial\psi_3}{\partial x_1}-\frac{\partial\psi_1}{\partial x_3}\Big)\frac{\partial\varphi_0}{\partial x_2}+\Big(\frac{\partial\varphi_1}{\partial x_2}-\frac{\partial\varphi_2}{\partial x_1}\Big)\frac{\partial\psi_0}{\partial x_3}+\Big(\frac{\partial\psi_1}{\partial x_2}-\frac{\partial\psi_2}{\partial x_1}\Big)\frac{\partial\varphi_0}{\partial x_3}\Big]\\
&+\beta(x)\Big(\frac{\partial\psi_1}{\partial x_1}+\frac{\partial\psi_2}{\partial x_2}+\frac{\partial\psi_3}{\partial x_3}\Big)\Big(\frac{\partial\varphi_1}{\partial x_1}+\frac{\partial\varphi_2}{\partial x_2}+\frac{\partial\varphi_3}{\partial x_3}\Big)\Big\}\mathrm{d}x.
\end{aligned}
$$

After differentiation by parts we have

$$
\begin{aligned}
&\int_\Omega\Big\{\alpha(x)\big(\psi_0\varphi_0+\psi_1\varphi_1+\psi_2\varphi_2+\psi_3\varphi_3\big)-\beta(x)(\Delta\psi_0)\varphi_0+\beta(x)\Big\{\Big[\frac{\partial}{\partial x_3}\Big(\frac{\partial\psi_3}{\partial x_1}-\frac{\partial\psi_1}{\partial x_3}\Big)-\frac{\partial}{\partial x_2}\Big(\frac{\partial\psi_1}{\partial x_2}-\frac{\partial\psi_2}{\partial x_1}\Big)\Big]\varphi_1\\
&\quad+\Big[\frac{\partial}{\partial x_1}\Big(\frac{\partial\psi_1}{\partial x_2}-\frac{\partial\psi_2}{\partial x_1}\Big)-\frac{\partial}{\partial x_3}\Big(\frac{\partial\psi_2}{\partial x_3}-\frac{\partial\psi_3}{\partial x_2}\Big)\Big]\varphi_2+\Big[\frac{\partial}{\partial x_2}\Big(\frac{\partial\psi_2}{\partial x_3}-\frac{\partial\psi_3}{\partial x_2}\Big)-\frac{\partial}{\partial x_1}\Big(\frac{\partial\psi_3}{\partial x_1}-\frac{\partial\psi_1}{\partial x_3}\Big)\Big]\varphi_3\Big\}\\
&+\beta(x)\Big\{\Big[-\frac{\partial}{\partial x_1}\Big(\frac{\partial\psi_2}{\partial x_3}-\frac{\partial\psi_3}{\partial x_2}\Big)-\frac{\partial}{\partial x_2}\Big(\frac{\partial\psi_3}{\partial x_1}-\frac{\partial\psi_1}{\partial x_3}\Big)-\frac{\partial}{\partial x_3}\Big(\frac{\partial\psi_1}{\partial x_2}-\frac{\partial\psi_2}{\partial x_1}\Big)\Big]\varphi_0\\
&\quad+\Big[\Big(\frac{\partial^2\psi_0}{\partial x_3\partial x_2}-\frac{\partial^2\psi_0}{\partial x_2\partial x_3}\Big)\varphi_1+\Big(\frac{\partial^2\psi_0}{\partial x_1\partial x_3}-\frac{\partial^2\psi_0}{\partial x_3\partial x_1}\Big)\varphi_2+\Big(\frac{\partial^2\psi_0}{\partial x_2\partial x_1}-\frac{\partial^2\psi_0}{\partial x_1\partial x_2}\Big)\varphi_3\Big]\Big\}\\
&+\beta(x)\Big[-\Big(\frac{\partial}{\partial x_1}\mathrm{div}\hat{\psi}\Big)\varphi_1-\Big(\frac{\partial}{\partial x_2}\mathrm{div}\hat{\psi}\Big)\varphi_2-\Big(\frac{\partial}{\partial x_3}\mathrm{div}\hat{\psi}\Big)\varphi_3\Big]\Big\}\mathrm{d}x=0.
\end{aligned}
$$

The last equation takes the form

$$
\begin{aligned}
&\int_\Omega\Big\{\alpha(x)\big(\psi_0\varphi_0+\psi_1\varphi_1+\psi_2\varphi_2+\psi_3\varphi_3\big)-\beta(x)\big(\Delta\psi_0\big)\varphi_0\\
&+\beta(x)\Big(\Big[\frac{\partial}{\partial x_3}\Big(\frac{\partial\psi_3}{\partial x_1}-\frac{\partial\psi_1}{\partial x_3}\Big)-\frac{\partial}{\partial x_2}\Big(\frac{\partial\psi_1}{\partial x_2}-\frac{\partial\psi_2}{\partial x_1}\Big)-\frac{\partial}{\partial x_1}\mathrm{div}\hat{\psi}\Big]\varphi_1\\
&\quad+\Big[\frac{\partial}{\partial x_1}\Big(\frac{\partial\psi_1}{\partial x_2}-\frac{\partial\psi_2}{\partial x_1}\Big)-\frac{\partial}{\partial x_3}\Big(\frac{\partial\psi_2}{\partial x_3}-\frac{\partial\psi_3}{\partial x_2}\Big)-\frac{\partial}{\partial x_2}\mathrm{div}\hat{\psi}\Big]\varphi_2\\
&\quad+\Big[\frac{\partial}{\partial x_2}\Big(\frac{\partial\psi_2}{\partial x_3}-\frac{\partial\psi_3}{\partial x_2}\Big)-\frac{\partial}{\partial x_1}\Big(\frac{\partial\psi_3}{\partial x_1}-\frac{\partial\psi_1}{\partial x_3}\Big)-\frac{\partial}{\partial x_3}\mathrm{div}\hat{\psi}\Big]\varphi_3\Big)\Big\}\mathrm{d}x=0.
\end{aligned}
$$

Equivalently we can write

$$
\begin{aligned}
&\int_\Omega\{[\alpha(x)\psi_0-\beta(x)\Delta\psi_0]\varphi_0+[\alpha(x)\psi_1-\beta(x)\Delta\psi_1]\varphi_1\\
&\quad+[\alpha(x)\psi_2-\beta(x)\Delta\psi_2]\varphi_2+[\alpha(x)\psi_3-\beta(x)\Delta\psi_3]\varphi_3\}\mathrm{d}x=0
\end{aligned}
\tag{A3}
$$

for all $\varphi\in C_0^\infty(\overline{\Omega},H)$. The Du Bois Reymond variational lemma [61] used for (A3) implies

$$\alpha(x)\psi_l-\beta(x)\Delta\psi_l=0 \text{ for } l=0,1,2,3.$$

In consequence, ψ must be a solution of the differential equation

$$\alpha(x)\psi-\beta(x)\Delta\psi=0. \tag{A4}$$

References

1. Home, D. *Conceptual Foundations of Quantum Physics: An Overview from Modern Perspectives*; Plenum Press: New York, NY, USA, 1997.
2. Bohm, D.; Hiley, B.J. *The Undivided Universe: An Ontological Interpretation of Quantum Theory*; Routledge: Philadelphia, PA, USA, 1993; p. 397.
3. Bohm, D.A. Suggested Interpretation of the Quantum Theory in Terms of Hidden Variables. I and II. *Phys. Rev.* **1952**, *85*, 166–193; [CrossRef]
4. Bell, J.S. On the Einstein Podolsky Rosen Paradox. *Physics* **1964**, *1*, 195–200. [CrossRef]
5. Bell, J.S. *Schrödinger–Centenary Celebration of a Polymath*; Kilmister, C.W., Ed.; Cambridge University Press: Cambridge, UK, 1987; pp. 41–52.

6. Ghirardi, G.C. The Interpretation of Quantum Mechanics: Where do we stand? *J. Phys. Conf. Ser.* **2009**, *174*, 012013. [CrossRef]
7. Bell, J.S. Bertlmann's Socks and the Nature of Reality. *J. Phys. Colloq.* **1981**, *42*, C2-41–C2-62. [CrossRef]
8. Feynman, R.P. *The Character of Physical Law*, 2nd ed.; The MIT Press: Cambridge, MA, USA, 2017.
9. Gell-Mann, M. *The Nature of the Physical Universe*; Huff, D., Prewett, O., Eds.; John Wiley & Sons: New York, NY, USA, 1979; p. 29.
10. Weinberg, S. Letter on 24 September 1996. Available online: https://bohmian-mechanics.net/weingold.html (accessed on 6 November 2020).
11. Hagar, A. Experimental Metaphysics: The Double Standard in the Quantum-Information Approach to the Foundations of Quantum Theory. *Stud. Hist. Philos. Mod. Phys.* **2007**, *38*, 906–919. [CrossRef]
12. Chen, P.; Kleinert, H. Deficiencies of Bohm Trajectories in View of Basic Quantum Principles. *Electr. J. Theor. Phys.* **2016**, *13*, 1–12.
13. Birkhoff, G.; von Neumann, J. The Logic of Quantum Mechanics. *Ann. Math.* **1936**, *37*, 823–843. [CrossRef]
14. Yang, C.N. High Energy Nuclear Physics. In Proceedings of the Seventh Annual Rochester Conference, Rochester, NY, USA, 15–19 April 1957; Midwestern Universities Research Association, distributed by Interscience Pubulishing, Inc.: New York, NY, USA, 1957; p. IX-26.
15. Finkelstein, D.; Jauch, J.M.; Schiminovich, S.; Speiser, D. Foundations of Quaternion Quantum Mechanics. *J. Math. Phys.* **1962**, *3*, 207–220. [CrossRef]
16. Brumby, S.P.; Joshi, G.C. Global Effects in Quaternionic Quantum Field Theory. *Found. Phys.* **1996**, *26*, 1591–1599. [CrossRef]
17. Brumby, S.P.; Joshi, G.C. Experimental Status of Quaternionic Quantum Mechanics. *Chaos Solitons Fract.* **1996**, *7*, 747–752; [CrossRef]
18. Panda1, S.; Muni, M.K.; Vasundhara, B.; Tripathy, L.K. Study the Relations for Different Components of Isospin with Quark States. *Int. J. Pure Appl. Phys.* **2016**, *12*, 61–69.
19. Bodurov, T. Solitary Waves Interacting with an External Field. *Int. J. Theor. Phys.* **1996**, *35*, 2489–2499. [CrossRef]
20. Bodurov, T. Derivation of the Nonlinear Schrödinger Equation from First Principles. *Ann. Fond. Louis Broglie* **2005**, *30*, 343–352.
21. Białynicki-Birula, I.; Mycielski, I.J. Nonlinear Wave Mechanics. *Ann. Phys.* **1976**, *100*, 62–93. [CrossRef]
22. Weng, Z.-H. Field Equations in the Complex Quaternion Spaces. *Adv. Math. Phys.* **2014**, *201*, 450262. [CrossRef]
23. Horodecki, R. De Broglie Wave and its Dual Wave. *Phys. Lett.* **1981**, *87A*, 95–97. [CrossRef]
24. Horodecki, R. Superluminal Singular Dual Wave. *Lett. Novo Cimento* **1983**, *36*, 509–511. [CrossRef]
25. Gantner, J. On the Equivalence of Complex and Quaternionic Quantum Mechanics. *Quantum Stud. Math. Found.* **2018**, *5*, 357–390. [CrossRef]
26. Adler, S.L. Quaternionic Quantum Field Theory. *Phys. Rev. Let.* **1985**, *55*, 783–786. [CrossRef]
27. Adler, S.L. *Mechanics and Quantum Fields*; Oxford University Press: New York, NY, USA, 1995.
28. Adler, S.L. *Quantum Theory as an Emergent Phenomenon*; Cambridge University Press: Cambridge, UK, 2004.
29. Nottale, L. Origin of Complex and Quaternionic Wavefunctions in Quantum Mechanics: The Scale-Relativistic View. *Adv. Appl. Clifford Algebr.* **2008**, *18*, 917–944; [CrossRef]
30. Arbab, A.I.; Al Ajmi, M. The Quaternionic Commutator Bracket and its Implications. *Symmetry* **2018**, *10*, 513. [CrossRef]
31. Thompson, S.P. *The Life of Lord Kelvin, Baron Kelvin of Largs, Volume II*; Macmillan: New York, NY, USA, 1910; p. 1070.
32. Deb, S.; Das, S.; Vagenas, E.C. Discreteness of Space from GUP in a Weak Gravitational Field. *Phys. Lett. B* **2016**, *755*, 17–23. [CrossRef]
33. Ali, A.F.; Das, S.; Vagenas, E.C. Discreteness of Space from the Generalized Uncertainty Principle. *Phys. Let. B* **2009**, *678*, 497–499. [CrossRef]
34. Kleinert, H. Gravity as Theory of Defects in a Crystal with Only Second–Gradient Elasticity. *Ann. Phys.* **1987**, *44*, 117–119. [CrossRef]
35. Danielewski, M. The Planck–Kleinert Crystal. *Z. Naturforsch.* **2007**, *62a*, 564–568. [CrossRef]
36. Danielewski, M.; Sapa, L. Diffusion in Cauchy Elastic Solid. *Diffus. Fundam.* **2020**, *33*, 1–14.

37. Danielewski, M.; Sapa, L. Nonlinear Klein–Gordon equation in Cauchy–Navier elastic solid. *Cherkasy Univ. Bull. Phys. Math. Sci.* **2017**, *1*, 22–29.
38. Gürlebeck, K.; Sprößig, W. *Quaternionic Analysis and Elliptic Boundary Value Problems*; Akademie-Verlag: Berlin, Germany, 1989.
39. Available online: http://en.wikipedia.org/wiki/Quaternion (accessed on 6 November 2020).
40. Maxwell, J.C. Remarks on the Mathematical Classification of Physical Quantities. *Proc. London Math. Soc.* **1869**, *3*, 224–233. [CrossRef]
41. Cauchy, A.L. Récherches sur L'équilibre et le Movement Intérieur des Corps Solides ou Fluides, Élastiques ou non Élastiques. *Bull. Sot. Philomath.* **1823**, *9*, 300–304.
42. Cauchy, A.L. De la Pression ou Tension dans un Corps Solide. *Exerc. Math.* **1827**, *2*, 60–81.
43. Poisson, D. Mémoire sur L'équilibre et le Mouvement des Corps Élastiques. *Mém. Acad. Sci. Paris* **1829**, *8*, 357–570.
44. Neumann, F. *Vorlesungen über die Theorie der Elasticität der Festen Körper und des Lichtäthers*; B.G. Teubner: Leipzig, Germany, 1885.
45. Duhem, P. Sur l'intégrale des équations des petits mouvements d'un solide isotrope. *Mém. Soc. Sci. Bordeaux, Ser. V* **1898**, *3*, 316.
46. Love, A.E.H. *Mathematical Theory of Elasticity*, 4th ed.; Dover Publications Inc.: New York, NY, USA, 1944; p. 8.
47. Todhunter, I. *A History of the Theory of Elasticity and of the Strength of Materials*; Pearson, K., Ed.; Cambridge University Press: Cambridge, UK, 2014. [CrossRef]
48. Kravchenko, V.V. *Applied Quaternionic Analysis, Research and Exposition in Mathematics*; Heldermann: Berlin, Germany, 2003.
49. Landau, L.D.; Lifshitz, E.M. *Theory of Elasticity*, 3rd ed.; Butterworth-Heinemann Elsevier Ltd.: Amsterdam, Holland, 1986.
50. Hamilton, W.R. On Quaternions, or on a New System of Imaginaries in Algebra. *London Edinb. Dublin Phil. Mag. J. Sci.* **1844**, *25*, 10–13. [CrossRef]
51. Maxwell, J.C. A Dynamical Theory of the Electromagnetic Field. *Phil. Trans. R. Soc. London* **1865**, *155*, 459–512; [CrossRef]
52. National Institute of Standards and Technology. Available online: http://physics.nist.gov (accessed on 10 November 2018).
53. Ulrych, S. Higher Spin Quaternion Waves in the Klein-Gordon Theory. *Int. J. Theor. Phys.* **2013**, *52*, 279–292. [CrossRef]
54. Weinberg, S. *The Quantum Theory of Fields, Volume 1*; Cambridge University Press: Cambridge, UK, 1995; pp. 7–8.
55. Sapa, L. Existence, uniqueness and estimates of classical solutions to some evolutionary system. *Opuscula Math.* **2015**, *35*, 935–956. [CrossRef]
56. Krasnoholovets, V. Reasons for the Gravitational Mass and the Problem of Quantum Gravity. In *Ether, Spacetime and Cosmology*; Duffy, M.J., Levy, J., Krasnoholovets, V., Eds.; PD Publications: Liverpool, UK, 2008; Volume 1, pp. 419–450.
57. Bodurov, T. Generalized Ehrenfest Theorem for Nonlinear Schrödinger Equations. *Int. J. Theor. Phys.* **1988**, *37*, 1299–1306. [CrossRef]
58. Popper, K. *The Logic of Scientific Discovery*; Hutchinson & Co.: London, UK; Basic Books Inc.: London, UK, 1959.
59. Hossenfelder, S. Testing Superdeterministic Conspiracy. *J. Phys. Conf. Ser.* **2014**, *504*, 012018; [CrossRef]
60. Brown, H.R.; Holland, P. Simple Applications of Noether's First Theorem in Quantum Mechanics and Electromagnetism. *Am. J. Phys.* **2004**, *72*, 34–39; [CrossRef]
61. Zeidler, E. *Nonlinear Functional Analysis and Its Applications II/A: Linear Monotone Operators*; Springer: New York, NY, USA, 1990; p. 18.

3

Probabilistic Resumable Quantum Teleportation of a Two-Qubit Entangled State

Zhan-Yun Wang [1], Yi-Tao Gou [2], Jin-Xing Hou [3,4], Li-Ke Cao [2,4] and Xiao-Hui Wang [2,4,*]

1 School of Electronic Engineering, Xi'an University of Posts and Telecommunications, Xi'an 710121, China; zywang@xupt.edu.cn
2 School of Physics, Northwest University, Xi'an 710127, China; 2014112142@stumail.nwu.edu.cn (Y.-T.G.); nwuclk@126.com (L.-K.C.)
3 Institute of Modern Physics, Northwest University, Xi'an 710127, China; jinxhou@163.com
4 Shaanxi Key Laboratory for Theoretical Physics Frontiers, Xi'an 710127, China
* Correspondence: xhwang@nwu.edu.cn

Abstract: We explicitly present a generalized quantum teleportation of a two-qubit entangled state protocol, which uses two pairs of partially entangled particles as quantum channel. We verify that the optimal probability of successful teleportation is determined by the smallest superposition coefficient of these partially entangled particles. However, the two-qubit entangled state to be teleported will be destroyed if teleportation fails. To solve this problem, we show a more sophisticated probabilistic resumable quantum teleportation scheme of a two-qubit entangled state, where the state to be teleported can be recovered by the sender when teleportation fails. Thus the information of the unknown state is retained during the process. Accordingly, we can repeat the teleportion process as many times as one has available quantum channels. Therefore, the quantum channels with weak entanglement can also be used to teleport unknown two-qubit entangled states successfully with a high number of repetitions, and for channels with strong entanglement only a small number of repetitions are required to guarantee successful teleportation.

Keywords: quantum teleportation; two-qubit entangled state; partially entangled state; local unitary operation; controlled-U gate

---

1. Introduction

Quantum teleportation (QT) is one of the most astonishing applications of quantum mechanics. This operable concept was originally presented by Bennett et al. in 1993 [1]. In this protocol, the sender (Alice) and the receiver (Bob) prearrange the sharing of an Einstein-Podolsky-Rosen (EPR) [2] correlated pair of particles. Alice makes a joint measurement on her EPR particle and the unknown quantum system; she then sends Bob the classical result of her measurement. Finally, Bob can convert the state of his EPR particle into an exact replica of the unknown state belonging to Alice by means of local operations and classical communication (LOCC). QT has been realized experimentally [3–5] and due to its fresh notion and latent applied prospects in the realm of quantum communication, various kinds of QT have been widely studied both theoretically [6–9] and experimentally [10–12].

From above researches, we can learn that a maximally entangled state as the quantum channel and two classical bits are the key ingredients for the deterministic teleportation with fidelity 1 [13,14]. However, in realistic situation, instead of the pure maximally entangled states, Alice and Bob usually share a mixed entangled state or a partially entangled state due to the decoherence. Teleportation using a mixed state as an entangled resource is, in general, equivalent to having a noisy quantum channel. As a mixed state can't be purified to a Bell state [15–17], a quantum channel of mixed states could never provide a teleportation with fidelity 1 [18,19]. Therefore, only pure entangled pairs should be considered if we prefer an exact teleportation, even if it is probabilistic. Li et al. [20] put forward a partially entangled quantum channel to probabilistically teleport the quantum state of a single particle and extended this scheme to a multi-particle system. Much attention has been paid to this direction [21–26].

Recently, the field of entanglement has become an intense research area due to its key role in many applications of quantum information processing, such as precise measurement, quantum communication, quantum network and quantum repeater, etc. It is therefore an interesting question how we can teleport a pair of entangled particles. In 1999, Gorbachev et al. [27] proposed their protocol, teleportation of a two-qubit entangled state (TTES), by using a three-particle maximally entangled state of the type Greenberger-Horne-Zeilinger (GHZ). Two schemes for generalized TTES (GTTES) have been reported soon. In one of them, the protocol is assisted with a generalized three-particle entangled state as quantum channel, which is based on GHZ [28]. In the other probabilistic scheme, teleportation is completed with an entanglement swapping process [29], which is carried out via two pairs of partially entangled channels (TPEC). In their schemes, since the fidelity cannot reach 1, the maximal probabilities of exact teleportation was provided. According to the no-cloning theorem [30,31] and the irreversibility of quantum measurements [32,33], the information of the states to be teleported will be lost if the processes fail. Obviously, their destructive protocols do not offer the chance to repeat the process if TTES fails.

Roa et al. [34] presented a scheme for teleporting probabilistically an unknown pure state with optimal probability and without losing the information of the state to be teleported, and its advantage is that the unknown state is recovered by the sender when teleportation fails. This property offers the chance to repeat the teleportation process as many times as one has available quantum channels.

This paper is organized as follow. In Section 2, we present the probabilistic quantum teleportation of a two-qubit entangled state protocol, which uses two pairs of partially entangled particles as quantum channel. We verify that the optimal probability of successful teleportation is determined by the smallest superposition coefficient of these partially entangled particles. We introduce an optimal scheme, probabilistic resumable quantum teleportation of a two-qubit entangled state (RTTES) in Section 3, which is assisted with TPEC and has the advantage that the unknown entangled state can be recovered by the sender when the process fails. That is to say, if the sender and receiver have more than one partially entangled quantum channel, then the sender is able to teleport many times until RTTES is successful because the sender still have the unknown state undisturbed. In Section 4, we discuss the success probability of probabilistic TTES process. Finally, the conclusions are summarized in Section 5.

2. Probabilistic Teleportation of a Two-Qubit Entangled State

Suppose Alice has an arbitrary partially entangled pair, consisting of particles (1,2), which can be described as

$$|\phi\rangle_{12} = (x|00\rangle + y|11\rangle)_{12}, \tag{1}$$

with $|x|^2 + |y|^2 = 1$, where $\{|0\rangle, |1\rangle\}$ are the eigenstates of the Pauli operator $\sigma_z = |0\rangle\langle 0| - |1\rangle\langle 1|$. Now Alice would like to teleport the unknown state $|\phi\rangle_{12}$ to Bob. She sets up two distant entangled pure states as quantum channel between herself (particles 3 and 5) and Bob (particles 4 and 6), which located in the following states, respectively:

$$\begin{cases} |\psi\rangle_{34} = & (a|00\rangle + b|11\rangle)_{34}, \\ |\psi\rangle_{56} = & (c|00\rangle + d|11\rangle)_{56}, \end{cases} \tag{2}$$

with $|a| \geqslant |b|$, $|a|^2 + |b|^2 = 1$, $|c| \geqslant |d|$, $|c|^2 + |d|^2 = 1$. Note that when $|a| = |b|$ and $|c| = |d|$, the quantum channel is composed of two EPR pairs, and the deterministic teleportation can be achieved. This is the special case of our scheme. We demonstrate that, by using entanglement swapping [29], Alice can successfully transmit state $|\phi\rangle_{12}$ to Bob with certain probability.

The state of the whole system is given by

$$\begin{aligned} |\Psi\rangle_{123456} &= |\phi\rangle_{12} \otimes |\psi\rangle_{34} \otimes |\psi\rangle_{56} \\ &= [x(ac|000000\rangle + ad|000011\rangle + bc|001100\rangle + bd|001111\rangle) \\ &\quad + y(ac|110000\rangle + ad|110011\rangle + bc|111100\rangle + bd|111111\rangle)]_{123456}. \end{aligned} \tag{3}$$

Now Alice firstly performs local Bell-state measurements [35] on particles $(1,3)$ and particles $(2,5)$, respectively. The particles $(4,6)$ belong to Bob will be projected to the corresponding quantum state, i.e., the above strategy is provided for the receiver to extract the quantum information by adopting a proper evolution. There are four outcomes in each Bell-state measurement, ($|\Phi^{\pm}\rangle = 1/\sqrt{2}(|00\rangle \pm |11\rangle), |\Psi^{\pm}\rangle = 1/\sqrt{2}(|01\rangle \pm |10\rangle)$), so there are sixteen specific results in total.

For example, here we analyse the case that the results of measurement are $|\Phi^{+}\rangle_{13}$ and $|\Psi^{-}\rangle_{25}$, respectively. The particles $(4,6)$ are collapsed into the following state

$$\begin{aligned} |\Phi\rangle_{46} &= \langle\Phi^{+}_{13}|\langle\Psi^{-}_{25}|\Psi\rangle_{123456} \\ &= \tfrac{1}{\sqrt{2}}(\langle 00| + \langle 11|)_{13}\tfrac{1}{\sqrt{2}}(\langle 01| - \langle 10|)_{25}|\Psi\rangle_{123456} \\ &= \tfrac{1}{2}(xad|01\rangle - ybc|10\rangle)_{46}. \end{aligned} \tag{4}$$

Next, Alice informs Bob of the results by the classical communication and Bob performs a unitary operation $(|0\rangle\langle 0| + |1\rangle\langle 1|)_4 \otimes (|0\rangle\langle 1| - |1\rangle\langle 0|)_6$ on Equation (4). Then its state changes to

$$\tfrac{1}{2}(xad|00\rangle + ybc|11\rangle)_{46}. \tag{5}$$

Without loss of generality, if the superposition coefficients satisfy $|a| \geqslant |c| \geqslant |d| \geqslant |b|$, we have $|ac| \geqslant |bd|$, $|ad| \geqslant |bc|$. In order to carry out the proper evolution, we need to introduce an auxiliary particle to Bob which initial state is $|0\rangle_{aux}$, and operate the following controlled unitary transformation under the basis $\{|0\rangle_4|0\rangle_{aux}, |1\rangle_4|0\rangle_{aux}, |0\rangle_4|1\rangle_{aux}, |1\rangle_4|1\rangle_{aux}\}$. The unitary transformation is described as the following controlled-U gate

$$U_1 = |0\rangle\langle 0| \otimes \hat{U}_1 + |1\rangle\langle 1| \otimes I, \tag{6}$$

with $\hat{U}_1$ being a rotation in a $\pi/2$ angle around the $\hat{n}_1$ direction, specifically,

$$\hat{U}_1 = e^{-i\frac{\pi}{2}}e^{i\frac{\pi}{2}\hat{n}_1\cdot\sigma}, \quad \hat{n}_1 = \left(\sqrt{1-\left(\tfrac{bc}{ad}\right)^2}, 0, \tfrac{bc}{ad}\right), \tag{7}$$

and $\sigma = (\sigma_x, \sigma_y, \sigma_z)$. Note that the amplitudes a, b, c and d of the quantum channel have to be known in order to apply the above unitary transformation. The collective unitary transformation U_1 transforms the un-normalized state Equation (5) to the result

$$|\Phi\rangle_{46aux} = \tfrac{1}{2}bc(x|00\rangle + y|11\rangle)_{46}|0\rangle_{aux} + \tfrac{1}{2}ad\sqrt{1-\left(\tfrac{bc}{ad}\right)^2}x|00\rangle_{46}|1\rangle_{aux}, \tag{8}$$

which is also un-normalized. Then we perform a measurement on the auxiliary particle. If the result of measurement is $|1\rangle_{aux}$, we can see that the teleportation fails with the state of qubits $(4,6)$ transformed to the state $1/2ad\sqrt{(1-bc/ad)^2}x|00\rangle_{46}$ and no information regarding the initial state $|\phi\rangle_{12}$ is left. On the contrary, if the result of measurement is $|0\rangle_{aux}$, the state of particles $(4,6)$ collapses to an exact replica of the teleported state $|\phi\rangle_{12}$. The teleportation is successfully accessed. The contribution of this un-normalized state can be expressed by the probabilistic amplitude of $(x|00\rangle + y|11\rangle)_{46}$ in Equation (8) as $|(1/2) \times bc|^2 = (1/4) \times |bc|^2$.

Similarly, if Alice's measurement results are $|\Psi^+\rangle_{13}$ and $|\Psi^-\rangle_{25}$, the particles (4, 6) are collapsed into the following state

$$\begin{aligned} |\Phi\rangle_{46} &= \langle\Psi^+_{13}|\langle\Psi^-_{25}|\Psi\rangle_{123456} \\ &= \tfrac{1}{2}(xbd|11\rangle - yac|00\rangle)_{46}. \end{aligned} \tag{9}$$

Bob operates a unitary operation $(|0\rangle\langle 1| + |1\rangle\langle 0|)_4 \otimes (|0\rangle\langle 1| - |1\rangle\langle 0|)_6$ on Equation (9) and changes it to

$$\tfrac{1}{2}(xbd|00\rangle + yac|11\rangle)_{46}. \tag{10}$$

It should be noted that the unitary operation here is different from Equation (6):

$$U_2 = |0\rangle\langle 0| \otimes I + |1\rangle\langle 1| \otimes \hat{U}_2, \tag{11}$$

with $\hat{U}_2$ being a rotation in a $\pi/2$ angle around the $\hat{n}_2$ direction, specifically,

$$\hat{U}_2 = e^{-i\frac{\pi}{2}} e^{i\frac{\pi}{2}\hat{n}_2\cdot\sigma}, \quad \hat{n}_2 = \left(\sqrt{1 - \left(\tfrac{bd}{ac}\right)^2}, 0, \tfrac{bd}{ac}\right). \tag{12}$$

The state of particles $(4,6)$ reduces to

$$|\Phi\rangle_{46aux} = \tfrac{1}{2}bd(x|00\rangle + y|11\rangle)_{46}|0\rangle_{aux} + \tfrac{1}{2}ac\sqrt{1 - \left(\tfrac{bd}{ac}\right)^2}\, y|11\rangle_{46}|1\rangle_{aux}. \tag{13}$$

So, for Equation (13), the probability of successful teleportation is $(1/4) \times |bd|^2$. Other measuring results can be discussed in the same way. The whole scheme is shown in Figure 1. We list all sixteen kinds of results, and show the corresponding operations respectively in Table 1, where the unitary operations $U_1' = |0\rangle\langle 0| \otimes I + |1\rangle\langle 1| \otimes \hat{U}_1$ and $U_2' = |0\rangle\langle 0| \otimes \hat{U}_2 + |1\rangle\langle 1| \otimes I$. Synthesizing all cases, we obtain the total probability of successful teleportation being

$$P = \tfrac{1}{4}|bc|^2 \times 8 + \tfrac{1}{4}|bd|^2 \times 8 = 2|b|^2. \tag{14}$$

Table 1. The specific situation of the measurement results and the corresponding unitary operations in GTTES.

Alice's Measurement Results	Probability	Bob's Unitary Operation
$\|\Phi^+\rangle_{13}\|\Phi^\pm\rangle_{25}$	$\frac{1}{4}\|bd\|^2$	$[I_4 \otimes (\|0\rangle\langle 0\| \pm \|1\rangle\langle 1\|)_6] \cdot U_2'$
$\|\Phi^-\rangle_{13}\|\Phi^\pm\rangle_{25}$	$\frac{1}{4}\|bd\|^2$	$[I_4 \otimes (\|0\rangle\langle 0\| \mp \|1\rangle\langle 1\|)_6] \cdot U_2'$
$\|\Phi^+\rangle_{13}\|\Psi^\pm\rangle_{25}$	$\frac{1}{4}\|bc\|^2$	$[I_4 \otimes (\|0\rangle\langle 1\| \pm \|1\rangle\langle 0\|)_6] \cdot U_1$
$\|\Phi^-\rangle_{13}\|\Psi^\pm\rangle_{25}$	$\frac{1}{4}\|bc\|^2$	$[I_4 \otimes (\|0\rangle\langle 1\| \mp \|1\rangle\langle 0\|)_6] \cdot U_1$
$\|\Psi^+\rangle_{13}\|\Phi^\pm\rangle_{25}$	$\frac{1}{4}\|bc\|^2$	$[(\|0\rangle\langle 1\| \pm \|1\rangle\langle 0\|)_4 \otimes I_6] \cdot U_1'$
$\|\Psi^-\rangle_{13}\|\Phi^\pm\rangle_{25}$	$\frac{1}{4}\|bc\|^2$	$[(\|0\rangle\langle 1\| \mp \|1\rangle\langle 0\|)_4 \otimes I_6] \cdot U_1'$
$\|\Psi^+\rangle_{13}\|\Psi^\pm\rangle_{25}$	$\frac{1}{4}\|bd\|^2$	$[(\|0\rangle\langle 1\| + \|1\rangle\langle 0\|)_4 \otimes (\|0\rangle\langle 1\| \pm \|1\rangle\langle 0\|)_6] \cdot U_2$
$\|\Psi^-\rangle_{13}\|\Psi^\pm\rangle_{25}$	$\frac{1}{4}\|bd\|^2$	$[(\|0\rangle\langle 1\| + \|1\rangle\langle 0\|)_4 \otimes (\|0\rangle\langle 1\| \mp \|1\rangle\langle 0\|)_6] \cdot U_2$

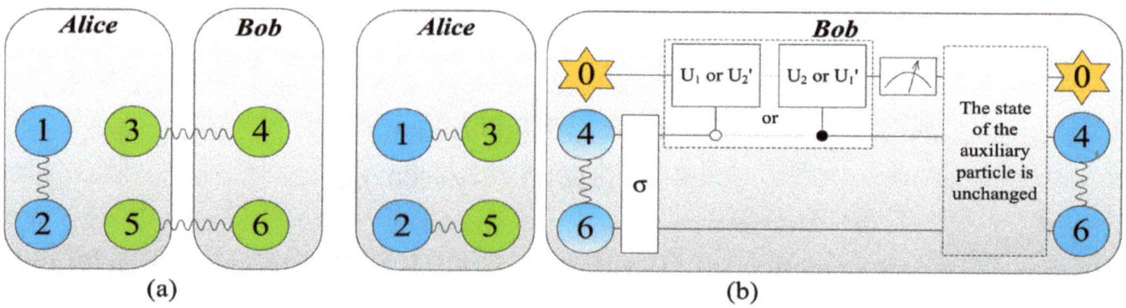

Figure 1. Colour online: blue represents the message to be teleported by Alice, and the information that needs to be extracted further is gradient. (**a**) System model for teleporting an arbitrary partially entangled pair consisting of particles (1, 2) by Alice. Particles (3, 4) and (5, 6) are partially entangled pairs. (**b**) By performing local Bell-state measurements on particles (1, 3) and (2, 5), the entanglement between particles (3, 4) and (5, 6) vanishes, while entanglement between particles (1, 3) and (2, 5) is built up. And then Bob extract the information from particles 4 and 6 via unitary operations and a von Neumann measurement. Eventually, the message has been teleported from Alice to Bob if and only if the state of auxiliary particle is unchanged after measurement.

3. Resumable Quantum Teleportation of a Two-Qubit Entangled Sstate

Now we consider an improved project of GTTES to teleport a two-qubit entangled state via weak entanglement quantum channels. Firstly, in order to implement the protocol it will be required to apply a series of joint unitary transformations known as the controlled-NOT gate [36,37].

$$U_{NOT}^{(ij)} = |0_i\rangle\langle 0_i| \otimes I_j + |1_i\rangle\langle 1_i| \otimes \sigma_x^{(j)}, \tag{15}$$

where I_j is the identity operator of target system j and i is the control system. We apply the $U_{NOT}^{(31)}$ and $U_{NOT}^{(52)}$ controlled-NOT gates onto the system, so the state (3) becomes

$$\begin{aligned} |\Gamma\rangle_{123456} &= U_{NOT}^{(31)} U_{NOT}^{(52)} |\Psi\rangle_{123456} \\ &= [x(ac|000000\rangle + ad|010011\rangle + bc|101100\rangle + bd|111111\rangle) \\ &\quad + y(ac|110000\rangle + ad|100011\rangle + bc|011100\rangle + bd|001111\rangle)]_{123456}. \end{aligned} \tag{16}$$

Now we introduce two extra auxiliary qubits m and n set into the state $|0\rangle_m$ and $|0\rangle_n$ to Alice. So Alice can apply the $U_{NOT}^{(1m)}$ and $U_{NOT}^{(2n)}$ gates. The process of taking the state $|\Gamma\rangle$ to the $|Y\rangle$ given by

$$\begin{aligned} |Y\rangle_{mn123456} &= U_{NOT}^{(1m)} U_{NOT}^{(2n)} |\Gamma\rangle_{mn123456} \\ &= [x(ac|00000000\rangle + ad|01010011\rangle + bc|10101100\rangle + bd|11111111\rangle) \\ &\quad + y(ac|11110000\rangle + ad|10100011\rangle + bc|01011100\rangle + bd|00001111\rangle)]_{mn123456}. \end{aligned} \tag{17}$$

To carry out our RTTES scheme, we shall apply the following controlled-U gate [38,39],

$$U^{(i,j)} = |00_i\rangle\langle 00_i| \otimes U_{00}^j + |01_i\rangle\langle 01_i| \otimes U_{01}^j + |10_i\rangle\langle 10_i| \otimes U_{10}^j + |11_i\rangle\langle 11_i| \otimes U_{11}^j, \tag{18}$$

where the superscript j means the target system (particles 1 and 2) and the subscript i is the control system (particles 3 and 5). We define that the unitary matrices of U_{00}, U_{01}, U_{10} and U_{11} below:

$$U_{00} = \begin{pmatrix} \frac{bd}{ac} & 0 & 0 & \sqrt{1-\frac{(bd)^2}{(ac)^2}} \\ 0 & 1 & 0 & 0 \\ 0 & 0 & 1 & 0 \\ \sqrt{1-\frac{(bd)^2}{(ac)^2}} & 0 & 0 & -\frac{bd}{ac} \end{pmatrix}, \tag{19}$$

$$U_{01} = \begin{pmatrix} 1 & 0 & 0 & 0 \\ 0 & \frac{bc}{ad} & \sqrt{1-\frac{(bc)^2}{(ad)^2}} & 0 \\ 0 & \sqrt{1-\frac{(bc)^2}{(ad)^2}} & -\frac{bc}{ad} & 0 \\ 0 & 0 & 0 & 1 \end{pmatrix}, \tag{20}$$

$$U_{10} = I, \tag{21}$$

$$U_{11} = I. \tag{22}$$

After applying the above $U^{(35,12)}$ gate, we can obtain the following state $|\Lambda\rangle$:

$$\begin{aligned} |\Lambda\rangle_{mn123456} &= U^{(35,12)}|\Upsilon\rangle_{mn123456} \\ &= x|00\rangle_{mn}(bd|00\rangle + \sqrt{(ac)^2-(bd)^2}|11\rangle)_{12}|0000\rangle_{3456} \\ &\quad +y|11\rangle_{mn}(-bd|11\rangle + \sqrt{(ac)^2-(bd)^2}|00\rangle)_{12}|0000\rangle_{3456} \\ &\quad +x|01\rangle_{mn}(bc|01\rangle + \sqrt{(ad)^2-(bd)^2}|10\rangle)_{12}|0011\rangle_{3456} \\ &\quad +y|10\rangle_{mn}(-bc|10\rangle + \sqrt{(ad)^2-(bd)^2}|01\rangle)_{12}|0011\rangle_{3456} \\ &\quad +x|10\rangle_{mn}bc|10\rangle_{12}|1100\rangle_{3456} \\ &\quad +y|01\rangle_{mn}bc|01\rangle_{12}|1100\rangle_{3456} \\ &\quad +x|11\rangle_{mn}bd|11\rangle_{12}|1111\rangle_{3456} \\ &\quad +y|00\rangle_{mn}bd|00\rangle_{12}|1111\rangle_{3456}. \end{aligned} \tag{23}$$

Next, we apply the $U_{NOT}^{(1m)}$ and $U_{NOT}^{(2n)}$ again, so the state (23) becomes

$$\begin{aligned} |\Pi\rangle_{mn123456} &= U_{NOT}^{(1m)}U_{NOT}^{(2n)}|\Lambda\rangle_{mn123456} \\ &= |11\rangle_{mn}[|0000\rangle_{3456}\sqrt{(ac)^2-(bd)^2}(x|11\rangle + y|00\rangle)_{12} \\ &\quad +|0011\rangle_{3456}\sqrt{(ad)^2-(bc)^2}(x|10\rangle + y|01\rangle)_{12}] \\ &\quad +|00\rangle_{mn}[bdx|00\rangle_{12}|0000\rangle_{3456} + bcx|01\rangle_{12}|0011\rangle_{3456} \\ &\quad +bcx|10\rangle_{12}|1100\rangle_{3456} + bdx|11\rangle_{12}|1111\rangle_{3456} \\ &\quad -bdy|11\rangle_{12}|0000\rangle_{3456} - bcy|10\rangle_{12}|0011\rangle_{3456} \\ &\quad +bcy|01\rangle_{12}|1100\rangle_{3456} + bdy|00\rangle_{12}|1111\rangle_{3456}]. \end{aligned} \tag{24}$$

In the following, we will analyse the different measurement results of the above final state.

If the qubits (m, n) are projected to $|11\rangle_{mn}$, after Alice performs a Bell-state measurement on her joint system consisting of particles (3, 5), she can recover the teleported entangled particle pair $|\phi\rangle_{12}$ of qubits (1, 2) by means of local operations. The detailed situations are summarized in Table 2, i.e., for the outcome $|11\rangle_{mn}$, the teleportation fails but the process performs the projection $|\phi\rangle_{12} \rightarrow |\phi\rangle_{12}$ on itself.

Table 2. The specific outcomes of measurement and the corresponding unitary operations.

Alice' Result	Probability	Operation
$\|00\rangle_{35}$	$\|(ac)^2) - (bd)^2\|$	$(\|0\rangle\langle 1\| + \|1\rangle\langle 0\|)_1 \otimes (\|0\rangle\langle 1\| + \|1\rangle\langle 0\|)_2$
$\|01\rangle_{35}$	$\|(ad)^2 - (bc)^2\|$	$(\|0\rangle\langle 1\| + \|1\rangle\langle 0\|)_1 \otimes (\|0\rangle\langle 0\| + \|1\rangle\langle 1\|)_2$

On the contrary, if the qubits (m, n) are projected to $|00\rangle_{mn}$, we can write out the system of residual particles as

$$\begin{aligned} |\Pi\rangle_{123456} &= bdx|00\rangle_{12}|0000\rangle_{3456} + bcx|01\rangle_{12}|0011\rangle_{3456} \\ &\quad + bcx|10\rangle_{12}|1100\rangle_{3456} + bdx|11\rangle_{12}|1111\rangle_{3456} \\ &\quad - bdy|11\rangle_{12}|0000\rangle_{3456} - bcy|10\rangle_{12}|0011\rangle_{3456} \\ &\quad + bcy|01\rangle_{12}|1100\rangle_{3456} + bdy|00\rangle_{12}|1111\rangle_{3456}. \end{aligned} \tag{25}$$

Then, we apply the controlled-NOT gates $U_{NOT}^{(31)}$ and $U_{NOT}^{(52)}$ on the system consisting of the residual particles, and the state $|\Pi\rangle$ becomes

$$\begin{aligned} |\Omega\rangle_{123456} &= U_{NOT}^{(31)} U_{NOT}^{(52)} |\Pi\rangle_{123456} \\ &= bdx|00\rangle_{12}|0000\rangle_{3456} + bcx|00\rangle_{12}|0011\rangle_{3456} \\ &\quad + bcx|00\rangle_{12}|1100\rangle_{3456} + bdx|00\rangle_{12}|1111\rangle_{3456} \\ &\quad - bdy|11\rangle_{12}|0000\rangle_{3456} - bcy|11\rangle_{12}|0011\rangle_{3456} \\ &\quad + bcy|11\rangle_{12}|1100\rangle_{3456} + bdy|11\rangle_{12}|1111\rangle_{3456}. \end{aligned} \tag{26}$$

Then Alice carry out the joint Bell-state measurements on the systems consisting of particles $(1, 3)$ and $(2, 5)$. For instance, here we assume that the results of the measurements are $|\Phi^+\rangle_{13}$ and $|\Psi^-\rangle_{25}$. The particles $(4, 6)$ will be projected to

$$\begin{aligned} |\Phi\rangle_{46} &= \langle\Phi_{13}^+|\langle\Psi_{25}^-|\Omega\rangle_{123456} \\ &= \tfrac{1}{\sqrt{2}}(\langle 00| + \langle 11|)_{13} \tfrac{1}{\sqrt{2}}(\langle 01| - \langle 10|)_{25}|\Omega\rangle_{123456} \\ &= \tfrac{1}{2} bc(x|01\rangle - y|10\rangle)_{46}. \end{aligned} \tag{27}$$

We can see that the projection of qubits (m, n) to $|00\rangle_{mn}$ allows one to achieve the RTTES process by means of LOCC. We list all 16 kinds of results and the corresponding operations in Table 3. The whole scheme is shown in Figure 2.

Table 3. The specific situation of the measurement results and the corresponding unitary operations in RTTES.

Alice's Measurement Results	the State of Particles (4, 6)	Bob's Unitary Operation
$\|\Phi^+\rangle_{13}\|\Phi^\pm\rangle_{25}$	$\frac{1}{2}bd(x\|00\rangle \pm y\|11\rangle)_{46}$	$I_4 \otimes (\|0\rangle\langle 0\| \pm \|1\rangle\langle 1\|)_6$
$\|\Phi^+\rangle_{13}\|\Psi^\pm\rangle_{25}$	$\frac{1}{2}bc(x\|01\rangle \pm y\|10\rangle)_{46}$	$I_4 \otimes (\|0\rangle\langle 1\| \pm \|1\rangle\langle 0\|)_6$
$\|\Phi^-\rangle_{13}\|\Phi^\pm\rangle_{25}$	$\frac{1}{2}bd(x\|00\rangle \mp y\|11\rangle)_{46}$	$I_4 \otimes (\|0\rangle\langle 0\| \mp \|1\rangle\langle 1\|)_6$
$\|\Phi^-\rangle_{13}\|\Psi^\pm\rangle_{25}$	$\frac{1}{2}bc(x\|01\rangle \mp y\|10\rangle)_{46}$	$I_4 \otimes (\|0\rangle\langle 1\| \mp \|1\rangle\langle 0\|)_6$
$\|\Psi^+\rangle_{13}\|\Phi^\pm\rangle_{25}$	$\frac{1}{2}bc(x\|10\rangle \mp y\|01\rangle)_{46}$	$(\|0\rangle\langle 1\| \mp \|1\rangle\langle 0\|)_4 \otimes I_6$
$\|\Psi^-\rangle_{13}\|\Phi^\pm\rangle_{25}$	$\frac{1}{2}bc(x\|10\rangle \pm y\|01\rangle)_{46}$	$(\|0\rangle\langle 1\| \pm \|1\rangle\langle 0\|)_4 \otimes I_6$
$\|\Psi^+\rangle_{13}\|\Psi^\pm\rangle_{25}$	$\frac{1}{2}bd(x\|11\rangle \mp y\|00\rangle)_{46}$	$(\|0\rangle\langle 1\| \mp \|1\rangle\langle 0\|)_4 \otimes (\|0\rangle\langle 1\| + \|1\rangle\langle 0\|)_6$
$\|\Psi^-\rangle_{13}\|\Psi^\pm\rangle_{25}$	$\frac{1}{2}bd(x\|11\rangle \pm y\|00\rangle)_{46}$	$(\|0\rangle\langle 1\| \pm \|1\rangle\langle 0\|)_4 \otimes (\|0\rangle\langle 1\| + \|1\rangle\langle 0\|)_6$

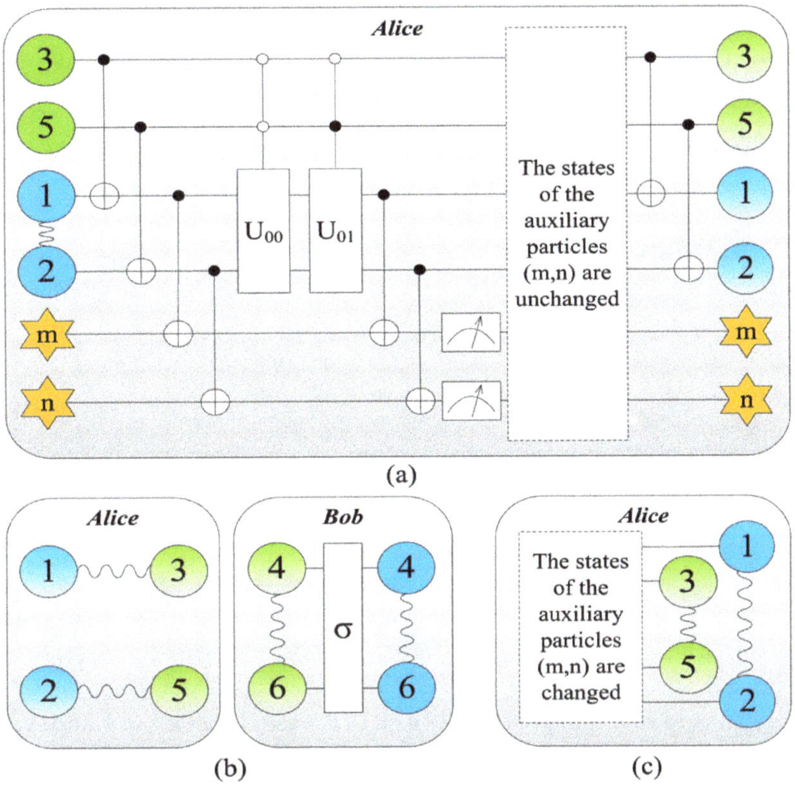

Figure 2. Colour online: blue represents the message to be teleported by Alice, and the information that needs to be extracted further is gradient. The initial state of the system is the same as that of (a) in Figure 1. (**a**) Alice introduces two extra auxiliary qubits m and n with the states $|0\rangle_m$ and $|0\rangle_n$ respectively to extract information stored in qubits 1 and 2. A series of joint unitary transformations known as the controlled-NOT gate and the controlled-U gate are performed by Alice. (**b**) Alice carry out the joint Bell-state measurements on the systems consisting of particles $(1,3)$ and $(2,5)$. The entanglement between particles $(3,4)$ and $(5,6)$ vanishes, while entanglement between particles $(1,3)$ and $(2,5)$ is built up. And then Bob extract the information from particles 4 and 6 via unitary operations and a von Neumann measurement. Eventually, the message has been teleported from Alice to Bob if and only if the states of auxiliary particles are unchanged after measurement. (**c**) The states of the auxiliary particles (m,n) are changed, the teleportation fails. Alice performs a Bell-state measurement on her joint system consisting of particles $(3,5)$, she can recover the teleported entangled particle pair $|\phi\rangle_{12}$ of qubits $(1,2)$ by means of local operations.

4. Discussion

We learn that the teleported two-qubit entangled state $|\phi\rangle_{12}$ can be recovered if RTTES process fails. Thus we can repeat a single RTTES process as many times as one has available quantum channels. And if $a = b = 1/\sqrt{2}$ and $c = d = 1/\sqrt{2}$ in Equation (2), the quantum channel reduces to maximally entangled states and hence the total probability equals 1 [27]. On the other hand, it is obvious that the RTTES is successful with probability $2\,|b|^2$, which is the same as Equation (14) in Section 2. Therefore, our RTTES scheme does not increase the probability of success by a single experiment, but provides a chance to repeat the RTTES process many times until this process is successful. It can be regarded as a Bernoulli experiment. After realizing N tries, the probability of having k successful events is in the form of binomial distribution $P_{N,k} = \binom{N}{k}[2b^2]^k[1-2b^2]^{N-k}$. Thus, we obtain the total probability of success as follows

$$P = \sum_{k=1}^{N} P_{N,k} = 1 - [1-2b^2]^N. \tag{28}$$

The coefficient b in Equation (2) represents the degree of entanglement of the quantum channel. When the quantum channel is a maximum entanglement state, i.e., $a = b = 1/\sqrt{2}$, $c = d = 1/\sqrt{2}$, see

Figure 3b, the RTTES process becomes deterministic, which is consistent with the result in Ref. [29]. Note that the success probability of probabilistic RTTES process will increase significantly for many tries. These results are summarized in Figure 3.

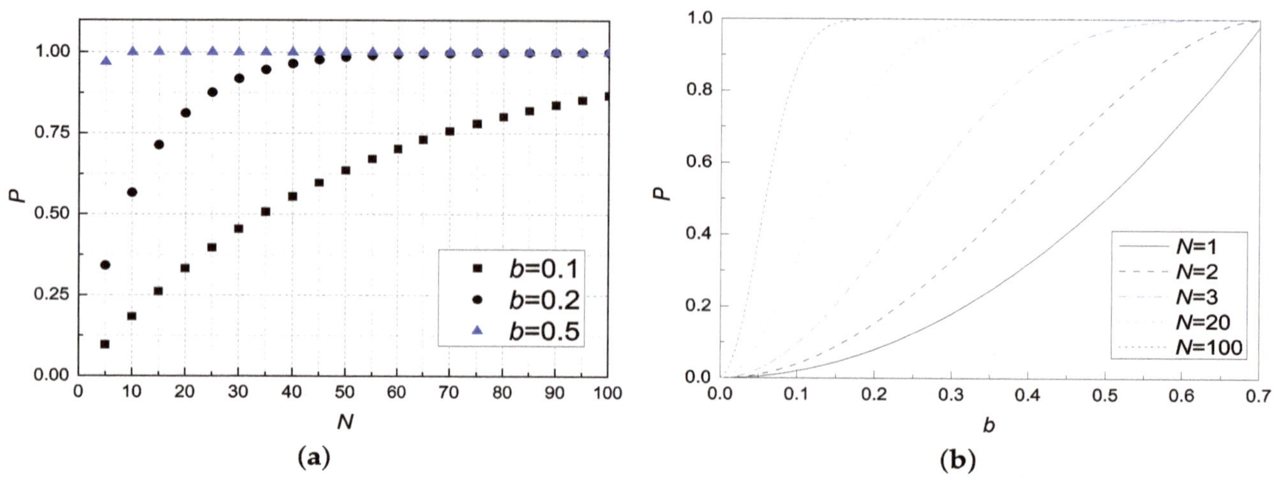

Figure 3. Colour online: total success probability of RTTES, as a function of b and N, here N means we performed RTTES N tries. (**a**) Total success probability of RTTES, as a function of the N for different values of b. The larger the coefficient b, the stronger the entanglement of quantum channel and the greater the probability of successful teleportation. (**b**) Total success probability of RTTES, as a function of b for different values of N. The probability increases significantly for N higher than 1. We can draw the conclusion that the quantum channels with weak entanglement ($b = 0.1$ or 0.2) can also be used to teleport successfully with a high number of repetitions, and for channels with strong entanglement ($b = 0.5$) only a small number of repetitions are required to guarantee successful teleportation.

5. Conclusions

In summary, we have proposed two teleportation schemes, the generalized probabilistic teleportation of a two-qubit entangled state (GTTES) and the probabilistic resumable teleportation of a two-qubit entangled state (RTTES), which use partially entangled pairs as quantum channel. In the standard deterministic protocol, a maximally entangled quantum channel is necessary for the success of teleportation. In real world, however, it is well known that the coupling between quantum systems and surrounding environment is inevitable [40], e.g., different kinds of decoherence, dephasing, and dissipation mechanisms reduce purity and entanglement of the channel. Therefore, sender and receiver may not shared a maximally entangled state but a partially entangled state. For this reason, GTTES is more general and practical. The differences between GTTES and TTES are that GTTES introduces an auxiliary particle, and need to perform local unitary operations before Bell-state measurements. We show that the optimal success probability of GTTES is only dependent on the smallest superposition coefficient of the partially entangled quantum channels. In other words, the success probability of GTTES cannot reach to 1. If GTTES fails, the state to be teleported will be destroyed. In addition, taking into account that an unknown state cannot be cloned, the above GTTES protocol do not offer the chance to repeat the process if GTTES fails. An improved scheme of GTTES (RTTES) is proposed. The advantage of this approach is that we are able to try repeatedly until the RTTES is successful. It is conformed to Bernoulli experiment, and total success probability of teleportation increases significantly by attempting many times. Finally, weak entanglement can be used to teleport a two-qubit entangled state effectively via RTTES. Our research also provides insights into the role of entanglement in quantum teleportation that it can be regarded as a key resource.

Author Contributions: Z.-Y.W., Y.-T.G., J.-X.H. and X.-H.W. initiated the research project and established the main results. L.-K.C. joined some discussions and provided suggestions. Z.-Y.W. and Y.-T.G. wrote the manuscript with advice from X.-H.W. and L.-K.C.

References

1. Bennett, C.H.; Brassard, G.; Crepeau, C.; Jozsa, R.; Peres, A.; Wootters, W.K. Teleporting an unknown quantum state via dual classical and Einstein-Podolsky-Rosen channels. *Phys. Rev. Lett.* **1993**, *70*, 1895–1899. [CrossRef] [PubMed]
2. Einstein, A.; Podolsky, B.; Rosen, N. Can quantum-mechanical description of physical reality be considered complete? *Phys. Rev.* **1935**, *47*, 777. [CrossRef]
3. Bouwmeester, D.; Pan, J.W.; Mattle, K.; Eibl, M.; Weinfurter, H.; Zeilinger, A. Experimental quantum teleportation. *Nature* **1997**, *390*, 575–579. [CrossRef]
4. Boschi, D.; Branca, S.; De Martini, F.; Hardy, L.; Popescu, S. Experimental realization of teleporting an unknown pure quantum state via dual classical and Einstein-Podolsky-Rosen channels. *Phys. Rev. Lett.* **1998**, *80*, 1121–1125. [CrossRef]
5. Furusawa, A.; Sørensen, J.L.; Braunstein, S.L.; Fuchs, C.A.; Kimble, H.J.; Polzik, E.S. Unconditional quantum teleportation. *Science* **1998**, *282*, 706–709. [CrossRef] [PubMed]
6. Banaszek, K. Optimal quantum teleportation with an arbitrary pure state. *Phys. Rev. A* **2000**, *62*, 024301. [CrossRef]
7. Yu, K.F.; Yang, C.W.; Liao, C.H.; Hwang, T. Authenticated semi-quantum key distribution protocol using Bell states. *Quantum Inf. Process.* **2014**, *13*, 1457–1465. [CrossRef]
8. Deng, F.G.; Li, C.Y.; Li, Y.S.; Zhou, H.Y.; Wang, Y. Symmetric multiparty-controlled teleportation of an arbitrary two-particle entanglement. *Phys. Rev. A* **2005**, *72*, 022338. [CrossRef]
9. Gottesman, D.; Chuang, I.L. Demonstrating the viability of universal quantum computation using teleportation and single-qubit operations. *Nature* **1999**, *402*, 390–393. [CrossRef]
10. Thapliyal, K.; Verma, A.; Pathak, A. A general method for selecting quantum channel for bidirectional controlled state teleportation and other schemes of controlled quantum communication. *Quantum Inf. Process.* **2015**, *14*, 4601–4614. [CrossRef]
11. Takeda, S.; Mizuta, T.; Fuwa, M.; van Loock, P.; Furusawa, A. Deterministic quantum teleportation of photonic quantum bits by a hybrid technique. *Nature* **2013**, *500*, 315–318. [CrossRef]
12. Kim, Y.H.; Kulik, S.P.; Shih, Y. Quantum teleportation of a polarization state with a complete Bell state measurement. *Phys. Rev. Lett.* **2001**, *86*, 1370. [CrossRef]
13. Prakash, H.; Verma, V. Minimum assured fidelity and minimum average fidelity in quantum teleportation of single qubit using non-maximally entangled states. *Quantum Inf. Process.* **2012**, *11*, 1951–1959. [CrossRef]
14. Li, G.; Ye, M.Y.; Lin, X.M. Entanglement fidelity of the standard quantum teleportation channel. *Phys. Lett. A* **2013**, *377*, 1531–1533. [CrossRef]
15. Linden, N.; Massar, S.; Popescu, S. Purifying noisy entanglement requires collective measurements. *Phys. Rev. Lett.* **1998**, *81*, 3279. [CrossRef]
16. Werner, R.F. Quantum states with Einstein-Podolsky-Rosen correlations admitting a hidden-variable model. *Phys. Rev. A* **1989**, *40*, 4277. [CrossRef]
17. Bennett, C.H.; Brassard, G.; Popescu, S.; Schumacher, B.; Smolin, J.A.; Wootters, W.K. Purification of noisy entanglement and faithful teleportation via noisy channels. *Phys. Rev. Lett.* **1996**, *76*, 722. [CrossRef]
18. Bowen, G.; Bose, S. Teleportation as a Depolarizing Quantum Channel, Relative Entropy, and Classical Capacity. *Phys. Rev. Lett.* **2001**, *87*, 267901–267904. [CrossRef]
19. Albeverio, S.; Fei, S.M.; Yang, W.L. Optimal teleportation based on bell measurements. *Phys. Rev. A* **2002**, *66*, 012301. [CrossRef]
20. Li, W.L.; Li, C.F.; Guo, G.C. Probabilistic teleportation and entanglement matching. *Phys. Rev. A* **2000**, *61*, 034301. [CrossRef]
21. Li, D.C.; Shi, Z.K. Probabilistic teleportation via entanglement. *Int. J. Theor. Phys.* **2008**, *47*, 2645–2654. [CrossRef]
22. Agrawal, P.; Pati, A.K. Probabilistic quantum teleportation. *Phys. Lett. A* **2002**, *305*, 12–17. [CrossRef]
23. An, N.B. Probabilistic teleportation of an M-qubit state by a single non-maximally entangled qubit-pair. *Phys. Lett. A* **2008**, *372*, 3778–3783. [CrossRef]

24. Gou, Y.T.; Shi, H.L.; Wang, X.H.; Liu, S.Y. Probabilistic resumable bidirectional quantum teleportation. *Quantum Inf. Process.* **2017**, *16*, 1–13. [CrossRef]
25. Liu, D.S.; Huang, Z.P.; Guo, X.J. Probabilistic Teleportation via Quantum Channel with Partial Information. *Entropy* **2015**, *17*, 3621–3630. [CrossRef]
26. Wang, K.; Yu, X.T.; Cai, X.F.; Zhang, Z.C. Probabilistic Teleportation of Arbitrary Two-Qubit Quantum State via Non-Symmetric Quantum Channel. *Entropy* **2018**, *20*, 238. [CrossRef]
27. Gorbachev, V.N.; Trubilko, A.I. Quantum teleportation of an Einstein-Podolsy-Rosen pair using an entangled three-particle state. *J. Exp. Theor. Phys.* **2000**, *91*, 894–898. [CrossRef]
28. Shi, B.S.; Jiang, Y.K.; Guo, G.C. Probabilistic teleportation of two-particle entangled state. *Phys. Lett. A* **2000**, *268*, 161–164. [CrossRef]
29. Lu, H.; Guo, G.C. Teleportation of a two-particle entangled state via entanglement swapping. *Phys. Lett. A* **2000**, *276*, 209–212. [CrossRef]
30. Wootters, W.K.; Zurek, W.H. A single quantum cannot be cloned. *Nature* **1982**, *299*, 802–803. [CrossRef]
31. Bužek, V.; Hillery, M. Universal optimal cloning of arbitrary quantum states: from qubits to quantum registers. *Phys. Rev. Lett.* **1998**, *81*, 5003. [CrossRef]
32. Brune, M.; Hagley, E.; Dreyer, J.; Maître, X.; Maali, A.; Wunderlich, C.; Raimond, J.M.; Haroche, S. Observing the progressive decoherence of the meter in a quantum measurement. *Phys. Rev. Lett.* **1996**, *77*, 4887. [CrossRef]
33. Ohya, M. Some aspects of quantum information theory and their applications to irreversible processes. *Rep. Math. Phys.* **1989**, *27*, 19–47. [CrossRef]
34. Roa, L.; Groiseau, C. Probabilistic teleportation without loss of information. *Phys. Rev. A* **2015**, *91*, 012344. [CrossRef]
35. Lütkenhaus, N.; Calsamiglia, J.; Suominen, K.A. Bell measurements for teleportation. *Phys. Rev. A* **1999**, *59*, 3295.
36. Bai, C.H.; Wang, D.Y.; Hu, S.; Cui, W.X.; Jiang, X.X.; Wang, H.F. Scheme for implementing multitarget qubit controlled-NOT gate of photons and controlled-phase gate of electron spins via quantum dot-microcavity coupled system. *Quantum Inf. Process.* **2016**, *15*, 1485–1498. [CrossRef]
37. De Martini, F.; Bužek, V.; Sciarrino, F.; Sias, C. Experimental realization of the quantum universal NOT gate. *Nature* **2002**, *419*, 815–818. [CrossRef] [PubMed]
38. DiVincenzo, D.P. Two-bit gates are universal for quantum computation. *Phys. Rev. A* **1995**, *51*, 1015. [CrossRef]
39. Sleator, T.; Weinfurter, H. Realizable universal quantum logic gates. *Phys. Rev. Lett.* **1995**, *74*, 4087.
40. Schlosshauer, M. Decoherence, the measurement problem, and interpretations of quantum mechanics. *Rev. Mod. Phys.* **2005**, *76*, 1267. [CrossRef]

4

Some Consequences of the Thermodynamic Cost of System Identification

Chris Fields

Independent Researcher, 23 rue des Lavandières, 11160 Caunes Minervois, France; fieldsres@gmail.com;

Abstract: The concept of a "system" is foundational to physics, but the question of how observers identify systems is seldom addressed. Classical thermodynamics restricts observers to finite, finite-resolution observations with which to identify the systems on which "pointer state" measurements are to be made. It is shown that system identification is at best approximate, even in a finite world, and that violations of the Leggett–Garg and Bell/CHSH (Clauser-Horne-Shimony-Holt) inequalities emerge naturally as requirements for successful system identification.

Keywords: Bell/CHSH inequality; coarse-graining; decoherence; Leggett–Garg inequality; LOCC protocol; observable; predictability sieve; system identification; thermodynamics

---

1. Introduction

The idea that all finite observers are characterized by uncertainty and must pay, in energetic currency, to reduce their uncertainty was introduced into classical physics by Boltzmann [1]. Shannon [2] showed that information obtained from observations can be naturally quantized into answers to yes/no questions and hence measured in bits. Landauer [3,4] then showed that such information has been "obtained" and is available for future use only after it has been irreversibly recorded on some physical medium. The resulting classical theory of *observation*—the exchange of energy for information—states that, for any finite, physically implemented observer O, each bit of irreversibly recorded uncertainty reduction (equivalently, each bit of information gain) costs $c^{(O)}k_B T$, where k_B is Boltzmann's constant, T is temperature, and $c^{(O)} \geq \ln 2$, is a measure of O's information-aquisition efficiency that can for simplicity be considered constant. As all classical observations in practice take place at $T > 0$, this energetic cost is always positive. This classical theory of observation has two familiar practical consequences: observations are limited to finite resolution and records of their outcomes to finite bit strings, and only some finite number of such finite-resolution observations can be made in any finite time.

The consequences of this classical, thermodynamic limitation to finite, finite-resolution observations have been investigated in both classical and quantum settings, particularly as they bear on issues of noise (i.e., uncontrolled degrees of freedom) and measurement uncertainty. It has been known since the pioneering work of Spekkens and colleagues [5,6], for example, that classical statistical mechanics reproduces wave-packet quantum theory in the special case in which wave packets are Gaussian. Jennings and Leifer [7] review this and other work, showing that classical statistical mechanics reproduces "quantum" features and behavior including the uncertainty principle, non-commutativity of measurements, state teleportation and the no-cloning theorem when a finite-resource restriction limiting the number and resolution of measurements is imposed. Krechmer [8] shows that "quantum" measurement disturbance and non-commutativity of observables result whenever two measurement devices are calibrated using the same physical standard.

My aim here is to investigate a different set of consequences of the classical thermodynamic restriction to finite energy resources and hence to finite, finite-resolution observations: its consequences for an observer's ability to *identify* the physical system being observed. The question of system identification has largely been neglected by theoretical physics, although it is of obvious practical relevance to experimental physics. Discussions of quantum measurement, for example, standardly examine the interaction between an observer and a fixed, well-defined system that is typically stipulated a priori by stipulating its Hilbert space (for reviews, see [9,10]). System identification has received more attention from engineers and computer scientists. Moore, for example, proved in 1956 that finite, finite-resolution observations cannot fully determine the state space of an otherwise-uncharacterized physical system; in particular, they cannot determine its state-space dimension d ([11] Theorem 2; see [12,13] for discussion). This result underlies the proven unsolvability of the halting problem in classical computer science [14]. Both of these results, however, rely on limits in which numbers of degrees of freedom become arbitrarily large, and neither considers the quantitative cost of system identification.

Here I characterize the thermodynamic cost of system identification in a general, operational framework covering both classical and quantum systems and investigate some of its consequences. The next section characterizes the system identification problem operationally as a search problem constrained by a finite-resource restriction. The consequences of this restriction for system identification and characterization are then discussed (Section 3), and the finite-resource restriction is shown to forbid the arbitrary refinement of state spaces of observed systems to assumed "objective" or "ontic" state spaces even when these are (Section 4). I then focus on two types of system-identification problems that regularly arise in practice: the identification of a single system at multiple, significantly separated times (Section 5) and the identification of a single system by multiple, spacelike-separated observers (Section 6). I show in each case that classical correlations of measurement outcomes are insufficient, in principle, for reliable system identification. A similar point regarding the second class of problems has been made previously by Grinbaum [15]. Violations of Leggett–Garg [16] and Bell [17] inequalities, respectively, thus arise naturally as requirements for reliable system identification in these settings. These results extend and elaborate on previous work of a more systems-theoretic nature [12,13,18,19]. The general theory of observation as a physical process, including the central role of the observer's memory as an encoding of observational outcomes, has also recently been discussed by Kupervasser [20].

2. Formalizing System Identification as a Search Process

Characterizing the thermodynamic cost of system identification requires redescribing observation in a way that makes the process of system identification explicit. Consider the standard, classical "picture" of observation shown in Figure 1. Here the "observer" is a physical system that interacts with a "system of interest" to obtain observational outcomes. Both observer and system are embedded in a surrounding environment, which can be regarded as "everything else" in the universe. This classical picture of observation is carried over unchanged into quantum theory, where the "observer" now terminates the von Neumann chain [21] by recording their outcome(s) in a thermodynamically irreversible way. It provides, by including the surrounding environment, the setting for environmental decoherence [22–28]. Tegmark has emphasized that the observer O in this setting comprises *only* the degrees of freedom that record observational outcomes, while the system S comprises only the "pointer" degrees of freedom that specify these outcomes; all other degrees of freedom are considered part of the "environment" E and traced over [29]. Tracing out the environment assures that information about the state of S reaches O only through the channel defined by the $O - S$ interaction, specified in Figure 1b by the Hamiltonian H_{OS}. The alternative channel via the environment E, given by the Hamiltonian $H_{SE} + H_E + H_{OE}$, contributes only classical noise. In the alternative "environment as witness" formulation of decoherence developed by Zurek and colleagues [30–33], O is assumed to be located sufficiently far from S that $H_{OS} \sim 0$. In this formulation, all information about S obtained by O flows through the channel $H_{SE} + H_E + H_{OE}$. The state $|E\rangle$ of the environment is regarded as "encoding" this information, with the encoding of information about the positions of macroscopic objects by the ambient photon field as the canonical example.

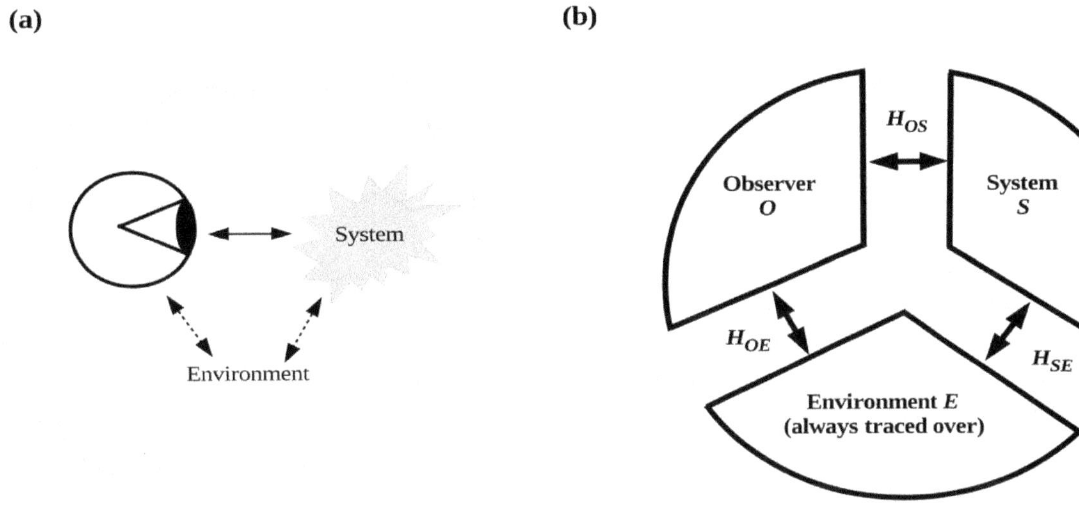

Figure 1. (**a**) A classical observer interacts with a system of interest; both are embedded in a surrounding environment. (**b**) Interactions between observer (O), system of interest (S) and environment (E) enabling environmental decoherence. The Hamiltonian H_{OS} transfers outcome information from S to O; H_{SE}, and H_{OE} decohere S and O respectively. Adapted from Figure 1 in ref. [29].

This conventional conception of observation, even when made precise using the formalism of decoherence, tells us nothing about how the observer *identifies* the system of interest. The system S is given a priori in Figure 1: the interactions H_{OS}, H_{OE}, H_{ES}, H_S, and H_E are all assumed to be given and well-defined. To include system identification in the picture, it is useful to describe it in operational terms. Suppose I want Alice to report the observational outcome registered by a particular macroscopic apparatus located in a laboratory filled with many other systems. How much information do I need to give Alice to assure that she reports the outcome from the *right* apparatus? In this scenario, the finite-resource restriction on Alice is clear: I can give Alice at most a finite description of the apparatus that I want her to report an outcome from. I could instruct her, for example, to locate a black laptop labeled "data 3," running linux, with a counter window open, and to report the outcome displayed in the counter window. I could add that "data 3" is connected to an ADC in the third rack from the right wall. Alice must then enter the laboratory and *look for*, using observational means at her disposal, an apparatus matching my finite description. The informational basis of this operational scenario can be made precise as follows:

Finite-resource restriction: No observer can employ more that a finite number of finite-resolution observational outcomes to identify a system of interest.

Classically, an observer subject to the finite-resource restriction has only a finite number of finite-resolution criteria for system identification; in quantum theory, this corresponds to a finite number of discrete-valued observables. Such criteria or observables can be considered to be binary without loss of generality.

It is obviously circular to assume that, when Alice enters the laboratory, she can identify the apparatus satisfying her finite criteria (or finite observables) without having to look at anything else: this is equivalent to assuming that the apparatus is given a priori and hence does not need to be identified. To *identify* the apparatus S, Alice must *distinguish it*, using her criteria/observables, from everything else in the laboratory, i.e., from E. Alice must, in other words, employ her criteria/observables to *search* the combined system $W = SE$ until she finds S. Hence, she is in the position illustrated in Figure 2b, not that of Figure 2a as is standardly assumed.

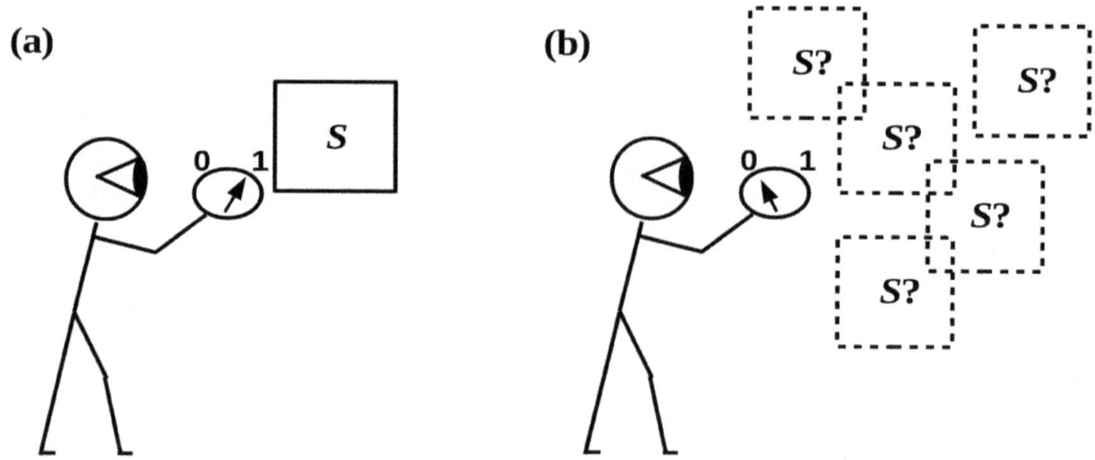

Figure 2. (**a**) An observer equipped with an observable (e.g., a meter reading) interacts with a pre-given system S. Adapted from Figure 1 in ref. [34]. (**b**) An observer with finite resources must *look for* the system of interest by probing the "world" W in which it is embedded.

To make this idea of searching W for S precise, suppose as above that an observer O and world W are given as collections of physical degrees of freedom, and assume for the present that they are quantum systems characterized by Hilbert spaces $\mathcal{H}_O$ and $\mathcal{H}_W$, respectively. Suppose further that O can perform n distinct (but not necessarily orthogonal) binary-outcome measurements M_i on W, that O's thermodynamic cost per bit of recorded outcome is $c^{(O)}k_BT$ as above, that deploying the M_i has no other energetic consequences, and that O's interaction with W consists entirely of deploying the M_i. In this case, each of the M_i can be regarded as extracting one bit of information from W and exhausting $c^{(O)}k_BT$ of waste heat into W. The operations M_i can be regarded informally as "questions to Nature" such as "is what I see before me a laptop?" or "is it black?" and formally as Hermitian operators on $\mathcal{H}_W$ in the usual way. No assumption need be made at this point about whether the M_i commute; this question is addressed in Section 3. For simplicity, suppose O deploys the M_i one at a time in the fixed order $i = 1, ..., n$, that each of the M_i is deployed for a fixed time $\Delta t^{(O)}$, the time required for O to record one bit, and that O makes m cycles of measurements. The total elapsed time during which O makes measurements on W is then $nm\Delta t^{(O)}$. Taking the $O - W$ interaction to be given by a Hamiltonian operator H_{OW} on $\mathcal{H}_O \otimes \mathcal{H}_W$, the total action is

$$\int_{t=0}^{nm\Delta t^{(O)}} dt\ H_{OW}(t) = nm\Delta t^{(O)}c^{(O)}k_BT. \tag{1}$$

To make $H_{OW}(t)$ explicit in the simplest case of sequential, equal-duration measurements, let $\Pi^{(i,m)}(t)$ be the rectangular Pi function with offset i, $0 \le i \le n-1$, duty cycle n, and the number of cycles m, i.e.,

$$\Pi^{(i,m)}(t) = \sum_{j=0}^{m-1} \Pi(t - (nj + i + 1/2)\Delta t^{(O)}) \tag{2}$$

where

$$\Pi(t) = \begin{cases} 0 & \text{if } |t| > 1/2 \\ 1/2 & \text{if } |t| = 1/2 \\ 1 & \text{if } |t| < 1/2 \end{cases}.$$

This $\Pi^{(i,m)}(t)$ is a sequence, starting at $t = i$, of m unit-height rectangular pulses with width $\Delta t^{(O)}$ and separation $n\Delta t^{(O)}$ as shown in Figure 3. In this case, we can write, for $0 \le t \le nm\Delta t^{(O)}$,

$$H_{OW}(t) = \sum_{i=0}^{n-1} \Pi^{(i,m)}(t) M_i, \tag{3}$$

with the heat dissipated by the action of the kth measurement operator during the first $j \le m$ cycles of measurement given by

$$(1/\Delta t^{(O)}) \int_{t=0}^{nj\Delta t^{(O)}} dt\, \Pi^{(k,j)}(t) M_k = jc^{(O)} k_B T. \tag{4}$$

If the requirement of a fixed sequence of equal-duration measurements is now dropped and O is simply assumed to make N total observations, Equation (3) can be generalized, for $0 \le t \le N\Delta t^{(O)}$, to

$$H_{OW}(t) = \sum_{i=1}^{n} \alpha_i(t) M_i, \tag{5}$$

subject to the constraints that, at all t,

$$\sum_{i=1}^{n} \alpha_i(t) = 1, \tag{6}$$

and, for any positive integer $k < N$,

$$(1/\Delta t^{(O)}) \sum_{i=1}^{n} \int_{t=k\Delta t^{(O)}}^{(k+1)\Delta t^{(O)}} dt\, \alpha_i(t) M_i = c^{(O)} k_B T. \tag{7}$$

Here the function $\alpha_i(t)$ is naturally interpreted as the probability of deploying the measurement M_i at t. The sequence of outcomes obtained will depend on the $\alpha_i(t)$; however, the incremental heat dissipation, expressed in Equation (7), of the measurements will not.

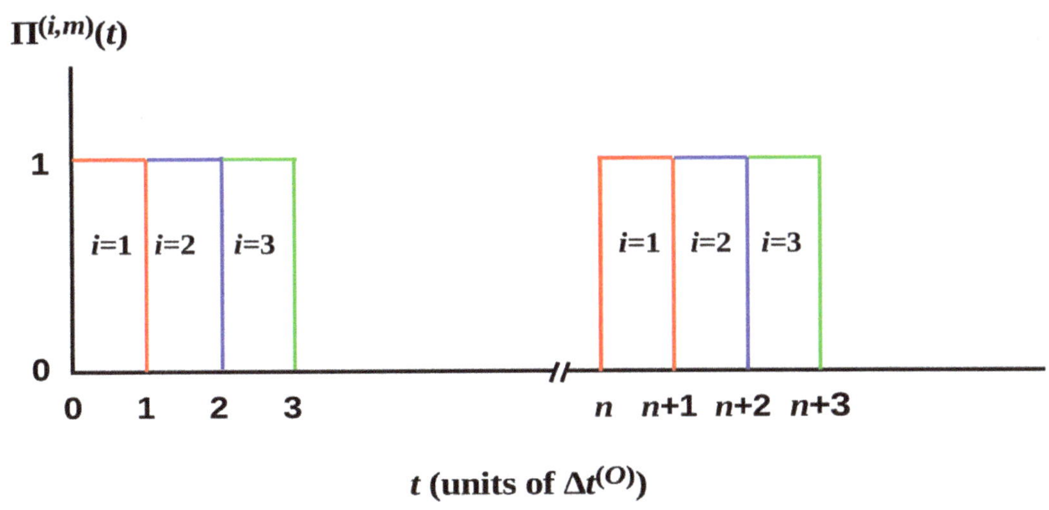

Figure 3. The first three components of $\Pi^{(i,m)}(t)$ of Equation (2) in the first and n^{th} cycles of deploying the M_i.

With $\Delta t^{(O)}$ finite, t can be treated as having only integer values $k\Delta t^{(O)}$ and hence regarded as a counter. This counter must be internal to O, as otherwise the values of t would be observational outcomes obtained from an external clock by some subset of the M_i and the above representation would be circular. The record of O's N observations can, in this case, be represented as Table 1 indexed by integer values of t:

Table 1. Sample record of O's observational outcomes from W, starting at $t = 1$.

Step t	Measure $M_i, i \leq n$	Outcome $x_i \in \{0, 1\}$
1	1	1
2	2	0
3	2	1
...	...	...
N	4	0

A table of this form contains all of the information about W available to O following N observations. The energetic cost of these data to O is $Nc^{(O)}k_BT$, which is dissipated into W as waste heat. The counter t can, alternatively, be regarded as counting sets of k simultaneously measurable outcomes obtained "in parallel" at a cost of $kc^{(O)}k_BT$; however, here we will maintain the convention that outcomes are obtained sequentially at discrete time steps.

As noted above, finite observations at finite resolution cannot fully determine the state space of an otherwise-uncharacterized system [11]. The only information about W available to O are the outcomes $x_1...x_N$ of N finite-resolution observations; hence, O cannot determine the state space of W, i.e., the assumed Hilbert space $\mathcal{H}_W$ or even its dimension d_W, and *ipso facto* can specify the measurements M_i being performed on W only operationally. The complete set of possible outcomes of the M_i are, however, fully specified: each action with M_i produces an outcome $x_i \in \{0, 1\}$. Associating each of these x_i with a unit basis vector $\vec{i}$ constructs a binary space $\mathcal{W}$ with dimension $d_{\mathcal{W}} \leq n$ (equality if the M_i are orthogonal and all are employed at least once), which we can call the *apparent* or *observable state space* of W for O. Each "observation" by O can, therefore, be thought of not as an action with some M_i on W but as an operation on $\mathcal{W}$ with a binary-valued POVM E_i that selects the same outcome x_i as M_i. The Hilbert spaces standardly employed in quantum theory are constructed in this way using possible outcomes as basis vectors and are hence "apparent" in this sense. The operators M_i are, similarly, standardly defined in terms of the outcomes they produce, i.e., as operators on such apparent state spaces; in this case, the relation $E_i = M_i^\dagger M_i$ can be viewed as operationally defining M_i. This standard practice justifies our starting assumption that O and W can be treated as quantum systems. The same formalism can be employed to represent finite, finite-resolution measurements of classical systems by requiring that all states be Gaussian [6,7].

3. Distinguishing Reference from Pointer Degrees of Freedom

We now turn to the question of commutativity requirements for the M_i. To be of empirical interest, a "system" S must (1) be distinguishable from its surroundings, (2) be sufficiently persistent in time to permit multiple observations (at minimum, "preparation" followed by observation), and (3) occupy more than one state. Determining the state of S at multiple times requires an ability to distinguish S from its surroundings, i.e., to *identify* S, at multiple times. Hence, any system S of empirical interest can be decomposed as $S = PR$, where the generalized "pointer" component P indicates the system's time-varying state, and the remaining "reference" component R permits, by remaining in a time-invariant state $|R\rangle$, re-identification of S at multiple times. For ordinary items of laboratory apparatus like voltmeters or oscilloscopes, size, shape, mass, and the layout of controls and displays on the surface are components of R and their fixed, system-identifying values are components of $|R\rangle$, while the position of the apparatus, what the leads are connected to, control settings, and what is indicated on the displays are components of P. The state $|S\rangle$ of S is then given by $|S\rangle = |R\rangle|P\rangle$ with $|R\rangle$ fixed and only $|P\rangle$ free to vary. Requiring S to be identifiable by observation is thus requiring $|S\rangle$ to be separable as $|R\rangle|P\rangle$. If my laptop's mass or the color of its exterior casing, for example, become entangled with what is displayed in one of its windows, I will no longer be able to identify it by observation.

This requirement of re-identifiability can be formulated using Zurek's notion of a "predictability sieve" [28], a criterion that allows the future state of a system, here the state $|R\rangle$ of the time-invariant reference component R, to be predicted with confidence. Predictability is only assured if, for all i,

$$[H_W + H_{OW}, M_i^{(R)}] = 0 \tag{8}$$

where the measurement operators $M_i^{(R)}$ act on R but not P (cf. [28]; Equation 4.41). In practice, it is sufficient that, for all i, $[H_W + H_{OW}, M_i^{(R)}] < \delta$ for some sufficiently small δ over the course of an experiment involving multiple observations. Given Equation (5), the predictability sieve condition expressed in Equation (8) requires that system identification using the $M_i^{(R)}$ does not disturb system identity and that pointer-state measurements using some set of measurement operators $M_j^{(P)}$ that act only on P do not disrupt system identification, i.e.,

$$[M_i^{(R)}, M_j^{(R)}] = 0 \quad \text{and} \quad [M_i^{(R)}, M_j^{(P)}] = 0 \tag{9}$$

for all i, j. Nothing, however, requires the pointer measurements $M_j^{(P)}$ to all mutually commute, and they do not, for example, if calibration is included ([8] or Section 5 below). With these definitions, system *identification* is distinct from system *preparation*; operations employed for preparation must preserve system identity and thus must commute with the $M_i^{(R)}$, but need not, and in general will not, commute with the $M_j^{(P)}$. Preparation and observation of the "pointer state" $|P\rangle$ of P will be considered equivalent in what follows.

In terms of the equivalent operators E_i defined on the apparent state space $\mathcal{W}$, an "observable system" S in W can now be operationally defined as

Definition 1. *An observable system S in W is a collection $(E_i^{(R)}, x_i^{(R)})$ of $1 < k < n$ mutually commuting POVMs $E_i^{(R)}$ defined on the apparent state space $\mathcal{W}$ with specified outcomes $x_i \in \{0,1\}$ that measure "reference" degrees of freedom of W that are fixed and no longer free in S and hence "identify" S, together with a distinct collection of $1 < l < (n-k)$ POVMs $E_j^{(P)}$ defined on $\mathcal{W}$ with unspecified binary outcomes $x_j^{(P)}$ that measure "pointer" degrees of freedom of W that remain free in S, where for each $E_j^{(P)}$, $[E_j^{(P)}, E_i^{(R)}] = 0$ for every $E_i^{(R)}$.*

Note that, while O and W are collections of degrees of freedom and are hence "ontic" entities, observable systems are collections of operations and outcomes and are hence in some sense "epistemic" entities. The notations "S," "R," "P," and, below, "E" will be maintained for consistency with the literature, and to recognize that in practice systems are standardly defined in terms of observational outcomes as noted above.

The l pointer degrees of freedom of S comprise its pointer P and their measured outcome values constitute its pointer states $|P\rangle = |x_1^{(P)} ... x_l^{(P)}\rangle$. While the $E_j^{(P)}$ selecting pointer outcomes are not required to commute, at least pairs of pointer outcomes must be compatible in any "interesting" system (an EPR/Bell experiment, for example, requires simultaneous measurement of two pointer states, the "measurement setting" and the outcome, by each observer (Section 6)). The collection $(E_i^{(R)}, x_i^{(R)})$ of k specified (POVM, outcome) pairs specifies the pointer-state independent reference component R and its time-invariant state $|R\rangle = |x_1^{(R)} ... x_k^{(R)}\rangle$. We require that $P \cap R = \emptyset$ and $PR = S$. For macroscopic systems such as laboratory apparatus, the number of pointer degrees of freedom $l << n$. Hence, the number of reference degrees of freedom $k \sim n$; this will be assumed in what follows.

Two observable systems S and S' are discernible in isolation only if they differ by at least one reference (POVM, outcome) pair. Every observable system S has a *complement* $\bar{S}$ that is the maximal observable system that does not overlap S. In the limit $n \to \infty$, $S\bar{S} \to W$, i.e., $\bar{S} \to E$ as defined above.

This limit cannot, clearly, be reached with finite observational resources; the consequences of this are considered in Section 4 below.

Several remarks are in order:

Remark 1. *Observable systems are defined here in terms of both the assumed partition of "the universe" into O and W and the operations employed by O to identify them. They are, therefore, observer-relative in the sense defined by Rovelli [35] for quantum states. However, as noted above, the present considerations apply to both classical and quantum systems provided the finite-resource restriction is respected. This observer-relativity naturally suggests counterfactual indefiniteness, i.e., that "unidentified systems have no states" (cf. [36]), regardless of the equations of motion they obey while being observed.*

Remark 2. *The "world" W is not an observable system. As $S \rightarrow W$ the notion of "system identification" loses any operational meaning.*

Remark 3. *The apparent state space $\mathcal{W}$ coarse-grains W. As will be made precise in the next section, unless $d_W >> n$ (hence effectively, $W >> O$), waste heat cannot be dissipated by O and commutativity of observables breaks down. This corresponds to the "large environment" assumption of decoherence.*

Remark 4. *No assumption is made that W exhibits objective classical randomness. The characterization of the energetic cost of observation as waste heat reflects O's objective uncertainty about the distribution of this energetic input across the degrees of freedom of W.*

Remark 5. *The requirement that every $E_j^{(P)}$ commutes with all $E_i^{(R)}$ enables repeated pointer measurements to have the same outcome, and hence enables "ideal measurements" as defined by Cabello [37], provided calibration procedures are implemented as discussed in Section 5.*

Remark 6. *The support of the $E_j^{(P)}$ and $E_i^{(R)}$ in $\mathcal{W}$ can be considered the apparent or observable state space $\mathcal{S}$ of S; again, this is the usual approach to defining state spaces for stipulated quantum systems. State transitions in $\mathcal{S}$ can be represented as actions of a discrete observed propagator $\mathcal{P}^S : |S\rangle|_t \mapsto |S\rangle|_{t+1}$. This $\mathcal{P}^S$ maps each observational outcome to its successor and so can be regarded as defining a computational process, regardless of whether the system S is classical or quantum, provided the finite-resource restriction is respected [19].*

4. System Identification Cannot Be Arbitrarily Refined

In contrast to the operational, observer-dependent conception of "systems" defined above, classical (or "effectively classical") macroscopic systems such as laboratory apparatus are standardly thought of as both observation- and observer-independent. They are, in particular, standardly viewed both as invariant under decompositions of "the universe" into alternative observer—world pairs—and as well-defined independently of any particular observer or observables (see [38] for an example of this "realist" position). Let us use the notation $\mathbb{S}$ to indicate an observer-independent ("objective" or "ontic") system, i.e., one that is considered well-defined in the absence of any observers, reserving S for "observed systems" defined operationally as above in terms of sets of observational outcomes. It is, for example, completely standard in classical physics to describe two observers interacting with or otherwise obtaining information about a single, observer-independent, macroscopic "object" $\mathbb{S}$. This assumption of observer-independence is often carried over into quantum theory. Extensions of the environment as witness formulation of decoherence to models of quantum Darwinism [32,33] or quantum-state broadcasting [39,40], for example, postulate that multiple observers can independently interact with separable, redundant encodings of the eigenvalues of a single, observer-independent interaction $H_{\mathbb{SE}}$ between an observer-independent quantum system $\mathbb{S}$ and its observer-independent environment $\mathbb{E}$. That such an encoding is redundant, i.e., that the multiple "copies" of the information are encoded by the single, objectively well-defined interaction $H_{\mathbb{SE}}$ must be assumed a priori, as it cannot be established by observation [41]. It is also commonly assumed, for example, in stating the Pusey–Barrett–Rudolph theorem [42] that multiple "copies" of a single quantum system can be acted upon (e.g., prepared and/or measured) independently by multiple, mutually distant observers. The copies in this case are assumed to objectively have all and only the same degrees of freedom, the same self-Hamiltonian, and the same interaction with their respective environments.

Here we consider whether, and to what extent, observers subject to the finite resource restriction imposed in Section 2 can identify, and hence either prepare or measure, a postulated "objective" system $\mathbb{S}$. We first consider, in this section, the case of a single observer O interacting during one time period with a single $\mathbb{S}$. We then consider two cases of practical interest: in Section 5, that of a single observer interacting with $\mathbb{S}$ during multiple time periods and, in Section 6, that of multiple observers interacting with $\mathbb{S}$ during a single time period. We show that violations of Leggett–Garg and Bell inequalities, respectively, can be interpreted as criteria for successful system identification in these two scenarios.

As noted above, the methods developed here apply equally to both classical and quantum systems provided the finite-resource restriction is respected. Let us now assume, as is typical in classical physics and as the simplest case, that W has an *observer-independent, classical* state $|W\rangle$, and first consider the finite case in which W can be described by a d_W-dimensional, classical, binary state space, e.g., a real Hilbert space. Let us also assume that an observer-independent, classical system $\mathbb{S}$ is embedded in W, that O obtains information specifically from $\mathbb{S}$ while dumping waste heat specifically into an observer-independent environment $\mathbb{E}$ defined by $\mathbb{SE} = W$, and that the dimension $d_{\mathbb{S}} << d_{\mathbb{E}}$. We assume that O interacts with $\mathbb{S}$ via a set of operators M_i as defined by Equation (5) above. The outcomes x_i of this interaction can be associated with unit vectors to construct the apparent state space $\mathcal{S}$ of O's observed system S as described above. In this case, O can, given a sufficient number (i.e., $n \geq d_{\mathbb{S}}$) of binary measurement operators, refine the observed S to the objective $\mathbb{S}$, i.e., the dimension $d_{\mathcal{S}} \to d_{\mathbb{S}} << d_W$, at a energetic cost of

$$H_{diss}^{(\mathbb{S})} = (1/\Delta t^{(O)}) \int_{t=0}^{\tau} H_{O\mathbb{S}} dt \to d_{\mathbb{S}} c^{(O)} k_B T \tag{10}$$

where t is a time coordinate associated with W, and $\tau \to d_{\mathbb{S}} \Delta t^{(O)}$ is the interval in t required by O to identify S at the given refinement. By dissipating $H_{diss}^{(\mathbb{S})}$ exclusively into $\mathbb{E}$, O assures that $\mathbb{S}$ remains undisturbed. It is this transfer of waste heat to a large, unobserved, observer-independent environment that enables the typical classical assumption of arbitrary measurement resolution and hence real-valued measurement outcomes.

If the assumption that O obtains information specifically from $\mathbb{S}$ is now dropped and O is required to *identify* $\mathbb{S}$ by observation as described above, O must search and therefore interact with, in the limit, all of W. In this case, refining the observed S to the objective $\mathbb{S}$ requires refining the apparent state space $\mathcal{W}$ to the full "ontic" state space of W. The number of measurement operators required is now $n \geq d_W$, and the energetic cost is now

$$H_{diss}^{(W)} = (1/\Delta t^{(O)}) \int_{t=0}^{\tau} H_{OW} dt \to d_W c^{(O)} k_B T \tag{11}$$

where now $\tau \to d_W \Delta t^{(O)}$. In this limit, $c^{(O)} k_B T$ is transferred, on average, to every binary degree of freedom of W. The environment $\mathbb{E}$ can no longer be treated as an unobserved "sink" for waste heat, as in the limit every degree of freedom of W must be examined to see whether it is a degree of freedom of the as-yet unidentified $\mathbb{S}$. Equation (11) does not depend in any way on W being classical but rather is a straightforward consequence of Equation (5); it is, therefore, completely general. Hence, we have the following.

Theorem 1. *An observed system S cannot be refined to an objective system $\mathbb{S}$ with finite resources.*

Proof. Consider the states $|W\rangle|_{t=1}$ and $|W\rangle|_{t=k\Delta t^{(O)}}$ acted on by measurement operators $M_1^{(R)}$ and $M_k^{(R)}$, respectively, for some $k >> 1$. As $d_{\mathcal{W}} \to d_W$, under increasing refinement, the maximum value of $k \to d_W$, and the energy difference between $|W\rangle|_{t=1}$ and $|W\rangle|_{t=k\Delta t^{(O)}}$ at maximum k, $\Delta H_{1,k} \to H_{diss}^{(W)}$. None of the M_i are, however, orthogonal to H_{OW}, so in this case $[M_1^{(R)}, M_k^{(R)}] \neq 0$. This violates the predictability sieve condition expressed in Equation (9), rendering $|R\rangle$ no longer invariant. Hence, S is, by definition, unidentifiable in this limit, and the desired refinement of S to $\mathbb{S}$ fails. □

Note that Equation (11) is independent of $d_{\mathbb{S}}$: the energy dissipation required for system identification increases with d_W even if $d_{\mathbb{S}} << d_W$. As $d_W \to \infty$ or becomes continuous, arbitrary refinement of $\mathcal{W}$ requires $H_{OW} \sim H_W$ and again commutativity of the (now infinitely or continuously many) $M^{(R)}$ fails. Theorem 1 thus provides a quantitative extension of Moore's qualitative result that finite, finite-resolution observations cannot fully determine the state space of an otherwise-uncharacterized system [11], and shows that it holds even in a finite "world" W.

System identification cannot, therefore, be arbitrarily refined to the limit of an "objective system" even in classical physics. The predictability sieve expressed in Equation (8) that allows system identification is only operable provided the measurement interaction $H_{OW} << H_W$ and the apparent state space dimension $d_{\mathcal{W}} << d_W$. Coarse-graining W is, therefore, required to identify any embedded system S, even if W is classical; if observer-independent "objective systems" exist in W, identifiable systems only approximate them. An observed S can, at best, only be associated with a set $\{\mathbb{S}\}$ of objective systems that could, in some theoretical model specifying some set of reference operators $\{M_i^{(R)}\}$, generate the observational outcomes $\{x_i^{(R)}\}$ that identify S. The dimensions of the elements of $\{\mathbb{S}\}$ are constrained only by d_W and d_S as upper and lower bounds, respectively. Hence, Theorem 1 rules out any confirmation by finite observations that two independently observed systems S and S', whether classical or quantum, are copies of a single objective $\mathbb{S}$.

In practice, observers search for systems only locally, effectively coupling a small, searched region of W to a large, unobserved reservoir—the rest of W—into which energy can be dissipated. If this coupling is weak and the dissipation constant $c^{(O)} >> 1$, the predictability sieve condition expressed in Equation (9) fails as search resolution increases, i.e. as $S \to \mathbb{S}$. Observers typically search even for macroscopic systems at low resolution and then refine the search slightly after plausible candidates have been identified. One may, for example, locate multiple systems of the right size and shape to be one's laptop and then refine the search by looking for identifying marks, checking the splash screen, etc. Refining the search toward an "objective" limit by examining every transistor, much less every atom, disrupts the commutativity of the M_i^R and is therefore infeasible.

5. System Identification at Multiple Times

Let $S|_t$ be the observed system identified when O deploys n measurement operators M_i during the interval between $(t - n\Delta t^{(O)})$ and t. Theorem 1 above show that S cannot be refined to some specific objective $\mathbb{S}$. However, $S|_t$ can be associated with a set $\{\mathbb{S}\}|_t$ of all objective systems for which the M_i would yield, at t, the outcomes obtained. For example, if $S|_t$ is identified by the two criteria of being red and having no linear dimension greater than 1 m, then the set $\{\mathbb{S}\}|_t$ contains all objective systems meeting these criteria at t. If O deploys the M_i at multiple times, a sequence $S|_t$, $S|_{t'}$, $S|_{t''}$, etc. is obtained, with corresponding sets of objective systems $\{\mathbb{S}\}|_t$, $\{\mathbb{S}\}|_{t'}$, $\{\mathbb{S}\}|_{t''}$, etc. The sequence $S|_t$, $S|_{t'}$, and $S|_{t''}$ identifies a single observed system S if there is a time-invariant set of reference outcome values $\{x_i^{(R)}\}$ that fixes a reference state $|R\rangle$ and hence a reference component $R \subset S$. However, O cannot determine by observation that $\{\mathbb{S}\}|_t = \{\mathbb{S}\}|_{t'}$ or even $\{\mathbb{S}\}|_t \cup \{\mathbb{S}\}|_{t'} \neq \emptyset$, as doing so requires determining the self-Hamiltonians of the $\mathbb{S}$, i.e., arbitrarily accurate refinement forbidden by Theorem 1. To continue the previous example, O cannot determine that each red thing will remain red or that each small thing will remain small without examining, for each thing, more degrees of freedom than color and size. Hence, even perfect correlation of each of the reference outcome values $x_i^{(R)}$ between all pairs of measurement times cannot gaurantee that O is interacting with the same objective system(s) at t, t', t'', etc. If the probability distributions over pointer outcome values $x_j^{(P)}$ are time-invariant, their time correlations are similarly insufficient to guarantee that O is interacting with the same objective system(s) at all measurement times. Hence, we have the following.

Theorem 2. *If for a set of measurements M_i and measurement times t_j and t_k, the two-time outcome correlation functions $C_{jk} = \langle x_i(t_j), x_i(t_k) \rangle$ satisfy the Leggett–Garg inequality, the observed system S identified by the M_i cannot be associated with any single element of the set $\{\mathbb{S}\}$ of objective systems associated with S.*

Proof. Mapping each binary outcome from $\{0,1\}$ to $\{-1,1\}$, the Leggett–Garg inequality can be written $C_{21} + C_{32} - C_{31} \leq 1$ for consecutive measurements at t_1, t_2, and t_3 [16]. The reference outcomes $x_i^{(R)}$ and hence the reference state $|R\rangle$ must remain fixed at all observation times to identify S; hence, the $x_i^{(R)}$ satisfy this inequality trivially. To see that the fixed $x_i^{(R)}$ cannot identify any particular element of the set $\{\mathbb{S}\}$ of objective systems associated with S, it is enough to note that obtaining $x_i^{(R)}$ from a measurement on $\mathbb{S}$ at t provides no evidence that $\mathbb{S}$ was in state $|R\rangle$ at $t-1$. Hence, if $\mathbb{S}$ is to be identified, it must be identified by correlations between the pointer outcomes $x_j^{(P)}$. If these satisfy the Leggett–Garg inequality, however, each measurement of the $x_j^{(P)}$ is independent of all previous as well as all future measurements. Hence, no measurement of the $x_j^{(P)}$ on $\mathbb{S}$ at t can provide information about the state of $\mathbb{S}$ at $t-1$. It is, therefore, consistent with both constant $x_i^{(R)}(t)$ and classically correlated $x_j^{(P)}(t)$ that outcomes have been obtained from a different element of $\{\mathbb{S}\}$ at each measurement time. □

Theorem 2 restates, in effect, the general principle that classical correlation does not imply joint causation; even perfectly correlated outcome values can have different causal sources. It shows, in the present context, that an observed system S cannot be associated with a particular objective $\mathbb{S}$ without violating the Leggett–Garg inequality. Violations of this inequality provide, therefore, *evidence* that a single objective system has been identified over time.

To assure violations of the Leggett–Garg inequality, O must choose pointer measurement operators $M_i^{(P)}$ such that, for $t_j < t_k$, $Prob(x_i^{(P)}(t_k) = 1|x_i^{(P)}(t_j) = 1) \neq Prob(x_i^{(P)}(t_k) = 1|x_i^{(P)}(t_j) = 0)$, i.e., the pointer state $|P\rangle$ must "remember" previous applications of $M_i^{(P)}$. Pointer states with this property are commonplace in classical systems; magnetic hysteresis and work hardening in metals are familiar examples. Direct measurements of such states are not non-disturbing. If the pointer states of a macroscopic apparatus "remember" disturbances caused by previous measurements in this way, the standard corrective is frequent recalibration. Calibrating an apparatus, i.e., using measurement of a designated standard to adjust (i.e., intentionally disturb), the pointer state of the apparatus, effectively erases the memory of the previous measurement-induced disturbance. By providing evidence that the Leggett–Garg inequality has been violated, a need for re-calibration provides evidence of previous use and hence evidence that a single objective system $\mathbb{S}$, i.e., the apparatus, has been identified.

System identification over time, therefore, requires a significant asymmetry between reference and pointer degrees of freedom. Measurements of reference degrees of freedom must be non-disturbing in order for the reference state $|R\rangle$ to remain fixed and the observed system S to be identifiable. If S is to be identified with an objective $\mathbb{S}$, however, consecutive pointer measurements cannot be non-disturbing. Re-preparing $\mathbb{S}$ between designated, non-consecutive "informative" measurements, i.e., re-calibration to erase the memory of previous measurements, allows the "informative" measurements to be mutually non-disturbing and hence ideal.

Quantum violations of the Leggett–Garg inequality can, clearly, only be observed if the pointer state component exhibiting the violation is not re-prepared by calibration between measurements. Observing quantum Leggett–Garg violations while maintaining a constant objective $\mathbb{S}$ requires at least one pointer state component that both exhibits memory of previous measurements and can be recalibrated between measurements. An apparatus control setting that is re-set, and hence re-prepared, between measurements satisfies this requirement. Here the "standard" to which the state of the control setting is effectively being calibrated is the observer who manipulates the setting. Note that such control settings cannot, while preserving their function of enabling re-preparation and hence re-identification, become entangled with other components of P that register the observational outcomes of interest. As in the case of R becoming entangled with P discussed in Section 3 above, entanglement between control and outcome-registering components of P can lead to system-identification failure.

6. Joint System Identification by Multiple Observers

Suppose Alice deploys measurement operators A_i with outcomes $a_i(t)$ to identify and obtain pointer-state outcomes from an observed system $S = RP$ and Bob, who is spacelike separated from Alice at each measurement time t, deploys measurement operators B_j with outcomes $b_j(t)$ to identify and obtain pointer-state outcomes from an observed system $S' = R'P'$. Under what conditions can Alice and Bob conclude, when later comparing their separate sequences of observations, that they were observing two "parts" of the same objective system $\mathbb{S}$? It is useful to consider this question from the perspective of an adversarial game; from this perspective, Alice and Bob determining that they share a single $\mathbb{S}$ is equivalent to Alice and Bob determining that they share a communication channel that cannot be, or at least has not been, manipulated by an adversary, Charlie. Suppose S and S' are connected by a classical, timelike communication channel C, such that $SCS' = \mathbb{S}$. Under what conditions can Alice and Bob conclude that their observations of S and S' are unaffected by Charlie breaching and manipulating the channel C? In particular, under what conditions can Alice and Bob conclude that their observational outcomes obtained from S and S' are not the result of Charlie breaching C and sending instructions to S and S' that determine the observational outcomes? This question has been extensively investigated in the guise of quantum communication security [43,44], and the answer is well known. Any pattern of classical correlations between $|S\rangle$ and $|S'\rangle$ can be undetectably produced by a manipulative Charlie; therefore, no pattern of classical correlations between Alice's and Bob's observational outcomes can demonstrate that the channel C is secure. Hence, we have the following.

Theorem 3. *If correlations between sequences $a_i(t)$ and $b_j(t)$ of observational outcomes obtained by spacelike-separated observers A and B are consistent with a deterministic hidden-variable theory, they cannot mutually identify a single jointly observed objective system $\mathbb{S}$.*

Proof. Any pattern of correlations between the $a_i(t)$ and the $b_j(t)$ that is consistent with a deterministic hidden-variable theory can be implemented by Charlie; hence, any such pattern of correlations is consistent with A and B observing separate systems, both of which are manipulated by Charlie. □

As the reference states $|R\rangle$ and $|R'\rangle$ remain fixed throughout Alice's and Bob's measurements to identify the observed systems S and S', respectively, and hence remain perfectly classically correlated, Theorem 3 effectively concerns patterns of correlations between the pointer-state outcomes $a_i^{(P)}(t)$ and the $b_j^{(P')}(t)$. Such pointer-state correlations only permit identification of a single jointly observed objective system $\mathbb{S}$ if they are inconsistent with any deterministic hidden-variable theory.

Patterns of pointer-state correlations that are inconsistent with any deterministic hidden-variable theory are well-known in the special case in which Alice and Bob perform a canonical EPR/Bell type experiment. In this case, their sets of pointer measurement operators $\{A_i^{(P)}\}$ and $\{B_j^{(P')}\}$, respectively, each comprise one "control setting" observable and two mutually noncommuting, two-valued "outcome" observables; the observed correlations between the "outcome" observables are inconsistent with any deterministic hidden-variable theory if and only if they violate at least one Bell/CHSH inequality [45]. Mermin [46] explicitly considers deterministic hidden variables as "instruction sets" carried by particles from a central source to spacelike-separated detectors in discussing such experiments. Correlations that violate one or more Bell/CHSH inequalities cannot be replicated by such instruction sets or by a manipulative Charlie, and so provide evidence that Alice and Bob are jointly observing a single objective system $\mathbb{S}$. Such correlations can, in particular, identify an entangled state of $\mathbb{S}$ that Alice and Bob share (for recent experimental demonstrations, see [47–49]). They cannot, however, by Theorem 1, specify the complete state space of $\mathbb{S}$. Therefore, they cannot identify the one system that Alice and Bob are guaranteed to share under any circumstances, viz. the system comprising everything in the universe except Alice and Bob.

Two features of the use of Bell/CHSH inequalities as an entanglement witness are of particular relevance to system identification. First, at least one of the A_i and one of the B_i must measure the state of a pointer observable not manipulable by Charlie. In the canonical EPR-type experimental setup, these observables correspond to the orientation settings for the polarization/spin measurements, which are assumed to be freely chosen by Alice and Bob, respectively, at each t. This free-choice assumption rules out super-determinism [50] by preventing Charlie from specifying correlations involving these settings. As seen in Section 5 above, the existence of at least one pointer observable controlled by the observer enables objective system identification over time. Hence, the free-choice assumption can also be viewed as the assumption that Alice and Bob can each, independently, identify their respective apparatus as objective. Second, Alice and Bob must, after their observations have been completed, exchange a classical message encoding their observational outcomes to compute the correlations observed. This separate, classical communication step (i.e., use of a LOCC (Local Operations, Classical Communication) protocol) is required for shared entanglement to serve as a communication resource [51]. It introduces a second system, the classical message, that Alice and Bob must share, but without the restriction of spacelike separation. The separate, local observations employed in a LOCC protocol can be regarded as detecting a Bell/CHSH inequality violation only if the joint identification of this later, classical message—in practice, Alice and Bob agreeing that they have securely shared reports of their outcomes—is regarded as unproblematic.

Theorem 3 shows that joint system identification by spacelike-separated observers is demonstrable empirically only within quantum theory; in classical theory it can at best be *assumed*. By showing that joint system identification requires use of a LOCC protocol, it suggests that all systems are equivalent to communication channels. This idea is implicit in operational reconstructions of quantum theory [52,53] and has been made explicit by Grinbaum [15].

Acknowledgments: Thanks to Mauro D'Ariano, Don Hoffman, Ken Krechmer, Antonino Marcianò, Chetan Prakash, Robert Prentner and participants in the Quantum Contextuality in Quantum Mechanics and Beyond 2018 workshop for relevant discussions, and to three anonymous referees and the academic editor for helpful comments.

Abbreviations

The following abbreviations are used in this manuscript:

ADC	Analog-to-Digital Converter
CHSH	Clauser-Horne-Shimony-Holt
EPR	Einstein-Podolsky-Rosen
LOCC	Local Operations, Classical Communication
POVM	Positive Operator-Valued Measure

References

1. Boltzmann, L. *Lectures on Gas Theory*; Dover Press: New York, NY, USA, 1995; First published 1896.
2. Shannon, C.E. A mathematical theory of communication. *Bell Syst. Tech. J.* **1948**, *27*, 379–423. [CrossRef]
3. Landauer, R. Irreversibility and heat generation in the computing process. *IBM J. Res. Dev.* **1961**, *5*, 183–195. [CrossRef]
4. Landauer, R. Information is a physical entity. *Phys. A* **1999**, *263*, 63–67. [CrossRef]

5. Spekkens, R.W. Evidence for the epistemic view of quantum states: A toy theory. *Phys. Rev. A* **2007**, *75*, 032110. [CrossRef]
6. Bartlett, S.D.; Rudolph, T.; Spekkens, R.W. Reconstruction of gaussian quantum mechanics from Liouville mechanics with an epistemic restriction. *Phys. Rev. A* **2012**, *86*, 012103. [CrossRef]
7. Jennings, D.; Leifer, M. No return to classical reality. *Contempl. Phys.* **2016**, *57*, 60–82. [CrossRef]
8. Krechmer, K. Relative measurement theory: The unification of experimental and theoretical measurements. *Measurement* **2018**, *116*, 77–82. [CrossRef]
9. Landsman, N.P. Between classical and quantum. In *Handbook of the Philosophy of Science: Philosophy of Physics*; Butterfield, J., Earman, J., Eds.; Elsevier: Amsterdam, The Netherlands, 2007; pp. 417–553.
10. Schlosshauer, M. *Decoherence and the Quantum to Classical Transition*; Springer: Berlin, Germany, 2007.
11. Moore, E.F. Gedankenexperiments on sequential machines. In *Autonoma Studies*; Shannon, C.W., McCarthy, J., Eds.; Princeton University Press: Princeton, NJ, USA, 1956; pp. 129–155.
12. Fields, C. Bell's theorem from Moore's theorem. *Int. J. Gen. Syst.* **2013**, *42*, 376–385. [CrossRef]
13. Fields, C. Building the observer into the system: Toward a realistic description of human interaction with the world. *Systems* **2016**, *4*, 32. [CrossRef]
14. Hopcroft, J.E.; Ullman, J.D. *Introduction to Automata Theory, Languages and Computation*; Addison-Wesley: Reading, MA, USA, 1979.
15. Grinbaum, A. How device-independent approaches change the meaning of physical theory. *Stud. Hist. Philos. Mod. Phys.* **2017**, *58*, 22–30. [CrossRef]
16. Emary, C.; Lambert, N.; Nori, F. Leggett–Garg inequalities. *Rep. Prog. Phys.* **2014**, *77*, 039501. [CrossRef]
17. Mermin, D. Hidden variables and the two theorems of John Bell. *Rev. Mod. Phys.* **1993**, *65*, 803–815. [CrossRef]
18. Fields, C. If physics is an information science, what is an observer? *Information* **2012**, *3*, 92–123. [CrossRef]
19. Fields, C. A model-theoretic interpretation of environment-induced superselection. *Int. J. Gen. Syst.* **2012**, *41*, 847–859. [CrossRef]
20. Kupervasser, O. *Application of New Cybernetics in Physics*; Elsevier: Amsterdam, The Netherlands, 2017.
21. Von Neumann, J. *The Mathematical Foundations of Quantum Mechanics*; Princeton University Press: Princeton, NJ, USA, 1955.
22. Zeh, D. On the interpretation of measurement in quantum theory. *Found. Phys.* **1970**, *1*, 69–76. [CrossRef]
23. Zeh, D. Toward a quantum theory of observation. *Found. Phys.* **1973**, *3*, 109–116. [CrossRef]
24. Zurek, W.H. Pointer basis of the quantum apparatus: Into what mixture does the wave packet collapse? *Phys. Rev. D* **1981**, *24*, 1516–1525. [CrossRef]
25. Zurek, W.H. Environment-induced superselection rules. *Phys. Rev. D* **1982**, *26*, 1862–1880. [CrossRef]
26. Joos, E.; Zeh, D. The emergence of classical properties through interaction with the environment. *Z. Phys. B Condens. Matter* **1985**, *59*, 223–243. [CrossRef]
27. Zurek, W.H. Decoherence, einselection and the existential interpretation (the rough guide). *Philos. Trans. R. Soc. A* **1998**, *356*, 1793–1821.
28. Zurek, W.H. Decoherence, einselection, and the quantum origins of the classical. *Rev. Mod. Phys.* **2003**, *75*, 715–775. [CrossRef]
29. Tegmark, M. How unitary cosmology generalizes thermodynamics and solves the inflationary entropy problem. *Phys. Rev. D* **2012**, *85*, 123517. [CrossRef]
30. Ollivier, H.; Poulin, D.; Zurek, W.H. Objective properties from subjective quantum states: Environment as a witness. *Phys. Rev. Lett.* **2004**, *93*, 220401. [CrossRef] [PubMed]
31. Ollivier, H.; Poulin, D.; Zurek, W.H. Environment as a witness: Selective proliferation of information and emergence of objectivity in a quantum universe. *Phys. Rev. A* **2005**, *72*, 042113. [CrossRef]
32. Blume-Kohout, R.; Zurek, W.H. Quantum Darwinism: Entanglement, branches, and the emergent classicality of redundantly stored quantum information. *Phys. Rev. A* **2006**, *73*, 062310. [CrossRef]
33. Zurek, W.H. Quantum Darwinism. *Nat. Phys.* **2009**, *5*, 181–188. [CrossRef]
34. Fuchs, C. QBism, the perimeter of Quantum Bayesianism. *arXiv* **2010**, arxiv:1003.5201v1.
35. Rovelli, C. Relational quantum mechanics. *Int. J. Theor. Phys.* **1996**, *35*, 1637–1678. [CrossRef]
36. Peres, A. Unperformed experiments have no results. *Am. J. Phys.* **1978**, *46*, 745–747. [CrossRef]
37. Cabello, A. A simple explanation of Born's rule. *arXiv* **2018**, arxiv:1801.06347.
38. Bell, J.S. Against measurement. *Phys. World* **1990**, *3*, 33–41. [CrossRef]

39. Chiribella, G.; D'Ariano, G.M. Quantum information becomes classical when distributed to many users. *Phys. Rev. Lett.* **2006**, *97*, 250503. [CrossRef] [PubMed]
40. Korbicz, J.K.; Horodecki, P.; Horodecki, R. Objectivity in a noisy photonic environment through quantum state information broadcasting. *Phys. Rev. Lett.* **2014**, *112*, 120402. [CrossRef] [PubMed]
41. Fields, C. Quantum Darwinism requires an extra-theoretical assumption of encoding redundancy. *Int. J. Theor. Phys.* **2010**, *49*, 2523–2527. [CrossRef]
42. Pusey, M.F.; Barrett, J.; Rudolph, T. On the reality of the quantum state. *Nat. Phys.* **2012**, *8*, 475–478. [CrossRef]
43. Ekert, A.K. Quantum cryptography based on Bell's theorem. *Phys. Rev. Lett.* **1991**, *67*, 661–663. [CrossRef] [PubMed]
44. Gisin, N.; Thew, R. Quantum communication. *Nat. Photonics* **2007**, *1*, 165–171. [CrossRef]
45. Fine, A. Hidden variables, joint probability, and the Bell inequalities. *Phys. Rev. Lett.* **1982**, *48*, 291–295. [CrossRef]
46. Mermin, N.D. Quantum mysteries revisited. *Am. J. Phys.* **1990**, *58*, 731–734. [CrossRef]
47. Hensen, B.; Bernien, H.; Dreau, A.E.; Reiserer, A.; Kalb, N.; Blok, J.; Ruitenberg, M.S.; Vermeulen, R.F.L.; Schouten, R.N.; Abellán, C.; et al. Loophole-free Bell inequality violation using electron spins separated by 1.3 kilometres. *Nature* **2015**, *526*, 682–686. [CrossRef] [PubMed]
48. Giustina, M.; Versteegh, M.A.M.; Wengerowsky, S.; Handsteiner, J.; Hochrainer, A.; Phelan, K.; Steinlechner, F.; Kofler, J.; Larsson, J.-A.; Abellán, C.; et al. A significant-loophole-free test of Bell's theorem with entangled photons. *Phys. Rev. Lett.* **2015**, *115*, 250401. [CrossRef] [PubMed]
49. Shalm, L.K.; Meyer-Scott, E.; Christensen, B.G.; Bierhorst, P.; Wayne, M.A.; Stevens, M.J.; Gerrits, T.; Glancy, S.; Hamel, D.R.; Allman, M.S.; et al. A strong loophole-free test of local realism. *Phys. Rev. Lett.* **2015**, *115*, 250402. [CrossRef] [PubMed]
50. Hofer-Szabó, G. How human and nature shake hands: The role of no-conspiracy in physical theories. *Stud. Hist. Philos. Mod. Phys.* **2017**, *57*, 89–97. [CrossRef]
51. Bartlett, S.D.; Rudolph, T.; Spekkens, R.W. Reference frames, superselection rules, and quantum information. *Rev. Mod. Phys.* **2007**, *79*, 555–609. [CrossRef]
52. Coecke, B. Quantum picturalism. *Contempl. Phys.* **2010**, *51*, 59–83. [CrossRef]
53. Chiribella, G.; D'Ariano, G.M.; Perinotti, P. Informational derivation of quantum theory. *Phys. Rev. A* **2011**, *84*, 012311. [CrossRef]

5

Tsirelson's Bound Prohibits Communication through a Disconnected Channel

Avishy Carmi [1,2,*] and Daniel Moskovich [1,2,*]

1 Center for Quantum Information Science and Technology, Ben-Gurion University of the Negev, Beersheba 8410501, Israel
2 Faculty of Engineering Sciences, Ben-Gurion University of the Negev, Beersheba 8410501, Israel
* Correspondence: avcarmi@bgu.ac.il (A.C.); dmoskovich@gmail.com (D.M.)

Abstract: Why does nature only allow nonlocal correlations up to Tsirelson's bound and not beyond? We construct a channel whose input is statistically independent of its output, but through which communication is nevertheless possible if and only if Tsirelson's bound is violated. This provides a statistical justification for Tsirelson's bound on nonlocal correlations in a bipartite setting.

Keywords: nonlocality; Bell inequality; Tsirelson's bound; no-signaling; information causality; Fisher information

---

1. Introduction

Some of the predictions made by quantum mechanics appear to be at odds with common sense. Yet quantum mechanics remains the most precisely tested and successful quantitative theory of nature. It is therefore believed that even if quantum mechanics is someday replaced, any successor will have to inherit at least some of its "preposterous" but highly predictive principles. Perhaps the most counter-intuitive quantum mechanical feature is *nonlocality* [1]: the correlations exhibited by remote parties may exceed those allowed by any local realistic model.

The mystery of nonlocality is not only why nature is as nonlocal as it is, but why nature is not *more* nonlocal than it is. There are alternative *Non-Signaling* theories which permit nonlocality beyond the quantum limit [2,3]; why doesn't nature choose one of these theories over quantum mechanics? In Section 1.1 we review several previously proposed explanations. This paper presents another explanation, from statistics.

In this paper we construct a protocol (a repeated oblivious transfer) which sends messages through a disconnected channel. We show that Alice can communicate nontrivial information to Bob via this protocol if and only if the maximal quantum mechanical violation of the Bell–CHSH inequality [1,4], *Tsirelson's bound* [5], is exceeded. We thus provide a statistical explanation of this bound that is independent of the mathematical formalism of quantum mechanics.

We briefly recall the setting for the Bell–CHSH experiment. Section 2 provides a more detailed account. A famous application of nonlocality is to construct an 1-2 *oblivious transfer protocol* between two distant agents (A)lice and (B)ob. Alice and Bob each hold a box. Alice's box might, for example, contain one half of a singlet state of spin-$\frac{1}{2}$ particles, with Bob's box containing the other half [1,4]. In addition, Alice possesses a pair of bits x_0 and x_1, each of which is a zero or a one. Using boolean algebra and her boxes (the protocol will be described later), Alice encodes her pair of bits into a single bit $x^{(1)}$ which she sends across a classical channel to Bob. Bob wants to know the value either of x_0 or of x_1, but Alice doesn't know which of these Bob wants to know. Bob uses the received bit $x^{(1)}$, his box, and some boolean algebra to construct an estimate y_i for his desired bit x_i. See Figure 2 later on.

What is the probability that Bob correctly estimates the bit he wishes to know? He has two possible sources of knowledge—the bit $x^{(1)}$ he received from Alice, and some mysterious "nonlocal" correlation between his box and Alice's. The strength of such a nonlocal coordination between two systems is captured by a parameter $c \in [-1, 1]$ called the *Bell–CHSH correlator*. Bob's probability of guessing the value of Alice's bit correctly is $(1 + |c|)/2$. The *Bell–CHSH inequality* states that $|c| \leq 1/2$ in a world governed by classical (non-quantum) mechanics [1,4]. *Nonlocality* is the state of affairs in which the Bell–CHSH inequality is violated. To the best of our knowledge, real world physics is nonlocal. Over the years, the violation of the Bell–CHSH inequality has been measured in increasingly accurate and loophole-free experiments, culminating in celebrated loophole-free verifications [6–8].

Thus, we know that $|c|$ can exceed $1/2$. How large can $|c|$ be? Tsirelson's bound tells us that $|c|$ cannot exceed $1/\sqrt{2}$ in a world described by quantum mechanics [5]. This quantum bound on nonlocality:

$$|c| \leq \frac{1}{\sqrt{2}}\,, \tag{1}$$

has been tested experimentally, with the current state of the art being an experiment which has achieved a value of c which is only 0.00084 ± 0.00051 distant from Tsirelson's bound [9]. Such experimental evidence supports the contention that Tsirelson's bound indeed holds true in the real world. Tsirelson's result as presented in the original paper is a specifically quantum mechanical fact, following from the Hilbert-space mathematical formalism for quantum mechanics, for which there has been no good conceptual physical explanation. How fundamental is Tsirelson's bound? Must this inequality also hold for any future theory which might someday supercede quantum mechanics [10]? We are led to the following question: *Can we identify a plausible physical principle, independent of quantum mechanics (or independent of functional analysis), which is necessary and sufficient to guarantee that* $|c| \leq 1/\sqrt{2}$?

1.1. Existing Principles

For the last two decades, people have searched for physical principles that bound nonlocality. It was initially expected that the physical principle of relativistic causality (no-signaling) itself restricts the strength of nonlocality [11–13]. But then it was discovered that no-signaling theories may exist for which $|c| > 1/\sqrt{2}$. This led to the device-independent formalism of *No-Signaling (NS)–boxes* [2,14] (see also [3]). In particular, maximum violation of the Bell–CHSH inequality is achieved by *Popescu–Rohrlich (PR)–boxes* which are consistent with relativistic causality.

So relativistic causality doesn't limit nonlocality after all; Why then does nature not permit (1) to be violated (as far as we know)? Several suggestions have been made. Superquantum correlations lead to violations of the Heisenberg uncertainty principle [15,16], which is another seemingly purely quantum result. PR–boxes would allow distributed computation to be performed with only one bit of communication [17], which looks unlikely but doesn't violate any known physical law. Similarly, in stronger-than-quantum nonlocal theories some computations exceed reasonable performance limits [18]. The principle of *Information Causality* [19] shows that no sensible measure of mutual information exists between pairs of systems in superquantum nonlocal theories. Our approach is most directly comparable with Information Causality, with a conceptual difference being that we use variance of an efficient estimator, therefore Fisher information, whereas information causality uses mutual information (Shannon information). The relationship between our approach and theirs is the topic of Section 6. Finally, it was shown that superquantum nonlocality does not permit local (non-nonlocal) physics to emerge in the limit of infinitely many microscopic systems [20,21].

1.2. Tsirelson's Bound from a Statistical No-Signaling Condition

Here we show that Tsirelson's bound follows from the following principle applied to a certain limiting Bell–CHSH setting:

Statistical No-Signaling: It is impossible to communicate a nontrivial message through a channel whose output is independent of its input.

Our strategy is to construct a channel whose input is a Bernoulli random variable X of mean θ and whose output is another Bernoulli random variable Y (Section 3.2). The construction of our channel is not new— it is a reinterpretation of the well-known van Dam protocol [17]. Through the channel, Alice sends 2^n samples $\mathcal{A} \stackrel{\text{def}}{=} \{x_0, x_1, \ldots, x_{2^n-1}\}$ from X, and at the other end Bob receives a set of values $\mathcal{B} \stackrel{\text{def}}{=} \{y_0, y_1, \ldots, y_{m-1}\}$.

We imagine $\theta \in [-1, 1]$ as encoding a message, perhaps in the digits of its binary expansion. Bob's task is to estimate θ. The following theorem states that he can do so if and only if Tsirelson's bound fails.

Theorem 1.1.

1. *The channel from X to Y we construct is described by the conditional probability $p(Y = x \mid X = x) = (1 + c^n)/2$, where c is the Bell–CHSH correlator. Its output satisfies:*

$$p(Y = 1 \mid \theta) = \frac{1}{2} + \frac{c^n \cdot \theta}{2} \quad .$$

In the $n \to \infty$ limit it disconnects for $p(Y \mid X) = p(Y)$ (i.e. we can arrange that $c < 1$).

2. *The unbiased estimator:*

$$\hat{\theta} \stackrel{\text{def}}{=} \frac{1}{2^n c^n} \sum_{i=0}^{2^n-1} y_i \ ,$$

for θ has variance:

$$\mathrm{Var}\left[\hat{\theta} \mid \theta\right] = \lim_{n\to\infty} \frac{1 - c^{2n}\theta^2}{(2c^2)^n} = \begin{cases} 0, & 2c^2 > 1 \text{ (signaling)} \\ 1, & 2c^2 = 1 \text{ (randomness)} \\ \infty, & 2c^2 < 1 \text{ (no-signaling)} \end{cases}$$

3. *The estimator $\hat{\theta}$ is efficient, i.e. it has the minimal variance of any estimator of θ constructed from Bob's set of samples $\mathcal{B}$ for all $n \in \mathbb{N}$.*

The theorem is visually summarized by Figure 1.

The theorem shows that failure of Tsirelson's bound leads to failure of the following consequence of Statistical No-Signaling—*Consequence of Statistical No-Signaling*—In the above notation, if X and Y are independent, then no estimator constructed from $\mathcal{B}$ has both mean θ and variance 0.

Section 5 shows that a violation of Uffink's inequality [22], a generalization of Tsirelson's bound, also leads to the failure of the same consequence of Statistical No-Signaling. Uffink's inequality is also known to be recovered by Information Causality [23].

Theorem 1.1 is formulated as an asymptotic construction, but in practice a finite number of samples suffices because for any experimental setup there exists a nonzero minimal possible environmental noise level $\epsilon > 0$. By Theorem 1.1, $p(Y = 1 \mid \theta)$ is physically indistinguishable from $1/2$ when the absolute value of $c^n\theta/2$ is less than ϵ. Since $|\theta| \le 1$, we need $n \ge \ln 2\epsilon / \ln c$ trials. As an example, for a photon pair where ϵ is greater than or equal to the reduced Planck constant $\hbar$, we find that $n \ge 244$ suffices to make $p(Y = 1 \mid \theta)$ physically indistinguishable from $1/2$ when $|c| \le 1/\sqrt{2}$. Thus, if we can still distinguish $p(Y = 1 \mid \theta)$ from $1/2$ for $n = 244$, we know that Tsirelson's bound has been violated, and if not then it holds.

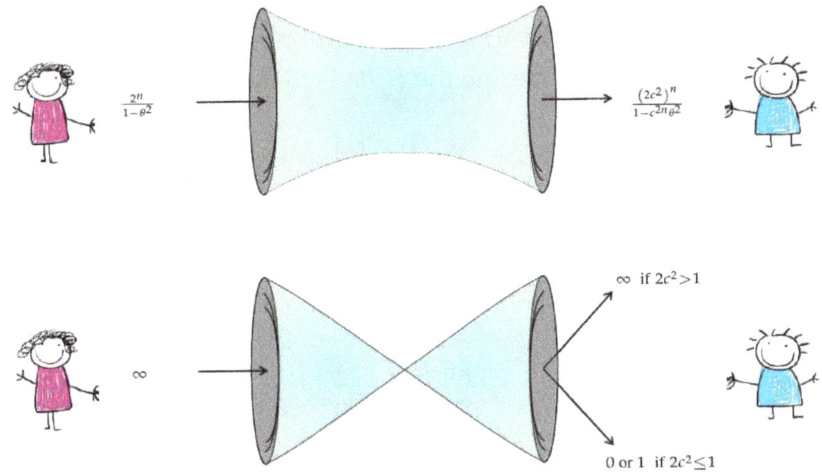

Figure 1. The Statistical No-Signaling condition. The van Dam protocol defines an underlying channel which becomes disconnected in the $n \to \infty$ limit. The upper illustration shows this channel and the Fisher information (one over the variance) of the maximum likelihood estimators for θ at its input and at its output. When the number of nonlocal resources increases unboundedly, the two ends of the channel become disconnected as illustrated by a vanishing bottleneck in the lower illustration. Statistical No-Signaling dictates that in this case no information can pass through. This occurs if and only if $2c^2 \leq 1$. The case of $2c^2 > 1$ leads to a physically unreasonable limit where Bob can fully read off the value of Alice's θ through a disconnected channel.

1.3. *Organization of This Paper*

Section 2 recalls the bipartite Bell experiment and exhibits the Bell–CHSH correlator c as the correlator of a certain noisy symmetric channel. Section 3 presents the van Dam protocol as an extension of the Bell–CHSH setup, and explain how it defines a noisy symmetric channel with correlator c^n. Section 4 computes the means and variance of an estimator $\hat{\theta}$ for θ, and proves that $\hat{\theta}$ is an efficient estimator. Section 5 extends Theorem 1.1 to recover Uffink's inequality [22,23] for anisotropic correlators from Statistical No-Signaling. Finally, Section 6 discusses the relationship of Statistical No-Signaling with Information Causality.

2. The Bipartite Bell Experiment as a Noisy Symmetric Channel

In this section we recall the definition of the Bell–CHSH correlator c and we formulate the Bell–CHSH inequality, establishing notation. We then exhibit c as the correlator of a symmetric binary channel.

2.1. *The Bell–CHSH Inequality*

Let us recall the classical bipartite Bell experiment [1]. Alice and Bob each hold one half of an EPR pair (a pair of particles with certain properties summarized below) such as a singlet state of spin–$\frac{1}{2}$ particles. They each possess two different measuring instruments. Alice measures her particle using one of the instruments, and Bob measures his particles using one of his. We write i for the index of the instrument used by Alice, and a for its reading. Similarly, we let j and b denote the index of an instrument chosen by Bob and its reading correspondingly. In the language of probability, a and b are ± 1–valued Bernoulli random variables. The choices of measuring instrument, i and j, may be either parameters or $0/1$–valued Bernoulli random variables.

Repeating the experiment for many different EPR pairs, Alice and Bob may compute the two-point correlator $E\left[ab \mid i,j\right]$ of their readings a and b for any given pair of indices i and j, where $E[\cdot]$ is the statistical expectation operator. We now define the *Bell–CHSH correlator* c by the formula:

$$c \stackrel{\text{def}}{=} \frac{1}{4}\left\{E\left[ab \mid 0,0\right] + E\left[ab \mid 0,1\right] + E\left[ab \mid 1,0\right] - E\left[ab \mid 1,1\right]\right\} . \tag{2}$$

In a theory in which both Alice and Bob's choices, and the readings of their measuring devices, are *local*, the Bell–CHSH inequality [4] holds:

$$|c| \leq \frac{1}{2} . \tag{3}$$

Operationally speaking, locality means that Alice's readings may only be affected by her own choices (and perhaps by other variables hidden locally at her site), and similarly for Bob's readings. Quantum mechanically, however, Alice and Bob may violate (3). Correlators violating (3) are said to be *nonlocal*.

2.2. The Bell–CHSH Correlator c as a Channel Correlator

Non-signaling (NS)–boxes provide an abstraction and an extension of the Bell–CHSH experiment [2,14]. This time, Alice and Bob each owns a box. Such a box may be thought of as a complete laboratory containing two measuring devices. Either participants inserts their choice of measuring device into their box. The box output is the respective reading of the chosen measuring device.

Alice and Bob share a pair of NS–boxes whose $0/1$–valued inputs are i and j and whose ± 1–valued outputs are Bernoulli random variables a and b. We will show that the Bell–CHSH correlator (2) represents the correlator of a symmetric binary channel whose input is the Bernoulli random variable $X \stackrel{\text{def}}{=} (-1)^{ij}$ and whose output is the Bernoulli random variable $Y \stackrel{\text{def}}{=} a \cdot b$.

Let $x \in \{-1, 1\}$. Define the *channel correlators* c_x as follows:

$$c_x \stackrel{\text{def}}{=} E\left[XY \mid X = x\right] = p(Y = x \mid X = x) - p(Y \neq x \mid X = x) = 2p(Y = x \mid X = x) - 1 . \tag{4}$$

With respect to a particular choice of measuring devices i and j and for $x = (-1)^{ij}$, (4) becomes:

$$c_x(i,j) = E\left[a \cdot b \cdot (-1)^{ij} \mid i,j\right] = 2p(a \cdot b = (-1)^{ij} \mid i,j) - 1 . \tag{5}$$

Assume the underlying channel is symmetric and therefore that $c_x(i,j)$ is fixed for all i,j. By (5) the Bell–CHSH correlator (2) may be written as:

$$c = \frac{1}{4}\left(c_1(0,0) + c_1(0,1) + c_1(1,0) + c_{-1}(1,1)\right) = c_x(i,j) = 2p(a \cdot b = ij \mid i,j) - 1 . \tag{6}$$

which is our promised interpretation of the Bell–CHSH correlator as a correlator of a noisy symmetric binary channel.

3. The Van Dam Protocol as a Noisy Symmetric Channel

In this section we recall the construction of the van-Dam protocol [17,19]. We then reinterpret this protocol as underlying a noisy symmetric binary channel, as a special case of the construction of Section 2. We compute its correlator, and establish the effect of noise on its classical component.

3.1. The Van Dam Protocol

The van Dam protocol realizes an *oblivious transfer protocol* by means of a classical channel and a collection of NS-boxes. Each of Alice's boxes has a corresponding box on Bob's side, and different pairs of boxes are statistically independent. Suppose that Alice has in her possession the bits $x_0, \ldots, x_{m-1}$ where $m = 2^n$, $n \geq 1$. Bob wishes to know the value of one of her bits. He may do so by specifying the address of the bit whose value he wishes to know via its binary address $j = j_{n-1}j_{n-2}\cdots j_0$. For example, if $n = 2$ then Bob may specify which of the bits x_0 to x_3 he wants by specifying a binary address, 00, 01, 10, or 11. Alice bits and Bob addresses are encoded into the inputs of $2^n - 1$ NS-boxes following a particular protocol which is described next.

Alice uses outputs of boxes and choices of measuring device to determine choices of measuring device for other boxes. Such a procedure is called *wiring*. The wiring of boxes on Alice side admits a recursive description which we now give. Let $a_i^{k,l}$ denote the output of Alice's lth box on the kth level for the input i. We follow the convention that box outputs for the van Dam protocol are 0/1–valued (rather than ±1–valued) random variables. Let also:

$$f^{k,l}(q_1, q_2) \stackrel{\text{def}}{=} q_1 \oplus a^{k,l}_{q_1 \oplus q_2} \ . \tag{7}$$

Suppose that Alice wishes to encode $m = 4$ bits with her boxes. To do so, she first picks two boxes and computes:

$$x_1^{(1)} \stackrel{\text{def}}{=} f^{1,1}(x_0, x_1), \quad x_2^{(1)} \stackrel{\text{def}}{=} f^{1,2}(x_2, x_3) \ . \tag{8}$$

This forms the first level in her construction. The second level then follows:

$$x^{(2)} \stackrel{\text{def}}{=} f^{2,1}\left(x_1^{(1)}, x_2^{(1)}\right) \ . \tag{9}$$

In this example there are only two levels and so $x^{(2)}$ is the bit which Alice transmits to Bob through the classical channel. In case where $m = 2^n$ there will be n levels and thus $x^{(n)}$ is the bit Bob will receive from Alice.

Unbeknownst to Alice, Bob now decides which bit x_j he would like to know the value of. He takes its binary address $j = j_{n-1}j_{i-2}\cdots j_0$, and inserts j_{k-1} into all of his boxes whose counterparts are on the k level on Alice's side. He then uses the values $b^{k,l}_{j_{k-1}}$ that he obtains, together with the bit $x^{(n)}$ he received from Alice, to construct the decoding function:

$$y_j \stackrel{\text{def}}{=} x^{(n)} \oplus b^{1,l_1}_{j_0} \oplus b^{2,l_2}_{i_1} \oplus \cdots \oplus b^{n,l_n}_{j_{n-1}} \ . \tag{10}$$

The values $l_1, \ldots, l_n$ (which boxes Bob uses) are determined by the binary address $j = j_{n-1}j_{n-2}\cdots j_0$ via the recursive formula $l_{h-1} = 2l_h - 1 + l_{h-1}$ for $h = 1, 2, \ldots n-1$ starting from $l_n = 1$.

The van Dam protocol we have described above is summarized in Figure 2.

The probability that Bob will decode the correct value of the bit he desires is governed by the NS–box correlator c. In general, decoding any bit out of 2^n possible bits involves using n pairs of NS boxes. Noting that an even number of errors, $a \oplus b \neq ij$, will cancel out in such a construction, we obtain the following expression [19]:

$$c^n = 2p(y_j = x_j \mid x_j) - 1 \ . \tag{11}$$

For example, for $n = 2$:

$$p(a_{i_1} \oplus b_{j_1} \oplus a_{j_2} \oplus b_{j_2} = i_1 j_1 \oplus i_2 j_2 \mid i_{1,2}, j_{1,2}, i_1 j_1 \oplus i_2 j_2) =$$
$$p(a_{i_1} \oplus b_{j_1} = i_1 j_1 \mid a_1, b_1) p(a_{i_2} \oplus b_{j_2} = i_2 j_2 \mid i_2, j_2) +$$
$$p(a_{i_1} \oplus b_{j_1} \neq i_1 j_1 \mid i_1, j_1) p(a_{i_2} \oplus b_{j_2} \neq i_2 j_2 \mid i_2, j_2) =$$
$$\frac{1}{2}(1+c) \cdot \frac{1}{2}(1+c) + \frac{1}{2}(1-c) \cdot \frac{1}{2}(1-c) = \frac{1}{2}(1+c^2) \ . \quad (12)$$

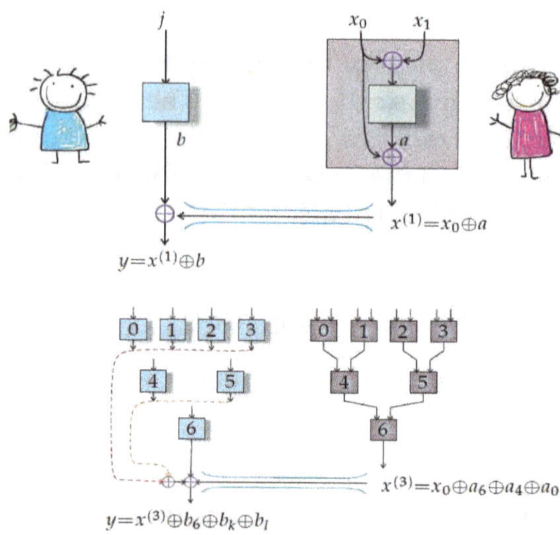

Figure 2. Distributed oblivious transfer (van Dam) protocol [17]. Its basic building block is on the left, where Alice inserts $x_0 \oplus x_1$ into her box, receives a, and sends $x_0 \oplus a$ to Bob. Bob decides that he wants to know the value of x_j, and he feeds j into his box, which outputs b. Bob's estimate of x_i is then $x^{(1)} \oplus b$. When there are multiple boxes, Alice concatenates (the process is called *wiring*). For example, with seven boxes, Alice begins with a collection of bits $x_0, x_1, \ldots, x_7$, and she inputs $x_{2i} \oplus x_{2i+1}$ into box i, where $i = 0, 1, 2, 3$, receiving a_0, a_1, a_2, a_3 correspondingly. The bits fed into the next level of boxes become $x_i^{(1)} \stackrel{\text{def}}{=} x_{2i} \oplus a_i$ with $i = 0, 1, 2, 3$. The final output $x^{(3)}$ is sent to Bob. Bob encodes the address of the bit he wants as the binary number $j_3 j_2 j_1$—for example, if he wants x_2, then he sets $j_3 = 0$, $j_2 = 1$, and $j_1 = 0$ because 10 is 2 in binary. This binary encoding describes a path in his binary tree from a root to a branch, where 0 means 'go left' and 1 means 'go right'. Bob inserts j_3 into the lowermost box to obtain b_6. Setting $k \stackrel{\text{def}}{=} 5 - (1 - j_3)$, he then inserts j_2 into box k to obtain b_k. Finally, setting $l \stackrel{\text{def}}{=} k - (3 - j_3) - (1 - j_2)$, Bob inserts j_1 into box l to obtain B_l. His final estimate for x_j is $y_j = x^{(3)} \oplus b_6 \oplus b_k \oplus b_l$.

3.2. Van Dam Protocol as a Symmetric Channel

This section describes the modification of the van Dam protocol that we use.

Alice has in her possession an information source that is a ± 1-valued Bernoulli random variable X whose mean is θ. Alice takes m iid samples, $\tilde{x}_0, \ldots, \tilde{x}_{m-1}$, from X and converts them into $0/1$-valued bits, $x_0, x_1, \ldots, x_{m-1}$ by mapping 0 to -1 and 1 to 1. Alice and Bob repeat the van Dam protocol m times, once for each of Alice's samples. Each time, Bob uses the protocol to estimate Alice's bit, first x_0, then x_1, and so on until x_{m-1}.

As in (12), the van Dam protocol has a *memoryless* property:

$$p(y_i = x_i \mid x_0, x_1, \ldots, x_{m-1}) = p(y_i = x_i \mid x_i) \ . \quad (13)$$

From this it follows that if Alice's inputs $x_0, x_1, \ldots, x_{m-1}$ are iid then Bob's outputs $y_0, y_1, \ldots, y_{m-1}$ are also iid. Therefore the set of $\tilde{y}_i \stackrel{\text{def}}{=} (-1)^{y_i}$ determines a Bernoulli random variable Y. In this way, the van Dam protocol may be viewed as a symmetric binary channel whose input is X and whose output is Y. By (11) the channel correlator is:

$$E\left[XY \mid X = \tilde{x}_i\right] = 2p(Y = \tilde{x}_i \mid X = \tilde{x}_i) - 1 = 2p(y_i = x_i \mid x_i) - 1 = c^n \ . \tag{14}$$

We generalize slightly, for the purpose of treating the $|c| = 1$ case in the next section. Suppose that Alice's bits are contaminated with noise and therefore might be flipped once injected into her boxes. Let $[1 - (c')^n]/2$ be the probability that the bit x_i is flipped where $|c'| \leq 1$. In this case the corresponding channel correlator (14) is $E\left[XY \mid X = \tilde{x}_i\right] = (cc')^n$, which follows from (4) and:

$$\begin{aligned} p(Y = \tilde{x}_i \mid X = \tilde{x}_i) = p(Y = \tilde{x}_i \mid X' = \tilde{x}_i)p(X' = \tilde{x}_i \mid X = \tilde{x}_i) + \\ p(Y = \tilde{x}_i \mid X' \neq \tilde{x}_i)p(X' \neq \tilde{x}_i \mid X = \tilde{x}_i) = \frac{1}{2}[1 + (cc')^n] \ , \end{aligned} \tag{15}$$

where $p(Y = \tilde{x}_i \mid X' = \tilde{x}_i) = [1 + c^n]/2$ underlies the channel defined by the ordinary van Dam protocol, and $p(X' \neq \tilde{x}_i \mid X = \tilde{x}_i) = [1 - (c')^n]/2$ is the probability of x_i having been flipped.

3.3. The Van Dam Channel Disconnects in the $n \to \infty$ Limit

If $|c| < 1$ or $|c'| < 1$ then it follows that:

$$E[XY] = 2p(Y = i \mid X = i) - 1 = (cc')^n \stackrel{n \to \infty}{\longrightarrow} 0 \ . \tag{16}$$

Therefore, in the $n \to \infty$ limit:

$$p(Y = i \mid X = i) = 1/2 \ . \tag{17}$$

But also:

$$p(Y = i) = p(Y = i \mid X = i)p(X = i) + p(Y = i \mid X \neq i)p(X \neq i) = \tfrac{1}{2}(p(X = i) + p(X \neq i)) = \tfrac{1}{2} \ . \tag{18}$$

Combining (17) with (18) gives:

$$p(Y \mid X) \stackrel{n \to \infty}{\longrightarrow} p(Y) \ . \tag{19}$$

Thus X and Y are statistically independent in the $n \to \infty$ limit, proving the first part of Theorem 1.1.

4. Bob's Estimator

4.1. Bob's Estimator

In Section 3 we used the van Dam protocol to construct a symmetric channel whose input is a ± 1–valued Bernoulli random variable X and whose output is another ± 1–valued Bernoulli random variable Y. The channel correlator is c^n.

Alice sends m iid random samples $\mathcal{X} \stackrel{\text{def}}{=} \{X_1, \ldots, X_m\}$ through the channel. Denote the set of respective outputs $\mathcal{Y} \stackrel{\text{def}}{=} \{Y_1, \ldots, Y_m\}$. Assume a prior distribution for X given by:

$$p(X = -1 \mid \theta) = \frac{1}{2}(1 + \theta) \ , \tag{20}$$

with parameter $\theta \in [-1, 1]$.

Bob attempts to estimate θ using the estimator:

$$\hat{\theta} \stackrel{\text{def}}{=} \frac{1}{2^n c^n} \sum_{i=0}^{2^n-1} Y_i \ . \tag{21}$$

We will show that Bob's estimator is unbiased, $E\left[\hat{\theta} \mid \theta\right] = \theta$. Note that

$$E\left[Y_i \mid \theta\right] = p(Y=1 \mid \theta) - p(Y=-1 \mid \theta) \ . \tag{22}$$

and

$$p(Y=-1 \mid \theta) = p(Y=-1 \mid X=-1)p(X=-1 \mid \theta) + p(Y=-1 \mid X=1)p(X=1 \mid \theta) = \tfrac{1+c^n\theta}{2} \ . \tag{23}$$

From (22) and (23) together, deduce:

$$E\left[Y_i \mid \theta\right] = c^n\theta \ . \tag{24}$$

and therefore, $E\left[\hat{\theta} \mid \theta\right] = \theta$.

As for variance, by (24):

$$\mathrm{Var}\left[Y_i \mid \theta\right] = E\left[Y_i^2 \mid \theta\right] - E\left[Y_i \mid \theta\right]^2 = 1 - c^{2n}\theta^2 \ . \tag{25}$$

Therefore:

$$\mathrm{Var}\left[\hat{\theta} \mid \theta\right] = \frac{1 - c^{2n}\theta^2}{(2c^2)^n} \ . \tag{26}$$

We have proved the second part of Theorem 1.1.

4.2. Bob's Estimator $\hat{\theta}$ is Efficient

We prove efficiency of $\hat{\theta}$ by calculating the Fisher information about θ contained in Bob's set of samples $\mathcal{B}$. The Cramer–Rao Theorem tells us that one over this Fisher information is a lower bound for the variance of an estimator for θ constructed from $\mathcal{B}$. By showing that $\hat{\theta}$ saturates this bound, we will have proven that it is efficient. In the derivation that follows, we assume that $|c| < 1$ by replacing c by cc' if necessary.

We compute the Fisher information. The *likelihood* of θ given the set $\mathcal{B}$ is given by the expression:

$$p(\mathcal{B} \mid \theta) = \left[p(Y=-1 \mid \theta)\right]^{\sum_{i=1}^{2^n} \mathbf{1}_{\{Y_i=-1\}}} \left[p(Y=1 \mid \theta)\right]^{\sum_{i=1}^{2^n} \mathbf{1}_{\{Y_i=1\}}} \ , \tag{27}$$

where the *indicator* random variable of a random event A is given as:

$$\mathbf{1}_A \stackrel{\text{def}}{=} \begin{cases} 1, & A \text{ occurred;} \\ 0, & \text{otherwise.} \end{cases} \tag{28}$$

According to (27) the log-likelihood is given by the expression:

$$\mathcal{L}(\theta) \stackrel{\text{def}}{=} \log p(\mathcal{B} \mid \theta) = \left[\sum_{i=1}^{2^n} \mathbf{1}_{\{Y_i=-1\}}\right] \log p(Y=-1 \mid \theta) + \left[\sum_{i=1}^{2^n} \mathbf{1}_{\{Y_i=1\}}\right] \log p(Y=1 \mid \theta) \ . \tag{29}$$

The *Fisher information* about θ contained in the set $\mathcal{B}$ is defined as:

$$\mathcal{I}_{\mathcal{B}}(\theta) \stackrel{\text{def}}{=} E\left[\left(\frac{\partial \mathcal{L}(\theta)}{\partial \theta}\right)^2\right] = -E\left[\frac{\partial^2 \mathcal{L}(\theta)}{\partial \theta^2}\right] . \tag{30}$$

Note that:

$$E\left[\sum_{i=1}^{2^n} \mathbf{1}_{\{Y_i=s\}}\right] = \sum_{i=1}^{2^n} E\left[\mathbf{1}_{\{Y_i=s\}}\right] = 2^n p(Y=s \mid \theta), \quad s=-1,1 . \tag{31}$$

Using this, (30) reads:

$$\mathcal{I}_{\mathcal{B}}(\theta) = \frac{(2c^2)^n}{1-c^{2n}\theta^2} . \tag{32}$$

Indeed the Fisher information about θ in $\mathcal{B}$ as given by Equation (32) equals one over the variance of $\hat{\theta}$ as given by Equation (26). Thus, by the Cramer–Rao Theorem, $\hat{\theta}$ is an efficient estimator for θ. Parenthetically, note that the minimum of $\mathcal{I}_{\mathcal{B}}(\theta)$ is obtained for $\theta = 0$ in which case $p(X \mid \theta) = 1/2$ and $\mathcal{I}_{\mathcal{B}}(0) = (2c^2)^n$. We have proved the final part of Theorem 1.1.

5. Uffink's Inequality from Statistical No-Signalling

The basic protocol in Section 3 assumes all box correlators are identical in absolute value. When this assumption is relaxed, Statistical No-Signaling leads to Uffink's inequality, which is a necessary condition for quantum mechanical Bell-CHSH correlators [22,23]. Our approach is based on evaluating the total Fisher information $\mathcal{I}_{\mathcal{B}}(\theta)$ gained by Bob in 2^n trials of the experiment.

Suppose that the mean of Alice's bits, x_i, is θ' for even i, and θ otherwise. Consider now a pair of NS-boxes with correlators, $c(i,j) \stackrel{\text{def}}{=} E[ab \mid i,j]$. The channel underlying the van Dam protocol in this case is described by

$$p(y_j = x_j \mid x_0, x_1) = p(a \oplus b = ij \mid j, i = x_0 \oplus x_1) = \left[1 + c(x_0 \oplus x_1, j)\right]/2, \tag{33}$$

where y_j is Bob's guess of Alice's bit x_j. It now follows that

$$\begin{aligned}
&p(y_j = 1 \mid \theta', \theta) = \\
&p(y_j = x_j \mid x_j = 1, x_{1-j} = 1)p(x_j = 1)p(x_{1-j} = 1) + p(y_j \neq x_j \mid x_j = 0, x_{1-j} = 0)p(x_j = 0)p(x_{1-j} = 0) + \\
&p(y_j = x_j \mid x_j = 1, x_{1-j} = 0)p(x_j = 1)p(x_{1-j} = 0) + p(y_j \neq x_j \mid x_j = 0, x_{1-j} = 1)p(x_j = 0)p(x_{1-j} = 1) = \\
&\tfrac{1}{2}\left[1 + \tfrac{1}{2}(c(0,j) + (-1)^j c(1,j))\theta' + \tfrac{1}{2}(c(0,j) - (-1)^j c(1,j))\theta\right] .
\end{aligned} \tag{34}$$

For simplicity, assume that $\theta' = 0$. It can now be verified that for a n-level construction in the van Dam protocol

$$p(y_{j_1,\ldots,j_n} = 1 \mid \theta) = \frac{1}{2}\left[1 + c_{j_1} c_{j_2} \cdots c_{j_n} \theta\right], \tag{35}$$

where $c_j \stackrel{\text{def}}{=} (c(0,j) - (-1)^j c(1,j))/2$. According to (32) the Fisher information about θ contained in $y_{j_1,\ldots,j_n}$ is

$$\mathcal{I}_{j_1,\ldots,j_n}(\theta) = \frac{\left(c_{j_1} \cdots c_{j_n}\right)^2}{1 - \left(c_{j_1} \cdots c_{j_n}\right)^2 \theta^2}. \tag{36}$$

Assuming $|c(i,j)| < 1$, Bob's total amount of information about θ in 2^n trials is

$$\mathcal{I}_{\mathcal{B}}(\theta) = \sum_{j_1=0,1} \cdots \sum_{j_n=0,1} \mathcal{I}_{j_1,\ldots,j_n}(\theta) \approx \sum_{j_1=0,1} \cdots \sum_{j_n=0,1} \left(c_{j_1} \cdots c_{j_n}\right)^2 = \left[c_0^2 + c_1^2\right]^n , \tag{37}$$

for large n. As before, the underlying channel asymptotically disconnects for $c_{j_1} \cdots c_{j_n} \to 0$ in the $n \to \infty$ limit. Statistical No-Signaling dictates that in this case the variance of Bob's estimator $\lim_{n\to\infty} \mathrm{Var}\left[\hat{\theta} \mid \theta\right] = \lim_{n\to\infty} \mathcal{I}_{\mathcal{B}}(\theta)^{-1} \geq 1$, which holds if and only if Uffink's inequality holds [22],

$$c_0^2 + c_1^2 = \frac{1}{4}\left[c(0,0) - c(1,0)\right]^2 + \frac{1}{4}\left[c(0,1) + c(1,1)\right]^2 \leq 1. \tag{38}$$

6. Relation to Information Causality

Of previous non-quantum justifications of Tsirelson's bound, Information Causality (IC) is perhaps the closest to Statistical No-Signalling [19]. IC is also stated as a limit on communication: *Information gain that Bob can reach about a previously unknown to him data set of Alice, by using all his local resources and m classical bits communicated by Alice, is at most m bits.*

IC is formally a restriction on the classical channel capacity. Detecting violation of this principle therefore requires the utilization of nonlocal resources, which the authors achieve through the application of IC to the van Dam protocol, that is the same communication protocol used in this paper.

The Information Causality quantity I is defined as the Shannon mutual information of Alice's input and Bob's output given the value of the single bit transmitted in the van Dam protocol. IC holds if $I \leq 1$ and is violated if $I > 1$. At the end of the supplementary section of [19], the following expression for the IC quantity is obtained:

$$I \geq \frac{1}{2\ln(2)}\left(c_1^2 + c_{-1}^2\right)^n , \tag{39}$$

where $c_i \stackrel{\text{def}}{=} E\left[XY \mid X = \tilde{i}\right]$ as in (4). In the symmetric setting, $c_1 = c_{-1} = c$, and for $\theta = 0$, Equations (39) and (32) combine to yield:

$$I \geq \frac{2^n c^{2n}}{2\ln(2)} = \frac{\left[1 - c^{2n}\theta^2\right]\mathcal{I}_{\mathcal{B}}(\theta)}{2\ln(2)} . \tag{40}$$

In particular, in the $n \to \infty$ limit, if $2c^2 > 1$ then $\mathcal{I}_{\mathcal{B}}(\theta) \to \infty$ implying that $I \to \infty$. Thus, violation of Statistical No-Signaling implies violation of IC. Conversely, as (39) is an inequality, it is unknown whether Tsirelson's bound being satisfied implies $I \leq 1$ (IC for the van Dam protocol), although, by our main theorem, it does imply $\mathcal{I}_{\mathcal{B}}(\theta) \leq 1$ (Statistical No-Signaling for the van Dam protocol).

7. Conclusions

We have formulated a *Statistical No-Signaling* principle which dictates that no information can pass through a disconnected channel. A violation of Tsirelson's bound, *i.e.* a value of $|c|$ greater that $1/\sqrt{2}$, allows us to violate Statistical No-Signalling by constructing a disconnected channel through which Bob can construct an unbiased estimator with variance 0 for Alice's parameter θ. Conversely, when Tsirelson's bound holds, then, through this channel, so does Statistical No-Signalling. Our construction thus provides a purely statistical justification for Tsirelson's bound, independent of quantum mechanics.

Acknowledgments: The authors thank Daniel Rohrlich for useful discussions. Avishy Carmi acknowledges support from Israel Science Foundation Grant No. 1723/16.

Author Contributions: Avishy Carmi and Daniel Moskovich have both written the text and worked out the mathematical proofs in this paper.

References

1. Bell, J.S. On the Einstein-Podolsky-Rosen paradox. *Physics* **1964**, *1*, 195–200.
2. Popescu, S.; Rohrlich, D. Quantum nonlocality as an axiom. *Found. Phys.* **1994**, *24*, 379–385.
3. Popescu, S. Nonlocality beyond quantum mechanics. *Nature Phys.* **2014**, *10*, 264–270.
4. Clauser, J.; Horne, M.; Shimony, A.; Holt, R. Proposed experiment to test local hidden—Variable theories. *Phys. Rev. Lett.* **1969**, *23*, 880–884.
5. Cirel'son, B. S. Quantum generalizations of Bell's inequality. *Lett. Math. Phys.* **1980**, *4*, 93–100.
6. Giustina, M.; Versteegh, M.A.M.; Wengerowsky, S.; Handsteiner, J.; Hochrainer, A.; Phelan, K.; Steinlechner, F.; Kofler, J.; Larsson, J.-A.; Abellán, C.; et al. Significant-loophole-free test of Bell's theorem with entangled photons. *Phys. Rev. Lett.* **2015**, *115*, 250401.
7. Hensen, B.; Bernien, H.;Dréau, A.E.; Reiserer, A.; Kalb, N.; Blok, M.S.; Ruitenberg, J.; Vermeulen, R.F.L.; Schouten, R.N.; Abellán, C.; et al. Experimental loophole-free violation of a Bell inequality using entangled electron spins separated by 1.3 km. *Nature* **2015**, *526*, 682–686.
8. Shalm, L. K.; Meyer-Scott, E.; Christensen, B.G.; Bierhorst, P.; Wayne, M.A.; Stevens, M.J.; Gerrits, T.; Glancy, S.; Hamel, D.R.; Allman, M.S.; et al. Strong loophole-free test of local realism. *Phys. Rev. Lett.* **2015**, *115*, 250402.
9. Poh, H.S.; Joshi, S.K.; Ceré, A.; Cabello, A.; Kurtsiefer, C. Approaching Tsirelson's bound in a photon pair experiment. *Phys. Rev. Lett.* **2015**, *115*, 180408.
10. Seife, C. Do deeper principles underlie quantum uncertainty and nonlocality? *Science* **2005**, *309*, 98.
11. Shimony, A. Controllable and Uncontrollable Non-Locality. In *Proceedings of the International Symposium on Foundations of Quantum Mechanics in the Light of New Technology*; Kamefuchi, S., Ed; Physical Society of Japan: Tokyo, Japan, 1984; pp. 225–230.
12. Shimony, A. Events and processes in the quantum world. In *Quantum Concepts in Space and Time*; Penrose, R., Isham, C.J, Eds.; Oxford University Press: Oxford, UK, 1986; pp. 182–203.
13. Aharonov, Y.; Rohrlich, D. Nonlocality and Causality. In *Quantum Paradoxes: Quantum Theory for the Perplexed*; Wiley-VCH: Weinheim, Germany, 2005.
14. Barrett, J.; Linden, N.; Massar, S.; Pironio, S.; Popescu, S.; Roberts, D. Non-local correlations as an information theoretic resource. *Phys. Rev. A* **2005**, *71*, 022101.
15. Wolf, M.; Garcia, D.P.; Fernandez, C. Measurements incompatible in quantum theory cannot be measured jointly in any other no-signaling theory. *Phys. Rev. Lett.* **2009**, *103*, 230402.
16. Oppenheim, J., Wehner, S. The uncertainty principle determines the non-locality of quantum mechanics. *Science* **2010** , *330*, 1072–1074.
17. van Dam, W. Implausible consequences of superstrong nonlocality. *Nat. Comput.* **2013**, *12*, 9–12.
18. Linden, N.; Popescu, S.; Short, A.J.; Winter, A. Quantum nonlocality and beyond: Limits from nonlocal computation. *Phys. Rev. Lett.* **2007**, *99*, 180502.
19. Pawlowski, M.; Paterek, T.; Kaszlikowski, D.; Scarani, V.; Winter, A.; Żukowski, M. Information causality as a physical principle. *Nature* **2009**, *461*, 1101–1104.
20. Rohrlich, D. PR-box correlations have no classical limit. In *Quantum Theory: A Two-Time Success Story*; Struppa, D.C., Tollaksen, J.M., Eds.; Springer: Berlin/Heidelberg, Germany, 2014; pp. 205–211.
21. Navascués, M.; Wunderlich, H. A glance beyond the quantum model. *Proc. R. Soc. A* **2010**, *466*, 881–890.
22. Uffink, J. Quadratic Bell inequalities as tests for multipartite entanglement. *Phys. Rev. Lett.* **2002**, *88*, 230406.
23. Allcock, J.; Brunner, N.; Pawlowski, M.; Scarani, V. Recovering part of the boundary between quantum and nonquantum correlations from information causality. *Phys. Rev. A* **2009**, *80*, 040103.

6

Image Thresholding Segmentation on Quantum State Space

Xiangluo Wang [1], Chunlei Yang [2,*], Guo-Sen Xie [2] and Zhonghua Liu [2]

[1] School of Information Technology, Luoyang Normal University, Luoyang 471934, China; kevinx_25@126.com
[2] School of Information Engineering, Henan University of Science and Technology, Luoyang 471023, China; xieguosen001@163.com (G.-S.X.); lzhlsh123@163.com (Z.L.)
* Correspondence: cleiyang@haust.edu.cn or cleiyang@126.com.

Abstract: Aiming to implement image segmentation precisely and efficiently, we exploit new ways to encode images and achieve the optimal thresholding on quantum state space. Firstly, the state vector and density matrix are adopted for the representation of pixel intensities and their probability distribution, respectively. Then, the method based on global quantum entropy maximization (GQEM) is proposed, which has an equivalent object function to Otsu's, but gives a more explicit physical interpretation of image thresholding in the language of quantum mechanics. To reduce the time consumption for searching for optimal thresholds, the method of quantum lossy-encoding-based entropy maximization (QLEEM) is presented, in which the eigenvalues of density matrices can give direct clues for thresholding, and then, the process of optimal searching can be avoided. Meanwhile, the QLEEM algorithm achieves two additional effects: (1) the upper bound of the thresholding level can be implicitly determined according to the eigenvalues; and (2) the proposed approaches ensure that the local information in images is retained as much as possible, and simultaneously, the inter-class separability is maximized in the segmented images. Both of them contribute to the structural characteristics of images, which the human visual system is highly adapted to extract. Experimental results show that the proposed methods are able to achieve a competitive quality of thresholding and the fastest computation speed compared with the state-of-the-art methods.

Keywords: image segmentation; thresholding; von Neumann entropy; density matrix

1. Introduction

Image segmentation is the task of dividing the image into different regions, each one of which ideally belongs to the same object or content. As a key step from image processing to computer vision, image segmentation is the target expression and has an important effect on the feature measurement, high-level image analysis and understanding [1,2]. Examples of image segmentation applications include medical imaging [3,4], document image analysis [5], object recognition [6,7] and quality inspection of materials [8,9]. In the last two decades, a wide variety of segmentation techniques have been developed, which conventionally fall into the following two categories [2]: layer-based and block-based segmentation methods [10,11]. Among all these techniques, the thresholding methods offer numerous advantages such as smaller storage space, fast processing and ease in manipulation.

In general, thresholding methods can be classified into parametric and nonparametric approaches [12]. Parametric approaches assume that the intensity distributions of images obey the Gaussian mixture (GM) model, which means the number and parameters of Gaussians in the mixture (the model selection) must be determined [13]. Although these problems have been traditionally solved by considering the expectation maximization (EM) algorithm [14] or gradient-based methods [15,16], the methods are time consuming. Nonparametric approaches find the thresholds that separate regions of an image in an optimal manner based on discriminating criteria such as the between-class variance [17], cluster distance [18], entropy [19–22], etc. Nonparametric methods have shown the advantage of dispensing with the modeling thresholding. However, they still suffer from the problem of high time consumption, although many techniques based on intelligent optimization algorithms (IOAs) [23–25] have been used to speed up the thresholding procedure.

Quantum computation and quantum information processing techniques have shown an immense potential and a revolutionary impact on the field of computer science, due to their remarkable resources: quantum parallelism, quantum interference and entanglement of quantum states. Information representing and processing in the framework of quantum theory is powerful for solving complex problems that are difficult or currently even impossible for conventional methods. The most significant works include Shor's quantum integer factoring algorithm, which can find the secret key encryption of the RSA algorithm in polynomial time [26], and Grover's quantum search algorithm for databases, which could achieve quadratic speedup [27]. In the recent years, quantum approaches have been introduced into the image processing field. Various quantum image representation models have been proposed, such as qubit lattice [28] and flexible representation of quantum images (FRQI) [29]. Meanwhile, several applications of quantum image processing have been researched including quantum image segmentation [30], quantum edge detection [31], quantum image recognition [32], quantum image watermarking [33] and quantum image reconstruction [34]. Though the research in quantum image processing still confronts fundamental aspects such as image representation on a quantum computer and the definition of basic processing operations, we still could be inspired to completely exploit new methods for some classical problems from a quantum information theoretical viewpoint.

In this paper, we address the thresholding problem on quantum state space. The proposed methods relate to the details of image representation by utilizing the density matrix, optimal threshold selection based on the criteria of the maximum von Neumann entropy, a novel image encoding scheme and the corresponding segmentation approaches, which can totally avoid the process of optimal solution searching. Specifically, the contributions of this paper mainly include the following aspects:

(1) We present an image thresholding method based on the criteria of global quantum entropy maximization (GQEM), which has an equivalent object function to Otsu's, but gives more explicit physical interpretation of image thresholding in the language of quantum mechanics.
(2) The quantum lossy-encoding based entropy maximization (QLEEM) approach is proposed to deal with the time consumption problem of thresholding. The QLEEM algorithm directly takes the eigenvalues of density matrices of lossy-encoded images as segmenting clues and then avoids the time-consuming process of searching for optimal thresholds. It can achieve the highest execution speed compared with the state-of-the-art methods.
(3) Due to the physical meaning of the lossy-encoding scheme and the unique procedure of optimal thresholding, a brand-new approach to determine the upper bound of the thresholding level automatically is offered in the proposed QLEEM algorithm. For most of the existing methods, this parameter is conventionally predetermined according to empirical knowledge.
(4) The QLEEM method provides the maximum inter-class separability with lower loss of intra-class information; thus, segmented images could keep more structural information. This feature is highly consistent with the way the human visual system (HVS) works.

The paper is organized as follows: Section 2 gives a brief description of the image thresholding and introduces some state-of-the-art thresholding methods including Otsu's between-class variance method [17], Kapur's entropy-criterion method [19], the quantum version of Kapur's method [35], and Tsallis entropy-based method [22]. Section 3 introduces the details of the proposed methods. Section 4 provides the experimental results and discussions about our method's performance. The conclusions of this study are drawn in the last part of this paper.

2. Related Works

Thresholding is a process in which a group of thresholds is selected under some criteria, and then, pixels of an image are divided into a series of sets or classes according to the rule of:

$$l \rightarrow C_i \; if \; th_{i-1} \leq l < th_i, \tag{1}$$

where $l \in [0, L-1]$ represents the intensity level of image pixels, $\{th_i \mid i = 1, 2, \cdots, M-1\}$ is the set of thresholds and $\{C_i \mid i = 1, 2, \cdots, M\}$ are classes labeling different groups of pixels.

Otsu's between-class variance method [17] selects the optimal thresholds by maximizing the following object function:

$$\sigma^{2^c} = \sum_{i,j} \omega_i \omega_j (\mu_i - \mu_j)^2. \tag{2}$$

Here, i and j index the intensity classes, and ω_i and μ_i are the probability of occurrence and the mean of a class, respectively. Such values are obtained as:

$$\omega_i = \sum_{j=th_{i-1}+1}^{th_i} p_j, \mu_i = \sum_{j=th_{i-1}+1}^{th_i} q_j j. \tag{3}$$

where p_j denotes the probability distribution of pixels and $q_j = p_j/\omega_i$. As we know, Otsu's method can achieve the best segmenting results if no contextual or semantic information is considered, but it suffers from the drawback of time-consuming searching for optimal thresholds.

Kapur presented another discriminant criterion based on maximum entropy [19]:

$$\arg\max_{TH} \sum_{i=0}^{M-1} H(C_i). \tag{4}$$

where $H(C_i)$ is the Shannon entropy corresponding to a specific class, which is defined as:

$$H(C_i) = -\sum_{j=th_{i-1}+1}^{th_i} q_j \log q_j. \tag{5}$$

Similarly, the quantum version of Kapur's method [35] determines the optimal thresholds by maximizing the von Neumann entropy:

$$\arg\max_{TH} \sum_{i=0}^{M-1} S(\rho_i). \tag{6}$$

where:

$$\rho_i = -\sum_{j=th_{i-1}+1}^{th_i} q_j \left|\theta_j\right\rangle \left\langle\theta_j\right| \tag{7}$$

is the density matrix representation of the i-th class and:

$$S(\rho_i) = -tr(\rho_i \log \rho_i). \tag{8}$$

Recently, the Tsallis entropy-based bi-level thresholding method was proposed [22], in which the optimal threshold is given by:

$$t^*(q) = \arg\max_t \left[S_T^A(t) + S_T^B(t) + (1-q) S_T^A(t) S_T^B(t)\right]. \tag{9}$$

Here, $S_T^A(t)$ and $S_T^B(t)$ represent the Tsallis entropy for object A and the background B, respectively, and the entropic index q can be calculated through q-redundancy maximization.

The effectiveness of these entropy-based methods has been proven. However, similar to Otsu's method, they also have the drawback of high computational complexity, which will affect the efficiency of the whole vision task.

3. Proposed Methods

In this section, we will start with a new method, which utilizes the criteria of global quantum entropy maximization to achieve optimal thresholding, and then propose a novel encoding scheme. Based on this scheme, the improved method for thresholding is derived, which can determine optimal thresholds with linear time complexity.

3.1. Thresholding Based on Global Quantum Entropy Maximization

For an image, we can represent its histogram with the following entangled state of a composite quantum system:

$$|I\rangle = \sum_{i=0}^{L-1} \sqrt{p_i}\,|\theta_i\rangle \otimes |i\rangle\,. \tag{10}$$

where we encode the i-th intensity level to the vector $|\theta_i\rangle = cos\theta_i\,|0\rangle + sin\theta_i\,|1\rangle$, which belongs to the state space of the first one-qubit subsystem (labeled as "A"), by establishing a bijective relationship between them, namely:

$$\theta_i = \frac{\pi}{2}\cdot\frac{i}{L-1}, i \in [0, L-1], \tag{11}$$

and $|i\rangle$ is the computational basis state of the second subsystem (labeled as "B"), which denotes the indices of pixel intensities. Though $|I\rangle$ is a pure state, the subsystem A or B is in a mixed state. Therefore, we describe these quantum systems in the language of the density matrix. Assuming $|I\rangle$ is rewritten as ρ^{AB}, then the reduced density matrix for the subsystem A can be defined by:

$$\begin{aligned}\rho &= tr_B(\rho^{AB}) \\ &= \sum_{i=0}^{L-1} p_i\,|\theta_i\rangle\,\langle\theta_i|\,.\end{aligned} \tag{12}$$

The density matrix ρ contains the information about the distance between any two intensities, as well as their probability distribution. This property will be very useful for thresholding.

If pixels of an image are divided into M classes by using M-1 thresholds, we represent the histogram of the segmented image with:

$$|I'\rangle = \sum_{i=0}^{M-1}\left(\sqrt{\omega_i}\,\left|\widetilde{\theta}_i\right\rangle \otimes \sum_{j=th_{i-1}+1}^{th_i} \sqrt{q_j}\,|i\rangle\right), \tag{13}$$

where $\widetilde{\theta}_i = \frac{\pi}{2}\cdot\frac{\mu_i}{L-1}$, ω_i and μ_i are defined in Equation (3). Then, the density matrix of the subsystem A becomes:

$$\rho' = \sum_{i=0}^{M-1} \omega_i \left|\widetilde{\theta}_i\right\rangle\left\langle\widetilde{\theta}_i\right|, \tag{14}$$

and the von Neumann entropy of ρ':

$$\begin{aligned}S(\rho') &= -tr(\rho' \log \rho') \\ &= -\lambda_1 \log \lambda_1 - \lambda_2 \log \lambda_2\end{aligned} \tag{15}$$

can quantify how much information is retained in the segmented image; where λ_1 and λ_2 are the eigenvalues of ρ'. As a result, we maximize it to determine the optimal thresholds:

$$TH_{op} = \arg\max_{TH} S(\rho'). \tag{16}$$

According to Equations (14) and (15), the following equation is established through simple algebraic computations:

$$\lambda_1\lambda_2 = \frac{1}{2}\sum_{i=0,j=0}^{M-1} \omega_i\omega_j sin^2(\widetilde{\theta}_i - \widetilde{\theta}_j), \tag{17}$$

where $\lambda_1 + \lambda_2 = 1$, as the restriction must be held.

It is worthwhile to note that Equation (17) can also be used to evaluate thresholding: when Equation (17) takes the maximum value, λ_1 and λ_2 will be most similar to each other, and then, $S(\rho')$ also reaches its best value. Meanwhile, Equation (17) indicates that the distance between intensities $sin^2(\widetilde{\theta}_i - \widetilde{\theta}_j)$, as well as the probability distribution (ω_i, ω_j) affect the thresholding results.

Different from Kapur's entropy-based method and its quantum version, our method has more explicit physical meaning for thresholding in terms of the following features:

(1) Encoding pixel intensities on the state space of a one-qubit system can be considered as a process in which independent intensities are squeezed into a two-dimensional space. The similarity between different state vectors, as well as its probability distribution, can be described with the density matrix. Both factors contribute to thresholding.
(2) According to the fundamental principles of information theory, the image segmenting process will causes the decrease of the information contained in images. Shannon entropy cannot directly be used to measure the information losses because it quantifies the amount of information on spaces with different dimensionality for original and segmented images. On the contrary, our method encodes the histograms of original and segmented images on the same quantum state space, which indicates that their entropies are comparable. As a result, the trivial solutions for segmentation, for example the thresholds equally dividing intensities into clusters with the same probability, could never appear since the entropy of the original image acts as the upper bound of our object function for all possible solutions.
(3) From Equation (17), we find that the object function of our method is very similar to Otsu's, described in Equation (2). The following experimental results will prove that they both achieve the best thresholding.

3.2. *Quantum Lossy-Encoding-Based Entropy Maximization Method*

As we have seen in Section 3.1, the proposed thresholding method derived from the viewpoint of quantum principles can achieve the best segmenting results similar to Otsu's. However, it still suffers from the efficiency problem of searching for optimal thresholds. In this subsection, we present another way for image thresholding on the quantum state space.

3.2.1. Quantum Lossy Encoding of Images

Different from the precedent method, we map the pixel intensities to quantum state vectors according to the following rules:

(1) Multiple qubits should be required for encoding intensity levels in accordance with the prospective number of thresholds. In other words, the state vectors are supposed to belong to an M-dimensional space if we want the M-level segmentation.
(2) The angle parameter of state vectors ranges from zero to $M \cdot \pi$ instead of $\pi/2$. Namely, $\theta_i = M\pi i/L$.
(3) After encoding, the terms contributing to density matrices should follow a π-periodic cyclical pattern. Namely, $|\theta\rangle\langle\theta| = |\theta+\pi\rangle\langle\theta+\pi|$.

Rule (1) provides the foundation for dividing pixel intensities into M classes, being linearly independent of each other. Rules (2) and (3) indicate that all state vectors representing pixel intensities are equally divided into M classes, and the corresponding density matrix:

$$\widetilde{\rho} = \sum_{i=0}^{N-1} \left(\sum_{j=0}^{M-1} p_{N \cdot j+i} \right) |\theta_i\rangle \langle\theta_i| \, , \theta_i \in [0, \pi], N = L/M, \tag{18}$$

only measures the information related to the local or intra-class uncertainty contributed by those adjoining intensity levels, but removes the global or inter-class information provided by those intensities far apart from each other.

According to the above rules, an alternative encoding scheme is given in the recursive form of:

$$\begin{aligned} |\theta_i\rangle^2 &= cos\theta_i\,|0\rangle + sin\theta_i\,|1\rangle \, , \theta_i = 2\pi i/L \\ |\theta_i\rangle^3 &= cos\theta_i cos2\theta_i\,|0\rangle + cos\theta_i sin2\theta_i\,|1\rangle + sin\theta_i\,|2\rangle \, , \theta_i = 3\pi i/L \\ &\cdots \\ |\theta_i\rangle^M &= cos\theta_i\,|2\theta_i\rangle^{M-1} + sin\theta_i\,|M-1\rangle \, , \theta_i = M\pi i/L \end{aligned} \tag{19}$$

where the superscript M is temporarily borrowed to label the dimensionality of state vectors and $i \in [0, L-1]$ denote pixel intensities. As an example, the traces of encoded state vectors in the 2D and 3D case are shown in Figure 1.

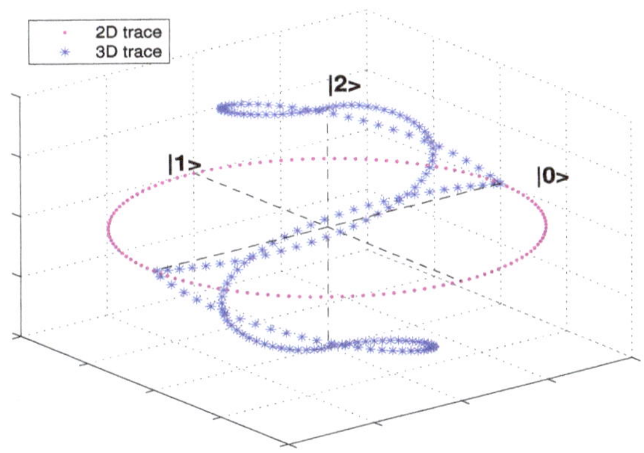

Figure 1. Traces of encoded state vectors on 2D and 3D space.

Differing from ordinary encoding practices, the proposed scheme records local information of images, but removes the global information. More precisely, the following evidence could be verified in the 2D case: we divide intensity levels into two classes equally and equivalently quantify the amount of information with the product of eigenvalues of $\widetilde{\rho}$:

$$\begin{aligned} \lambda_1\lambda_2 &= \frac{1}{2} \sum_{i=0,j=0}^{L-1} p_i p_j sin^2(\theta_i - \theta_j) \\ &= \frac{1}{2} \left(\sum_{i=0,j=0}^{L/2-1} p_i p_j sin^2(\theta_i - \theta_j) + \sum_{i=L/2,j=L/2}^{L-1} p_i p_j sin^2(\theta_i - \theta_j) \right) + \sum_{i=0}^{L/2-1} \sum_{j=L/2}^{L-1} p_i p_j sin^2(\theta_i - \theta_j). \end{aligned} \tag{20}$$

We note that the first term on the right of Equation (20) measures the local information (intra-class uncertainty) contributed by intensities in the same class, and the second term counts the global information (inter-class uncertainty) provided by intensities in different classes. Meanwhile, it is easy to verify that the values of the two terms will increase and decrease respectively when θ covers $[0, 2\pi]$ instead of $[0, \pi/2]$.

3.2.2. The QLEEM Method

Intuitively, the intensities far apart from each other and their probability distribution provide the evidence of thresholding. Therefore, we rewrite the density matrix of the given histogram in a decomposed form:

$$\rho = v_1\rho_1 + v_2\rho_2. \tag{21}$$

where ρ_1 and ρ_2 describe the probability distributions of local and remote intensity levels (that is, intra-class and inter-class uncertainty), respectively. Meanwhile, as there is no more knowledge about v_1 and v_2 except $v_1 + v_2 = 1$, we assume $v_1 = v_2 = 1/2$ according to the foundational principle of the entropy theory.

Now, we substitute ρ_1 with $\widetilde{\rho}$ given by the proposed lossy encoding scheme, since it contains the information contributed by local uncertainty of intensity vectors, and maximize the von Neumann entropy of Equation (21) for determining optimal thresholds:

$$TH_{op} = \arg\max S(\frac{1}{2}\widetilde{\rho} + \frac{1}{2}\widehat{\rho}). \tag{22}$$

Here, we adopt orthogonal state vectors in M-dimensional space representing M classes after thresholding, since we want these intensity classes to be as independent as possible. Let:

$$\widetilde{\rho} = \sum_{i=0}^{M-1} \lambda_{1,i} |\theta_{1,i}\rangle \langle\theta_{1,i}| , \widehat{\rho} = \sum_{i=0}^{M-1} \lambda_{2,i} |\theta_{2,i}\rangle \langle\theta_{2,i}| \tag{23}$$

be orthonormal decompositions for the states $\widetilde{\rho}$ and $\widehat{\rho}$, then for any one eigenvector of $\widetilde{\rho}$ denoted with $|\theta_{1,j} >$, there must exist an eigenvector of $\widehat{\rho}$ named $|\theta_{2,i} >$ satisfying the relationship of $|\theta_{2,i} >= \pm|\theta_{1,j} >$ when $S((\widetilde{\rho} + \widehat{\rho})/2)$ takes the max value. Meanwhile, the eigenvalues of the state can be determined according to the following equation:

$$\lambda_{2,i} = \frac{2}{M} - \lambda_{1,j}, if\ |\theta_{2,i}\rangle = \pm |\theta_{1,j}\rangle . \tag{24}$$

For the sake of representation, here we give the evidence of the above conclusion for the 2D situation. Assuming λ_1 and λ_2 are eigenvalues of the state $(\widetilde{\rho} + \widehat{\rho})/2$, its entropy will take the maximum value if we equivalently maximize:

$$\begin{aligned}\lambda_1\lambda_2 =& \lambda_{1,0}\lambda_{2,0}sin^2(\theta_{1,0} - \theta_{2,0}) + \lambda_{1,0}\lambda_{2,1}sin^2(\theta_{1,0} - \theta_{2,1}) + \lambda_{1,1}\lambda_{2,0}sin^2(\theta_{1,1} - \theta_{2,0}) \\ &+ \lambda_{1,1}\lambda_{2,1}sin^2(\theta_{1,1} - \theta_{2,1}) + \lambda_{1,0}\lambda_{1,1} + \lambda_{2,0}\lambda_{2,1}\end{aligned} \tag{25}$$

Notice that $\langle\theta_{1,0}|\theta_{1,1}\rangle = 0$ and $\langle\theta_{1,0}|\theta_{1,1}\rangle = 0$ must hold. Then:

$$\begin{aligned}\lambda_1\lambda_2 =& (\lambda_{1,0}\lambda_{2,0} + \lambda_{1,1}\lambda_{2,1})sin^2(\theta_{1,0} - \theta_{2,0}) + (\lambda_{1,0}\lambda_{2,1} + \lambda_{1,1}\lambda_{2,0})cos^2(\theta_{1,0} - \theta_{2,0}) \\ &+ \lambda_{1,0}\lambda_{1,1} + \lambda_{2,0}\lambda_{2,1}\end{aligned} \tag{26}$$

will take the extremum when $\langle\theta_{1,0}|\theta_{1,1}\rangle = 0\ or\ 1$. In other words,

$$\begin{cases} |\theta_{2,0}\rangle = \pm |\theta_{1,0}\rangle \\ |\theta_{2,1}\rangle = \pm |\theta_{1,1}\rangle \end{cases} or: \begin{cases} |\theta_{2,0}\rangle = \pm |\theta_{1,1}\rangle \\ |\theta_{2,1}\rangle = \pm |\theta_{1,0}\rangle \end{cases} \tag{27}$$

must hold. Without loss of generality, we adopt the first case of Equation (27) for the succeeding discussions. Then:

$$\begin{aligned}\lambda_1\lambda_2 &= \lambda_{1,0}\lambda_{2,1} + \lambda_{1,1}\lambda_{2,0}) + \lambda_{1,0}\lambda_{1,1} + \lambda_{2,0}\lambda_{2,1} \\ &= -\lambda_{2,0}^2 + (1 + \lambda_{1,0} - \lambda_{1,1})\lambda_{2,0} + \lambda_{1,0}\lambda_{1,1} + \lambda_{1,1}\end{aligned} \tag{28}$$

will reach its maximum value when:

$$\begin{aligned}\lambda_{2,0} &= (1+\lambda_{1,1}-\lambda_{1,0})/2 \\ &= \frac{2}{M}-\lambda_{1,2} \\ &= \lambda_{1,1}\end{aligned} \quad . \tag{29}$$

The above conclusions have the instructive function for thresholding, which can be seen in two aspects:

(1) Based on the proposed lossy-encoding scheme, we can directly calculate the eigenvalues of $\widehat{\rho}$ according to Equation (24), which represent the probability distribution of intensity classes after thresholding, and then determine the optimal threshold values.

(2) As the probability with which any one intensity class occurs must be greater than zero, according to Equation (24), all eigenvalues of the density matrix $\widetilde{\rho}$ would satisfy the condition of $\lambda_{1,i} < 2/M$. Otherwise,$\lambda_{1,i} \geq 2/M$ indicates that there exist meaningless and unnecessary classes for segmentation. In summary, the upper bound of the thresholding level can be determined using our method. This feature implies that our method is more feasible than the most of the other existing ones, since the thresholding level, as a hyperparameter, is often predetermined empirically.

Finally, the optimal thresholds $TH = th_1, th_2, \cdots, th_{M-1}$ can be determined according to the following relationships:

$$\begin{cases} \sum\limits_{i=0}^{th_1} p_i \leq \lambda_0 < \sum\limits_{i=0}^{th_1+1} p_i \\ \cdots \\ \sum\limits_{i=th_{M-2}+1}^{th_{M-1}} p_i \leq \lambda_{M-1} < \sum\limits_{i=th_{M-2}+1}^{th_{M-1}+1} p_i \end{cases} \tag{30}$$

where $\lambda_0, \lambda_1, \cdots, \lambda_{M-1}$ is the sequence taken from the eigenvalue set of $\widehat{\rho}$, and the corresponding sequence $|\theta_0\rangle, |\theta_1\rangle, \cdots, |\theta_{M-1}\rangle$ belongs to the circular permutation of all eigenvectors, which satisfy the following rules:

$$\begin{cases} |\theta_i\rangle = \arg\max\limits_j |\langle\theta_j|0\rangle| \\ \left|\theta_{(i+1)modM}\right\rangle = \arg\max\limits_j |\langle\theta_j|1\rangle| \\ \cdots \\ \left|\theta_{(i+M-1)modM}\right\rangle = \arg\max\limits_j |\langle\theta_j|M-1\rangle| \end{cases} \quad . \tag{31}$$

According to the methods mentioned above, the framework of the QLEEM algorithm is given in Algorithm 1.

Algorithm 1 The framework of the QLEEM algorithm

Input: The original image I, the thresholding level M
Output: The optimal thresholds
Init: Compute the histogram of the input image;
Step 1: Obtain density matrix $\widetilde{\rho}$ by using the lossy-encoding scheme;
Step 2: Calculate the eigenvalues and eigenvectors of $\widetilde{\rho}$ and then $\widehat{\rho}$
Step 3: Enumerate all possible M circular sequences of the eigenvalues of $\widehat{\rho}$, and then get M groups of thresholds;
Step 4: loop over the M groups of thresholds, and select the optimal one based on which the entropy denoted in Equation (15) takes the maximum value.

3.2.3. Time Complexity of the QLEEM Algorithm

For the problem of M-level thresholding segmentation of images containing L-level intensities, the time of calculating the density matrix $\widetilde{\rho}$ is $O(L)$; computing eigenvalues and eigenvectors of $\widetilde{\rho}$ needs $O(M^3)$; the time for performing Step 3 is $O(M! + L)$; and the loop in Step 4 consumes $O(M * 2^3)$ time. Since $M << L$ is satisfied in general cases, the optimal performance time of the QLEEM algorithm is achieved by $T = O(L)$, which notably outperforms Otsu's $T = O(A_{L-1}^{M-1}/2^{M-2})$.

4. Experiments and Comparisons

4.1. Datasets and Settings

To evaluate the performance of the proposed methods, a set of standard test images was obtained from the Berkeley segmentation dataset [36]. All of the test images are 8-bit in depth, with a size of 481 × 321 pixels. The algorithms used for comparison are Otsu's between-class variance method [17], Kapur's entropy criterion method [19], the quantum version of Kapur's [35] and our GQEM and QLEEM methods. These algorithms are implemented with MathWorks MATLAB 2014a on a Thinkpad notebook with an Intel Core-i5 2.2-GHz processor, 16 GB RAM and Ubuntu 14.04.

Threshold levels, quality of segmented images and time complexity are the most important indicators for evaluating the performance of image thresholding algorithms. Here, we evaluate the quality of segmented images by using the peak signal-to-noise ratio (PSNR) and structural similarity (SSIM). In addition, four measures: the Dice similarity coefficient (DICE) [37], the probabilistic rand index (PRI) [38], the global consistency error (GCE) [36] and the variation of information (VI) [39], are used to assess segmentations against ground truth data. Time complexity is measured by the execution time required in these methods. In particular, except for the proposed QLEEM, all the other exhaustive-search-based methods used in our experiments are sped up with the harmony search multithresholding algorithm (HSMA) [25].

4.2. Experimental Results and Comparisons

We applied these algorithms to all 300 pictures contained in the standard test dataset for assessing their performance. For the sake of representation, only five images, which are presented in Figure 2, have been used to show the bi-level segmented results. In Figure 3, the thresholding quality of the outcomes is analyzed considering the complete set, where the PSNR and SSIM scores are calculated under different thresholding levels, and we take the average values on the whole dataset.

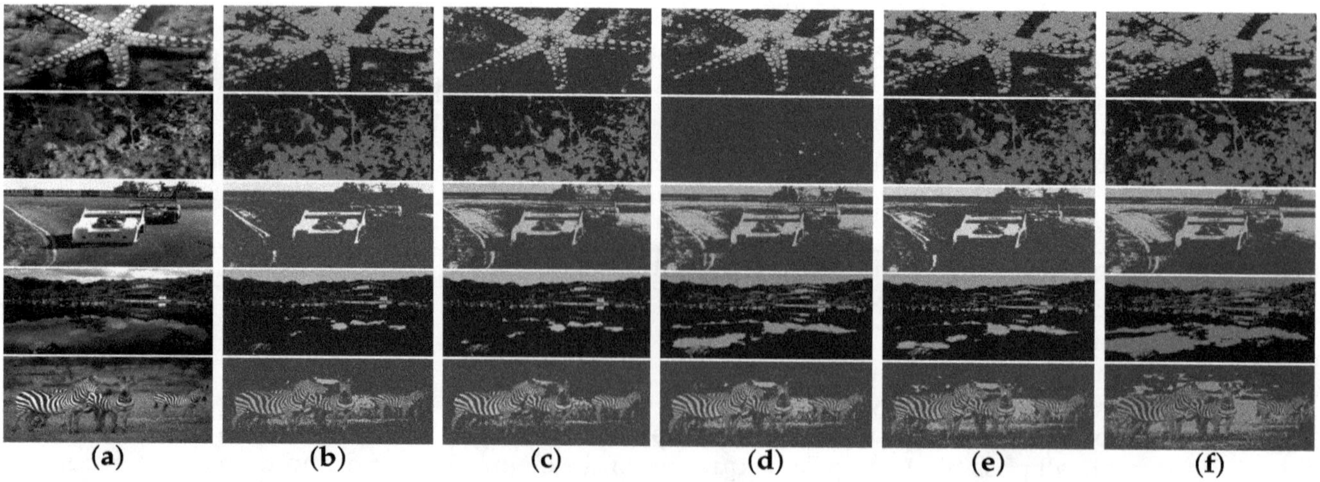

Figure 2. Visual comparison of (**a**) original images and bi-level segmented ones by using the (**b**) Otsu, (**c**) Kapur, (**d**) quantum version of Kapur's method (QKapur), (**e**) global quantum entropy maximization (GQEM) and (**f**) quantum lossy-encoding-based entropy maximization (QLEEM) methods, respectively.

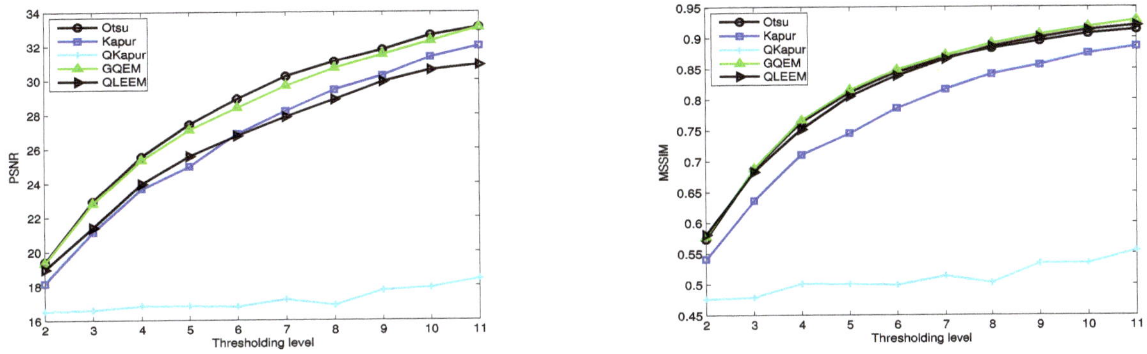

(a) (b)

Figure 3. Quality assessment of the segmented images in terms of (**a**) peak signal-to-noise ratio (PSNR) and (**b**) structural similarity (SSIM).

Meanwhile, we recorded the CPU time consumed by these algorithms, and the average values for all the test images under different thresholding levels are depicted in Figure 4. As an example, the experimental results in terms of thresholding level, thresholds and CPU time are tabulated in Table 1 for a randomly selected image.

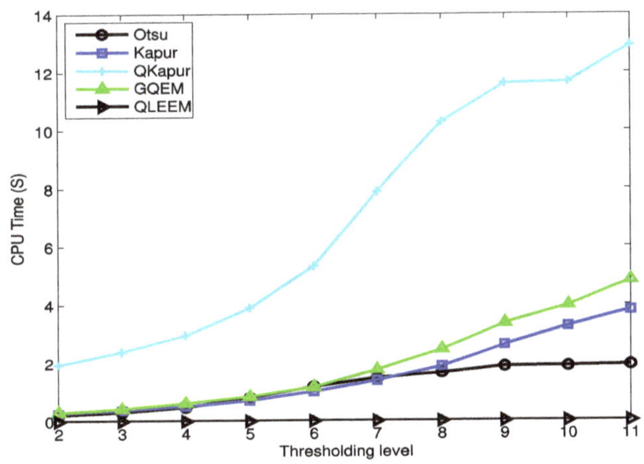

Figure 4. The comparison of the time consumption of different methods under different thresholding levels.

Table 1. Performance comparison in terms of thresholding level (M), thresholds and computation time.

Method	M	Thresholds	CPU Time (s)
Otsu	2	116	0.233523
	3	85-157	0.313405
	4	69-120-178	0.348805
	5	60-101-138-187	0.666153
	6	52-85-117-150-193	1.293793
Kapur	2	155	0.256437
	3	91-170	0.395228
	4	75-130-183	0.473902
	5	66-113-160-203	1.222006
	6	56-93-132-170-209	1.181965
QKapur	2	147	2.007029
	3	10-147	2.12888
	4	10-17-147	3.924715
	5	10-17-147-252	3.114482
	6	10-17-147-251-252	4.59602
GQEM	2	114	0.338808
	3	84-147	0.410247
	4	70-117-168	0.666514
	5	62-99-133-176	0.682051
	6	54-86-114-143-182	0.985941
QLEEM	2	107	0.001661
	3	86-135	0.002079
	4	62-106-153	0.002549
	5	53-90-121-160	0.003043
	6	49-83-106-133-166	0.003673

From Figure 2, we find that the segmentations obtained by using GQEM, QLEEM and Otsu are visually indistinguishable, which means these three methods have a similar performance. This conclusion can be further confirmed in Figure 3: the GQEM method obtains almost the same PSNR score as Otsu's in spite of very little computational error; meanwhile, both GQEM and QLEEM outperform the others in terms of SSIM. The experimental results can be explained with the criteria of maximizing quantum entropy and the lossy-encoding scheme proposed in our methods, because they emphasize the weight of between-class variance and retain the local information, respectively. This feature is highly consistent with the SSIM method, which assesses the perceived quality of images based on structural similarity indicators, such as contrast and local inter-dependencies of pixels.

Examining Figure 4 and Table 1, we can see that the proposed QLEEM algorithm achieves the fastest execution speed (at least 100-times faster than Otsu in the case of bi-level thresholding and up to 350-times when the number of thresholds increases to five). In addition, the time consumption of QLEEM was insensitive to increments of the threshold level, since the complexity of our algorithm was mainly correlated with the total intensity level, instead of the amount of thresholds.

On the other hand, the upper bounds of the thresholding level recommended by the proposed QLEEM algorithm were tested. We found that the maximum possible amount of thresholds was lower than 10 for about 40 images in the test set. Our algorithm would terminate when we try to apply more thresholds to them. Figure 5 lists two groups of images and corresponding histograms, for which the proposed algorithm gave one and two thresholds, respectively. According to the visual observation, it is reasonable to believe that the suggested amounts of thresholds are feasible, as there are no more than three distinct peaks in their histograms.

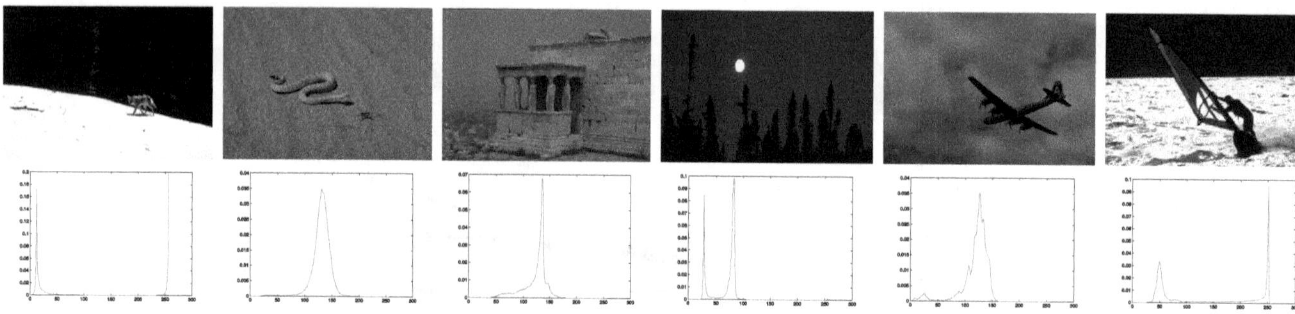

(a)

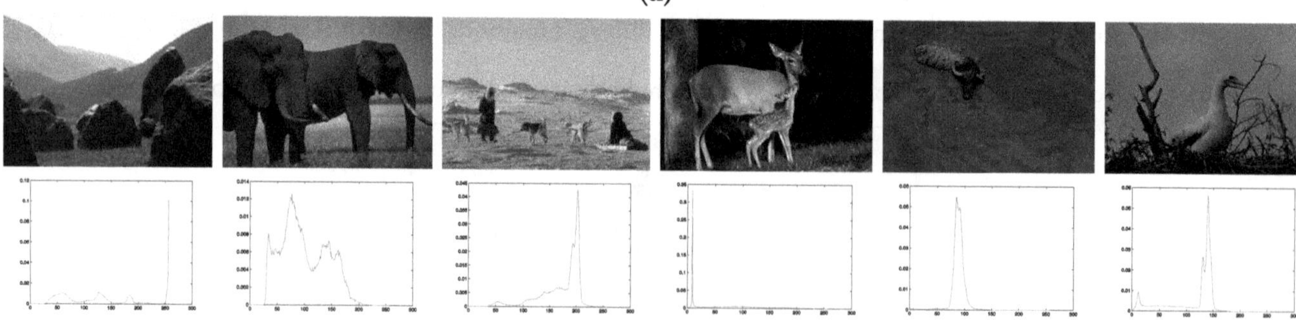

(b)

Figure 5. Two groups of images in the test dataset, to which the QLEEM algorithm suggests applying (**a**) bi-level and (**b**) tri-level thresholding, respectively.

Finally, we evaluate segmentations against the ground truth data. The first experiment is performed on a synthetic image corrupted by Gaussian noise (the mean value is zero, and the variance is 0.03), which is utilized for testing the efficiency and robustness of the proposed methods. Figure 6 shows the noisy image and segmentation results obtained by different algorithms. In addition, the performance indexes: the DICE ratio, PRI, GCE and VI scores, are used to assess the robustness of these algorithms. The corresponding scores are listed in Table 2.

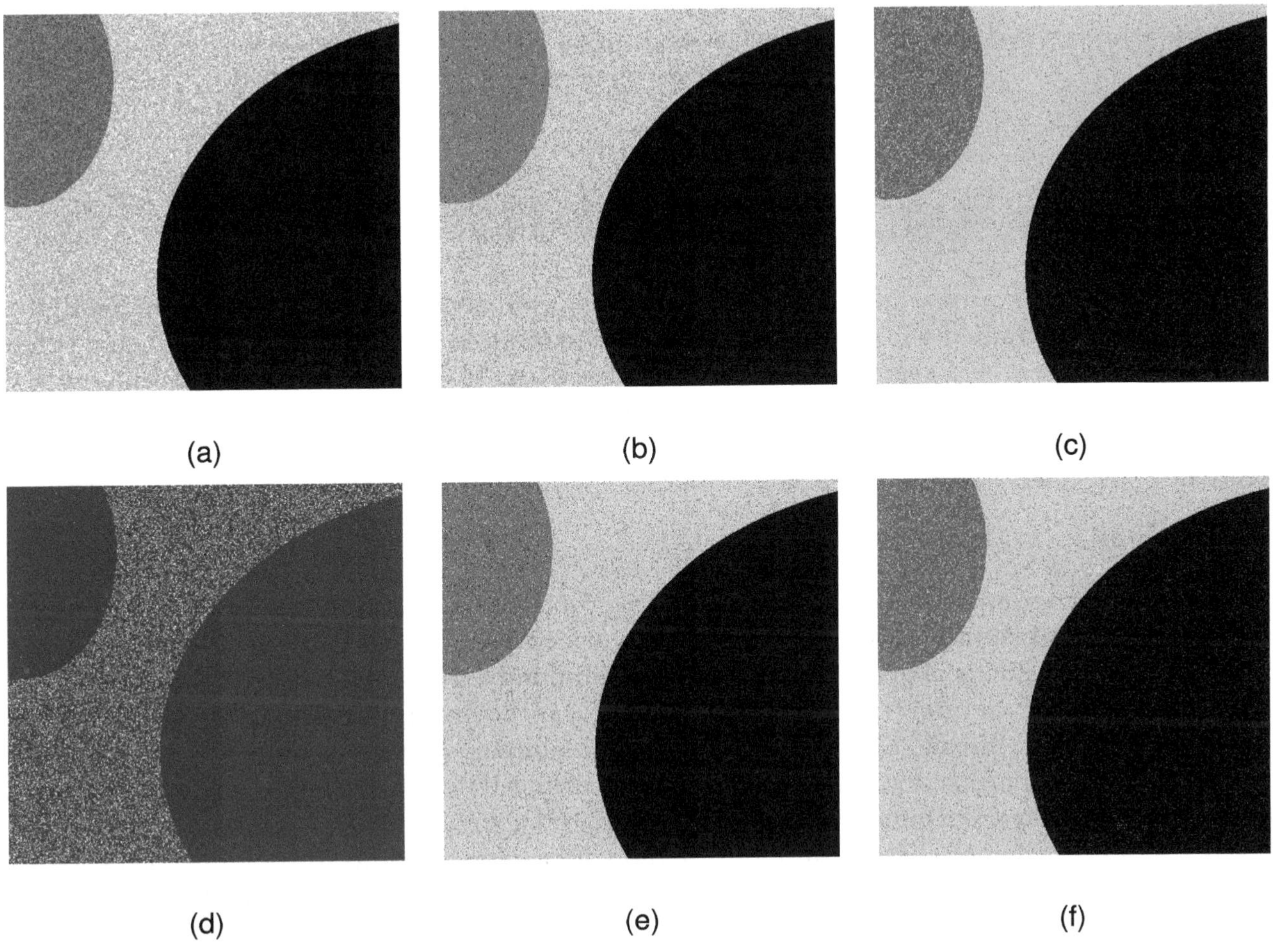

Figure 6. Comparison of segmentation results on a synthetic image. (a) noisy image (Gaussian noise with zero mean and 3% variance); (b) Otsu result; (c) Kapur result; (d) QKapur result; (e) GQEM result; (f) QLEEM result.

Table 2. Performance of different algorithms on a noisy image (the best values are highlighted). DICE, Dice similarity coefficient; PRI, probabilistic Rand index; GCE, global consistency error; VI, variation of information.

Algorithm	DICE	PRI	GCE	VI
Otsu	0.889787	0.934784	0.09807	0.54778
Kapur	0.908592	0.946141	0.093275	0.532568
QKapur	0.472366	0.426367	0.084447	1.570079
GQEM	***0.921509***	***0.955491***	***0.078646***	***0.45552***
QLEEM	0.908281	0.948501	0.097511	0.580201

The visual comparison in Figure 6 shows that the proposed GQEM and QLEEM algorithms produce clearer and more accurate segmentation results. From Table 2, we can confirm this conclusion: our GQEM clearly outperformed the others on the DICE, PRI, GCE and VI values. The robustness of the proposed GQEM for noisy images can be explained by comparing the object function of GQEM and Otsu. Considering the last term in Equations (2) and (17), both of them measure the distance between pixel intensities, but our GQEM method scaled the range $[0, L-1]$ of this parameter down to $[0, 1]$. This feature is helpful for suppressing the high contrast caused by noise, and then, our GQEM algorithm partly played the role of a low-pass filter in segmentation tasks.

In the second experiment, we performed thresholding segmentation on BSDS300 dataset and compared the results with the ground truth segmentations in terms of the DICE, PRI, GCE and VI indexes. The average scores of these indicators obtained by different algorithms are presented in Table 3.

Table 3. Average performance of different algorithms on BSDS300 dataset (the best values are highlighted).

Algorithm	DICE	PRI	GCE	VI
Otsu	0.411934	0.613044	0.385938	2.825647
Kapur	0.400079	***0.64313***	0.366348	2.49384
QKapur	0.363979	0.542463	***0.1704***	***1.802242***
GQEM	***0.412396***	0.611379	0.384827	2.892085
QLEEM	0.405824	0.614035	0.386781	2.931183

From Table 3, we can see that all the listed algorithms obtained lower scores compared with those that have been well trained with the manually-labeled dataset. In general, thresholding segmentation is a form of unsupervised segmentation, which cannot use any a priori knowledge involving the ground truth of a training set of images. Furthermore, the proposed GQEM and QLEEM along with the others used for comparison are all histogram-based algorithms. They achieve optimal segmentation by merely utilizing the probability distribution of colors, instead of the spatial and texture information.

5. Conclusions

In this paper, we address the image thresholding problem on quantum state space. The proposed GQEM and QLEEM methods follow a different way to represent images and determine the optimal thresholds in the language of quantum mechanics. In summary, the contributions of this paper mainly include the following aspects: (1) To our knowledge, this is the first application of the global quantum entropy criteria to the thresholding problem. The von Neumann entropy is more powerful for image segmentations than the Shannon entropy, because it measures the distance between pixel intensities, as well as the probability distribution. (2) Compared with other state-of-the-art approaches, our QLEEM algorithm tends to retain more structural information after segmentations. It is highly consistent with the way in which the HVS works. (3) The proposed QLEEM algorithm has the lowest consumption of execution times known to us, even compared with others that are sped up with some intelligent optimization techniques.

Author Contributions: Conceptualization, X.W. Formal analysis, X.W. Methodology, X.W. and C.Y. Project administration, C.Y. Validation, G.-S.X. Writing, original draft, X.W. Writing, review and editing, C.Y. and Z.L.

References

1. Bhat, M. Digital image processing. *Int. J. Sci. Technol.* **2014**, *3*, 272–276.
2. Zaitoun, N.M.; Aqel, M.J. Survey on image segmentation techniques. *Procedia Comput. Sci.* **2015**, *65*, 797–806. [CrossRef]
3. Saha, S.; Bandyopadhyay, B. Automatic MR brain image segmentation using a multiseed based multiobjective clustering approach. *Appl. Intell.* **2011**, *35*, 411–427. [CrossRef]
4. Iakovidis, D.; Savelonas, M.; Karkanis, S. A genetically optimized level set approach to segmentation of thyroid ultrasound images. *Appl. Intell.* **2007**, *27*, 193–203. [CrossRef]
5. Kamel, M.; Zhao, A. Extraction of binary character/graphics images from grayscale document images. *Graph. Models Image Process.* **1993**, *55*, 203–217. [CrossRef]
6. Vijayalakshmi, S.; Durairaj, D.C. Use of multiple thresholding techniques for moving object detection and tracking. *Int. J. Comput. Appl.* **2013**, *80*, 1–7. [CrossRef]
7. Valova, I.; Milano, G.; Bowen, K.; Gueorguieva, N. Bridging the fuzzy, neural and evolutionary paradigms for automatic target recognition. *Appl. Intell.* **2011**, *35*, 211–225. [CrossRef]
8. Yang, M.-D.; Su, T.-C.; Pan, N.-F.; Yang, Y.-F. Systematic image quality assessment for sewer inspection. *Expert Syst. Appl.* **2011**, *38*, 1766–1776. [CrossRef]
9. Goumas, S.K.; Dimou, L.N.; Zervakis, M.E. Combination of multiple classifiers for post-placement quality inspection of components: A comparative study. *Inf. Fusion* **2010**, *11*, 149–162. [CrossRef]
10. Sasirekha, D.; Chandra, D.E. Enhanced techniques for PDF image segmentation and text extraction. *Int. J. Comput. Sci. Inf. Secur.* **2012**, *10*, 1–5.

11. Yang, Y.; Hallman, S.; Ramanan, D.; Fowlkes, C.C. Layered object models for image segmentation. *IEEE Trans. Pattern Anal. Mach. Intell.* **2012**, *34*, 1731–1743. [CrossRef] [PubMed]
12. Ma, Z.; Tavares, J.; Jorge, R. A review of algorithms for medical image segmentation and their applications to the female pelvic cavity. *Comput. Methods Biomech. Biomed. Eng.* **2010**, *13*, 235–246. [CrossRef] [PubMed]
13. Cuevas, E.; Zaldivar, D.; Sossa, H. A multi-threshold segmentation approach based on Artificial Bee Colony optimization. *Appl. Intell.* **2012**, *37*, 321–336. [CrossRef]
14. Zhang, Z.; Chen, C.; Sun, J. EM algorithms for Gaussian mixtures with split-and-merge operation. *Pattern Recognit.* **2003**, *36*, 1973–1983. [CrossRef]
15. Lu, Z. Entropy regularized likelihood learning on Gaussian mixture: two gradient implementations for automatic model selection. *Neural Process. Lett.* **2007**, *25*, 17–30. [CrossRef]
16. Ma, J.; Wang, T.; Xu, L. A gradient BYY harmony learning rule on Gaussian mixture with automated model selection. *Neurocomputing* **2004**, *56*, 481–487. [CrossRef]
17. Otsu, N. A threshold selection method from gray level histograms. *IEEE Trans. Syst. Man Cybern.* **1979**, *9*, 62–66. [CrossRef]
18. Farshi, T.P.; Demirci, R.; Feizi-Derakhshi, M.R. Image clustering with optimization algorithms and color space. *Entropy* **2018**, *20*, 296. [CrossRef]
19. Kapur, J.; Sahoo, P.; Wong, A. A new method for gray-level picture thresholding using the entropy of the histogram. *Comput. Vis. Gr. Image Process.* **1985**, *29*, 273–285. [CrossRef]
20. Pare, S.; Kumar, A.; Bajaj, V. An efficient method for multilevel color image thresholding using cuckoo search algorithm based on minimum crossentropy. *Appl. Soft Comput.* **2017**, *61*, 570–592. [CrossRef]
21. Liang, Y.C.; Cuevas, J.R. An automatic multilevel image thresholding using relative entropy and meta-heuristic algorithms. *Entropy* **2013**, *15*, 2181–2209. [CrossRef]
22. Ramirez-Reyes, A.; Hernandez-Montoya, A.R.; Herrera-Corral, G.; Dominguez-Jimenez, I. Determining the entropic index q of Tsallis entropy in images through redundancy. *Entropy* **2016**, *18*, 299. [CrossRef]
23. Ye, Z.; Yin, H.; Ye, Y. Comparative analysis of two leading evolutionary intelligence approaches for multilevel thresholding. *Int. J. Signal Imaging Syst. Eng.* **2018**, *11*, 20–30. [CrossRef]
24. Dehshibi, M.M.; Sourizaei, M. A hybrid bio-inspired learning algorithm for image segmentation using multilevel thresholding. *Multimedia Tools Appl.* **2017**, *76*, 15951–15986. [CrossRef]
25. Oliva, D.; Cuevas, E.; Pajares, G. Multilevel thresholding segmentation based on harmony search optimization. *J. Appl. Math.* **2013**, *2013*, 1–24. [CrossRef]
26. Shor, W.P. Algorithms for quantum computation: discrete logarithms and factoring. In Proceedings of the 35th Annual Symposium Foundations of Computer Science, Santa Fe, NM, USA, 20–22 November 1994; pp. 124–134.
27. Grover, L. A fast quantum mechanical algorithm for database search. In Proceedings of the 28th Annual ACM Symposium on the Theory of Computing, Philadelphia, PA, USA, 22–24 May 1996; pp. 212–219.
28. Venegas-Andraca, S.E.; Bose, S. Storing, processing and retrieving an image using quantum mechanics. In Proceedings of the SPIE Conference Quantum Information and Computation, Orlando, FL, USA, 21–25 April 2003; pp. 137–147.
29. Le, P.Q.; Dong, F.; Hirota, K. A flexible representation of quantum images for polynomial preparation, image compression, and processing operations. *Quantum Inform. Process.* **2011**, *10*, 63–84. [CrossRef]
30. Caraiman, S.; Manta, V.I. Image segmentation on a quantum computer. *Quantum Inf. Process.* **2015**, *14*, 1693–1715. [CrossRef]
31. Yangguang, S. Quantum statistical edge detection using path integral monte carlo simulation. Bio-Inspired Computing-Theories and Applications. *Commun. Comput. Inf. Sci.* **2014**, *472*, 430–434.
32. Yan, F.; Iliyasu, A.M.; Fatichah, C.; Tangel M.L. Quantum image searching based on probability distributions. *J. Quantum Inf. Sci.* **2012**, *2*, 55–60. [CrossRef]
33. Ning, W.; Song, L. A watermarking strategy for quantum image based on least significant bit. *Chin. J. Quantum Electron.* **2015**, *32*, 263–269.
34. Feng, S.; Xiang, L.; Huabao, L. Sampling number of reconstruction arithmetic based on quantum correlated imaging. *Chin. J. Quantum Electron.* **2015**, *32*, 144–149.
35. Du, S.; Wu, G.; Ma, L. Maximum quantum entropy based optimal threshold selecting criterion for thresholding image segmentation. *J. Comput. Inf. Syst.* **2014**, *10*, 3359–3366.
36. Martin, D.; Fowlkes, C.; Tal, D.; Malik, J. A database of human segmented natural images and its application

to evaluating segmentation algorithms and measuring ecological statistics. *Proc. Int. Conf. Comput. Vis.* **2001**, *2*, 416–423.

37. Dice, L.R. Measures of the amount of ecologic association between species. *Ecology* **1945**, *26*, 297–302. [CrossRef]
38. Unnikrishnan, R.; Pantofaru, C.; Hebert, M. Toward objective evaluation of image segmentation algorithms. *IEEE Trans. Pattern Anal. Mach. Intell.* **2007**, *29*, 929–944. [CrossRef] [PubMed]
39. Arbelaez, P.; Maire, M.; Fowlkes, C. Contour detection and hierarchical image segmentation. *IEEE Trans. Pattern Anal. Mach. Intell.* **2011**, *33*, 898–916. [CrossRef] [PubMed]

7

The Entropic Dynamics Approach to Quantum Mechanics

Ariel Caticha

Physics Department, University at Albany-SUNY, Albany, NY 12222, USA; ariel@albany.edu

Abstract: Entropic Dynamics (ED) is a framework in which Quantum Mechanics is derived as an application of entropic methods of inference. In ED the dynamics of the probability distribution is driven by entropy subject to constraints that are codified into a quantity later identified as the phase of the wave function. The central challenge is to specify how those constraints are themselves updated. In this paper we review and extend the ED framework in several directions. A new version of ED is introduced in which particles follow smooth differentiable Brownian trajectories (as opposed to non-differentiable Brownian paths). To construct ED we make use of the fact that the space of probabilities and phases has a natural symplectic structure (i.e., it is a phase space with Hamiltonian flows and Poisson brackets). Then, using an argument based on information geometry, a metric structure is introduced. It is shown that the ED that preserves the symplectic and metric structures—which is a Hamilton-Killing flow in phase space—is the linear Schrödinger equation. These developments allow us to discuss why wave functions are complex and the connections between the superposition principle, the single-valuedness of wave functions, and the quantization of electric charges. Finally, it is observed that Hilbert spaces are not necessary ingredients in this construction. They are a clever but merely optional trick that turns out to be convenient for practical calculations.

Keywords: quantum mechanics; entropic dynamics; symplectic geometry; information geometry

---

1. Introduction

Quantum mechanics has been commonly regarded as a generalization of classical mechanics with an added element of indeterminism. The standard quantization recipe starts with a description in terms of the system's classical coordinates and momenta $\{q, p\}$ and then proceeds by applying a series of more or less ad hoc rules that replace the classical $\{q, p\}$ by self-adjoint linear operators $\{\hat{q}, \hat{p}\}$ acting on some complex Hilbert space [1]. The Hilbert space structure is given priority while the probabilistic structure is relegated to the less fundamental status of providing phenomenological rules for how to handle those mysterious physical processes called measurements. The result is a dichotomy between two separate and irreconcilable modes of wave function evolution: one is the linear and deterministic Schrödinger evolution and the other is the discontinuous and stochastic wave function collapse [2,3]. To put it bluntly, the dynamical and the probabilistic aspects of quantum theory are incompatible with each other. And furthermore, the dichotomy spreads to the interpretation of the quantum state itself [4–8]. It obscures the issue of whether the wave function describes the *ontic* state of the system or whether it describes an *epistemic* state about the system [9].

In the Entropic Dynamics (ED) approach these problems are resolved by placing the probabilistic aspects of QM at the forefront while the Hilbert space structure is relegated to the secondary role of a convenient calculational tool [10–12]. ED tackles QM as an example of entropic inference, a framework designed to handle insufficient information [13–18]. The starting point is to specify the subject matter, the ontology—are we talking about the positions of particles or the configurations of fields? Once this decision is made our inferences about these variables are driven by entropy subject to information expressed by constraints. The main effort is directed towards choosing those constraints since it is through them that the "physics" is introduced.

From the ED perspective many of the questions that seemed so urgent in other approaches are successfully evaded. For example, when quantum theory is regarded as an extension of classical mechanics any deviations from causality demand an explanation. In contrast, in the entropic approach uncertainty and probabilities are the norm. Indeterminism is just the inevitable consequence of incomplete information and no deeper explanation is needed. Instead, it is the certainty and determinism of the classical limit that require explanations. Another example of a question that has consumed an enormous effort is the problem of deriving the Born rule from a fundamental Hilbert space structure. In the ED approach this question does not arise and the burden of explanation runs in the opposite direction: how do objects such as wave functions involving complex numbers emerge in a purely probabilistic framework? Yet a third example concerns the interpretation of the wave function itself. ED offers an uncompromising and radically epistemic view of the wave function Ψ. This turns out to be extremely restrictive: in a fully epistemic interpretation there is no logical room for "quantum" probabilities obeying alternative rules of inference. Not only is the probability $|\Psi|^2$ interpreted as a state of knowledge but, in addition, the epistemic significance of the phase of the wave function must be clarified and made explicit. Furthermore, it is also required that all *updates of* Ψ, which include *both* its unitary time evolution *and* the wave function collapse during measurement, must be obtained as a consequence of entropic and Bayesian updating rules [19–24].

There is a large literature on reconstructions of quantum mechanics (see e.g., [25–31] and references therein) and there are several approaches based on information theory (see e.g., [32–46]). What distinguishes ED is a strict adherence to Bayesian and entropic methods and a central concern with the nature of time. The issue here is that any discussion of dynamics must inevitably include a notion of time but the rules for inference do not mention time—they are totally atemporal. One can make inferences about the past just as well as about the present or the future. This means that any model of dynamics based on inference must also include assumptions about time, and those assumptions must be explicitly stated. In ED "entropic" time is a book-keeping device *designed* to keep track of changes. The construction of entropic time involves several ingredients. One must introduce the notion of an 'instant'; one must show that these instants are suitably ordered; and finally, one must define a convenient measure of the duration or interval between the successive instants. It turns out that an arrow of time is generated automatically and entropic time is intrinsically directional.

This paper contains a review of previous work on ED and extends the formalism in several new directions. In [10–12] the Schrödinger equation was derived as a peculiar non-dissipative diffusion in which the particles perform an irregular Brownian motion that resembles the Einstein–Smoluchowski (ES) process [47]. The trajectories are continuous and non-differentiable so their velocity is undefined. Since the expected length of the path between any two points is infinite this would be a very peculiar motion indeed. Here we exhibit a new form of ED in which the Brownian motion resembles the much smoother Oernstein–Uhlenbeck (OU) process [47]. The trajectories have finite expected lengths; they are continuous and differentiable. On the other hand, although the velocities are well defined and continuous, they are not differentiable [25,48].

We had also shown that the irregular Brownian motion at the "microscopic" or sub-quantum level was not unique. One can enhance or suppress the fluctuations while still obtaining the same emergent Schrödinger behavior at the "macroscopic" or quantum level [49,50]. A similar phenomenon is also found in the smoother ED developed here. In both the ES and the OU cases the special limiting case in which fluctuations are totally suppressed turns out to be of particular interest because the particles evolve deterministically along the smooth lines of probability flow. This means that ED includes the Bohmian or causal form of quantum mechanics [51–53] as a limiting case.

ED consists of the entropic updating of probabilities through information supplied by constraints. The main concern is how these constraints are chosen including, in particular, how the constraints themselves are updated. In [54] an effective criterion was found by adapting Nelson's seminal insight that QM is a non-dissipative diffusion [55]. This amounts to updating constraints in such a way that a certain energy functional is conserved. Unfortunately, this criterion, while fully satisfactory in a non-relativistic setting, fails in curved space-times where the concept of a globally conserved energy may not exist.

The second contribution in this paper is a geometric framework for updating constraints that does not rely on the notion of a conserved energy. Our framework draws inspiration from two sources: one is the fact that QM has a rich geometrical structure [56–64]. The authors of [56–62] faced the task of unveiling geometric structures that, although well hidden, are already present in the standard QM framework. Our goal runs in the opposite direction: we impose these natural geometric structures as the foundation upon which we reconstruct the QM formalism.

The other source of inspiration is the connection between QM and information geometry [17,65–68] that was originally suggested in the work of Wootters [32]. This connection has been explored in the context of quantum statistical inference [69], in the operational description of quantum measurements [37,39], and in the reconstruction of QM [43,44]. Our previous presentation in [12] has been considerably streamlined by recognizing the central importance of symmetry principles when implemented in conjunction with concepts of information geometry.

In ED, the degrees of freedom are the probability densities $\rho(x)$ and certain "phase" fields $\Phi(x)$ that represent the constraints that control the flow of probabilities. Thus, we are concerned not just with the "configuration" space of probabilities $\{\rho\}$ but with the larger space of probabilities and phases $\{\rho,\Phi\}$. The latter has a natural symplectic structure, i.e., $\{\rho,\Phi\}$ is a phase space. Imposing a dynamics that preserves this symplectic structure leads to Hamiltonian flows, Poisson brackets, and so much of the canonical formalism associated with mechanics. To single out the particular Hamiltonian flow that reproduces QM we extend the information geometry of the configuration space $\{\rho\}$ to the full phase space. This is achieved by imposing a symmetry that is natural in a probabilistic setting: we extend the well-known spherically symmetric information geometry of the space $\{\rho\}$ to the full phase space $\{\rho,\Phi\}$. This construction yields a derivation of the Fubini–Study metric. A welcome by-product is that the joint presence of a symplectic and a metric structure leads to a complex structure. This is the reason QM involves complex numbers.

The dynamics that preserves the metric structure is a Killing flow. We propose that the desired geometric criterion for updating constraints is a dynamics that preserves both the symplectic and the metric structures. Thus, in the final step of our reconstruction of QM we show that the Hamiltonians that generate Hamiltonian-Killing flows lead to an entropic dynamics described by the linear Schrödinger equation.

We conclude with some comments exploring various aspects of the ED formalism. We show that despite the arrow of entropic time, the resulting ED is symmetric under time reversal. We discuss the connections between linearity, the superposition principle, the single-valuedness of wave functions, and the quantization of charge. We also discuss the classical limit and the Bohmian limit in which fluctuations are suppressed and particles follow deterministic trajectories. Finally, we discuss the introduction of Hilbert spaces. We argue that while strictly unnecessary in principle, Hilbert spaces are extremely convenient for calculational purposes.

This paper focuses on the derivation of the Schrödinger equation but the ED approach has been applied to a variety of other topics in quantum theory. These include: the quantum measurement problem [70,71]; momentum and uncertainty relations [50,72] (see also [73–76]); the Bohmian limit [49,50] and the classical limit [77]; extensions to curved spaces [78]; to relativistic fields [79,80]; and the ED of spin [81].

2. The ED of Short Steps

We deal with N particles living in a flat 3-dimensional space $\mathbf{X}$ with metric δ_{ab}. For N particles the configuration space is $\mathbf{X}_N = \mathbf{X} \times \ldots \times \mathbf{X}$. We assume that the particles have definite positions x_n^a

and it is their unknown values that we wish to infer [82]. (The index $n = 1, \ldots, N$ denotes the particle and $a = 1, 2, 3$ the spatial coordinates.)

In ED positions play a very special role: they define the ontic state of the system. This is in contradiction with the standard Copenhagen notion that quantum particles acquire definite positions only as a result of a measurement. For example, in the ED description of the double slit experiment the particle definitely goes through one slit or the other but one might not know which. The wave function, on the other hand, is a purely epistemic notion and, as it turns out, all other quantities, such as energy or momentum, are epistemic too. They do not reflect properties of the particles but properties of the wave function [70–72].

Having identified the microstates $x \in \mathbf{X}_N$ we tackle the dynamics. The main dynamical assumption is that the particles follow trajectories that are continuous. This represents an enormous simplification because it implies that a generic motion can be analyzed as the accumulation of many infinitesimally short steps. Therefore, the first task is to find the transition probability $P(x'|x)$ for a short step from an initial x to an unknown neighboring x' and only later we will determine how such short steps accumulate to yield a finite displacement.

The probability $P(x'|x)$ is found by maximizing the entropy

$$\mathcal{S}[P, Q] = -\int dx'\, P(x'|x) \log \frac{P(x'|x)}{Q(x'|x)} \tag{1}$$

relative to the joint prior $Q(x'|x)$ subject to constraints given below. (In multidimensional integrals such as (1) the notation dx' stands for $d^{3N}x'$.)

The prior. The choice of prior $Q(x'|x)$ must reflect the state of knowledge that is common to all short steps. (It is through the constraints that the information that is specific to any particular short step will be supplied.) We adopt a prior that carries the information that the particles take infinitesimally short steps and reflects the translational and rotational invariance of the Euclidean space $\mathbf{X}$ but is otherwise uninformative. In particular, the prior expresses total ignorance about any correlations. Such a prior can itself be derived from the principle of maximum entropy. Indeed, maximize

$$S[Q] = -\int dx'\, Q(x'|x) \log \frac{Q(x'|x)}{\mu(x')}\,, \tag{2}$$

relative to the uniform measure $\mu(x')$ [83], subject to normalization, and subject to the N independent constraints

$$\langle \delta_{ab} \Delta x_n^a \Delta x_n^b \rangle = \kappa_n\,, \quad (n = 1 \ldots N)\,, \tag{3}$$

where κ_n are small constants and $\Delta x_n^a = x_n'^a - x_n^a$. The result is a product of Gaussians,

$$Q(x'|x) \propto \exp -\frac{1}{2} \sum_n \alpha_n \delta_{ab} \Delta x_n^a \Delta x_n^b\,, \tag{4}$$

where, to reflect translational invariance and possibly non-identical particles, the Lagrange multipliers α_n are independent of x but may depend on the index n. Eventually we will let $\alpha_n \to \infty$ to implement infinitesimally short steps. Next we specify the constraints that are specific to each particular short step.

The drift potential constraint. In Newtonian dynamics one does not need to explain why a particle perseveres in its motion in a straight line; what demands an explanation—that is, a force—is why the particle deviates from inertial motion. In ED one does not require an explanation for why the particles move; what requires an explanation is how the motion can be both directional and highly correlated. This physical information is introduced through one constraint that acts simultaneously on all particles. The constraint involves a function $\phi(x) = \phi(x_1 \ldots x_N)$ on configuration space $\mathbf{X}_N$ that we call the "drift" potential. We impose that the displacements Δx_n^a are such that the expected change of the drift potential $\langle\Delta\phi\rangle$ is constrained to be

$$\langle\Delta\phi\rangle = \sum_{n=1}^{N} \langle\Delta x_n^a\rangle \frac{\partial\phi}{\partial x_n^a} = \kappa'(x)\,, \tag{5}$$

where $\kappa'(x)$ is another small but for now unspecified function. As we shall later see this information is already sufficient to construct an interesting ED. However, to reproduce the particular dynamics that describes quantum systems we must further require that the potential $\phi(x)$ be a multi-valued function with the topological properties of an angle—ϕ and $\phi + 2\pi$ represent the same angle [84].

The physical origin of the drift potential $\phi(x)$ is at this point unknown so how can one justify its introduction? The idea is that identifying the relevant constraints can represent significant progress even when their physical origin remains unexplained. Indeed, with the *single* assumption of a constraint involving a drift potential we will explain and coordinate *several* features of quantum mechanics such as entanglement, the existence of complex and symplectic structures, the actual form of the Hamiltonian, and the linearity of the Schrödinger equation.

The gauge constraints. The single constraint (5) already leads to a rich entropic dynamics but by imposing additional constraints we can construct even more realistic models. To incorporate the effect of an external electromagnetic field we impose that for each particle n the expected displacement $\langle\Delta x_n^a\rangle$ will satisfy

$$\langle\Delta x_n^a\rangle A_a(x_n) = \kappa_n'' \quad \text{for} \quad n = 1\ldots N\,, \tag{6}$$

where the electromagnetic vector potential $A_a(x_n)$ is a field that lives in the 3-dimensional physical space ($x_n \in \mathbf{X}$). The strength of the coupling is given by the values of the κ_n''. These quantities could be specified directly but, as is often the case in entropic inference, it is much more convenient to specify them indirectly in terms of the corresponding Lagrange multipliers.

The transition probability. An important feature of the ED model can already be discerned. The central object of the discussion so far, the transition probability $P(x'|x)$, codifies information supplied through the prior and the constraints which makes no reference to anything earlier than the initial position x. Therefore ED must take the form of a Markov process.

The distribution $P(x'|x)$ that maximizes the entropy $\mathcal{S}[P,Q]$ in (1) relative to (4) and subject to (5), and (6), and normalization is

$$P(x'|x) = \frac{1}{Z} \exp - \sum_n \left(\frac{\alpha_n}{2} \delta_{ab} \Delta x_n^a \Delta x_n^b - \alpha' \left[\partial_{na}\phi - \beta_n A_a(x_n)\right] \Delta x_n^a \right) \tag{7}$$

where α' and β_n are Lagrange multipliers. This is conveniently written as

$$P(x'|x) = \frac{1}{Z} \exp - \sum_n \frac{\alpha_n}{2} \delta_{ab} \left(\Delta x_n^a - \Delta\bar{x}_n^a\right) \left(\Delta x_n^b - \Delta\bar{x}_n^b\right)\,, \tag{8}$$

with a suitably modified normalization and

$$\Delta\bar{x}_n^a = \frac{\alpha'}{\alpha_n} \left[\partial_{na}\phi - \beta_n A_a(x_n)\right] = \langle\Delta x_n^a\rangle\,. \tag{9}$$

A generic displacement is expressed as a drift plus a fluctuation,

$$\Delta x_n^a = \langle\Delta x_n^a\rangle + \Delta w_n^a\,, \tag{10}$$

where

$$\langle\Delta w_n^a\rangle = 0\,, \quad \text{and} \quad \langle\Delta w_n^a \Delta w_{n'}^b\rangle = \frac{1}{\alpha_n} \delta_{nn'} \delta^{ab}\,, \tag{11}$$

The fact that the constraints (5) and (6) are not independent—both involve the same displacements $\langle \Delta x_n^a \rangle$—has turned out to be significant. We can already see in (7) and (9) that it leads to a gauge symmetry. As we shall later see the vector potential A_a will be interpreted as the corresponding gauge connection field and the multipliers β_n will be related to the electric charges through $\beta_n = q_n/\hbar c$.

3. Entropic Time

The task of iterating the short steps described by the transition probability (8) to predict motion over finite distances leads us to introduce a book-keeping parameter t, to be called time, in order to keep track of the accumulation of short steps. The construction of time involves three ingredients: (a) we must specify what we mean by an 'instant'; (b) these instants must be ordered; and finally; (c) one must specify the interval Δt between successive instants—one must define 'duration'.

Since the foundation for any theory of time is the theory of change, i.e., the dynamics, the notion of time constructed below will reflect the inferential nature of entropic dynamics. Such a construction we will call "entropic" time [10]. Later we will return to the question of whether and how this "entropic" time is related to the "physical" time that is measured by clocks.

3.1. Time as an Ordered Sequence of Instants

ED consists of a succession of short steps. Consider, for example, the ith step which takes the system from $x = x_{i-1}$ to $x' = x_i$. Integrating the joint probability, $P(x_i, x_{i-1})$, over x_{i-1} gives

$$P(x_i) = \int dx_{i-1} P(x_i, x_{i-1}) = \int dx_{i-1} P(x_i|x_{i-1}) P(x_{i-1}) \,. \tag{12}$$

No physical assumptions were involved in deriving this equation; it follows directly from the laws of probability. To establish the connection to time and dynamics we will make the physical assumption that if $P(x_{i-1})$ is interpreted as the probability of different values of x_{i-1} at one "instant" labelled t, then we will interpret $P(x_i)$ as the probability of values of x_i at the next "instant" labelled t'. More explicitly, if we write $P(x_{i-1}) = \rho_t(x)$ and $P(x_i) = \rho_{t'}(x')$ then we have

$$\rho_{t'}(x') = \int dx \, P(x'|x) \rho_t(x) \,. \tag{13}$$

This equation defines the notion of "instant": if the distribution $\rho_t(x)$ refers to one instant t, then the distribution $\rho_{t'}(x')$ generated by $P(x'|x)$ defines what we mean by the "next" instant t'. Iterating this process defines the dynamics.

This construction of time is intimately related to information and inference. An instant is an informational state that is complete in the sense that it is specified by the information—codified into the distributions $\rho_t(x)$ and $P(x'|x)$—that is sufficient for predicting the next instant. Thus, the present is defined through a sufficient amount of information such that given the present, the future is independent of the past.

In the ED framework the notions of instant and of simultaneity are intimately related to the distribution $\rho_t(x)$. To see how this comes about consider a single particle at the point $\vec{x} = (x^1, x^2, x^3)$. It is implicit in the notation that x^1, x^2, and x^3 occur simultaneously. When we describe a system of N particles by a single point $x = (\vec{x}_1, \vec{x}_2, \ldots \vec{x}_N)$ in $3N$-dimensional configuration space it is also implicitly assumed that all the $3N$ coordinate values refer to the same instant; they are simultaneous. The very idea of a *point* in configuration space assumes simultaneity. And furthermore, whether we deal with one particle or many, a distribution such as $\rho_t(x)$ is meant to describe our uncertainty about the possible configurations x of the system at the given instant. Thus, a probability distribution $\rho_t(x)$ provides a criterion of simultaneity [85].

3.2. The Arrow of Entropic Time

The notion of time constructed according to Equation (13) is intrinsically directional. There is an absolute sense in which $\rho_t(x)$ is prior and $\rho_{t'}(x')$ is posterior. Indeed, the same rules of probability that led us to Equation (13) can also lead us to the time-reversed evolution,

$$\rho_t(x) = \int dx' \, P(x|x')\rho_{t'}(x') . \tag{14}$$

Note, however, that there is a temporal asymmetry: while the distribution $P(x'|x)$, Equation (7), is a Gaussian derived using the maximum entropy method, its time-reversed version $P(x|x')$ is related to $P(x'|x)$ by Bayes' theorem,

$$P(x|x') = \frac{\rho_t(x)}{\rho_{t'}(x')} P(x'|x) , \tag{15}$$

which in general will not be Gaussian.

The puzzle of the arrow of time (see e.g., [86,87]) arises from the difficulty in deriving a temporal asymmetry from underlying laws of nature that are symmetric. The ED approach offers a fresh perspective on this topic because it does not assume any underlying laws of nature—whether they be symmetric or not. The asymmetry is the inevitable consequence of constructing time in a dynamics driven by entropic inference.

From the ED point of view the challenge does not consist of explaining the arrow of time—*entropic time itself only flows forward*—but rather in explaining how it comes about that despite the arrow of time some laws of physics, such as the Schrödinger equation, turn out to be time reversible. We will revisit this topic in Section 9.

3.3. Duration and the Sub-Quantum Motion

We have argued that the concept of time is intimately connected to the associated dynamics but at this point neither the transition probability $P(x'|x)$ that specifies the dynamics nor the corresponding entropic time have been fully defined yet. It remains to specify how the multipliers α_n and α' are related to the interval Δt between successive instants.

The basic criterion for this choice is convenience: *duration is defined so that motion looks simple.* The description of motion is simplest when it reflects the symmetry of translations in space and time. In a flat space-time this leads to an entropic time that resembles Newtonian time in that it flows "equably everywhere and everywhen." Referring to Equations (9) and (11) we choose α' and α_n to be independent of x and t, and we choose the ratio $\alpha'/\alpha_n \propto \Delta t$ so that there is a well-defined drift velocity. For future convenience the proportionality constants will be expressed in terms of some particle-specific constants m_n,

$$\frac{\alpha'}{\alpha_n} = \frac{\hbar}{m_n}\Delta t , \tag{16}$$

where $\hbar$ is an overall constant that fixes the units of the m_ns relative to the units of time. As we shall later see, the constants m_n will eventually be identified with the particle masses while the constant $\hbar$ will be identified as Planck's constant. Having specified the ratio α'/α_n it remains to specify α_n (or α'). It turns out that the choice is not unique. There is a variety of motions at the sub-quantum "microscopic" level that lead to the same quantum mechanics at the "macroscopic" level.

In previous work [10–12] we chose α_n proportional to $1/\Delta t$. This led to an ED in which the particles follow the highly irregular non-differentiable Brownian trajectories characteristic of an Einstein–Smoluchowski process. The first new contribution of this paper is to explore the consequences of choosing $\alpha_n \propto 1/\Delta t^3$,

$$\alpha_n = \frac{m_n}{\eta \Delta t^3} , \tag{17}$$

where a new constant η is introduced.

It is convenient to introduce a notation tailored to configuration space. Let $x^A = x_n^a$, $\partial_A = \partial/\partial x_n^a$, and $\delta_{AB} = \delta_{nn'}\delta_{ab}$, where the upper case indices $A, B, \ldots$ label both the particles $n, n', \ldots$ and their coordinates $a, b, \ldots$. Then the transition probability (8) becomes

$$P(x'|x) = \frac{1}{Z} \exp\left[-\frac{1}{2\eta\Delta t} m_{AB} \left(\frac{\Delta x^A}{\Delta t} - v^A\right)\left(\frac{\Delta x^B}{\Delta t} - v^B\right)\right], \tag{18}$$

where we used (9) to define the drift velocity,

$$v^A = \frac{\langle \Delta x^A \rangle}{\Delta t} = m^{AB}\left[\partial_B \Phi - \bar{A}_B\right]. \tag{19}$$

The drift potential is rescaled into a new variable

$$\Phi = \hbar\phi \tag{20}$$

which will be called *the phase*. We also introduced the "mass" tensor and its inverse,

$$m_{AB} = m_n \delta_{AB} = m_n \delta_{nn'}\delta_{ab} \quad \text{and} \quad m^{AB} = \frac{1}{m_n}\delta^{AB}, \tag{21}$$

and $\bar{A}_A$ is a field in configuration space with components,

$$\bar{A}_A(x) = \hbar\beta_n A_a(x_n), \tag{22}$$

A generic displacement is then written as a drift plus a fluctuation,

$$\Delta x^A = v^A \Delta t + \Delta w^A, \tag{23}$$

and the fluctuations Δw^A are given by

$$\langle \Delta w^A \rangle = 0 \quad \text{and} \quad \langle \Delta w^A \Delta w^B \rangle = \eta m^{AB} \Delta t^3, \tag{24}$$

or

$$\left\langle \left(\frac{\Delta x^A}{\Delta t} - v^A\right)\left(\frac{\Delta x^B}{\Delta t} - v^B\right)\right\rangle = \eta m^{AB}\Delta t. \tag{25}$$

It is noteworthy that $\langle \Delta x^A \rangle \sim O(\Delta t)$ and $\Delta w^A \sim O(\Delta t^{3/2})$. This means that for short steps the fluctuations are negligible and the dynamics is dominated by the drift. The particles follow trajectories that are indeterministic but differentiable. Since $\Delta w^A \sim O(\Delta t^{3/2})$ the limit

$$V^A = \lim_{\Delta t \to 0} \frac{\Delta x^A}{\Delta t} = v^A \tag{26}$$

is well defined. In words: the actual velocities of the particles coincide with the expected or drift velocities. From Equation (19) we see that these velocities are continuous functions. The question of whether the velocities themselves are differentiable or not is trickier.

Consider two successive displacements $\Delta x = x' - x$ followed by $\Delta x' = x'' - x'$. The velocities are

$$V^A = \frac{\Delta x^A}{\Delta t} \quad \text{and} \quad V'^A = \frac{\Delta x'^A}{\Delta t}. \tag{27}$$

The change in velocity is given by a Langevin equation,

$$\Delta V^A = \langle \Delta V^A \rangle_{x''x'} + \Delta U^A, \tag{28}$$

where $\langle\cdot\rangle_{x''x'}$ denotes taking the expectations over x'' using $P(x''|x')$, and then over x' using $P(x'|x)$, and ΔU^A is a fluctuation. It is straightforward to show that

$$\langle\Delta V^A\rangle_{x''x'} = (\partial_t v^A + v^B\partial_B v^A)\,\Delta t\,, \tag{29}$$

so that the expected acceleration is given by the convective derivative of the velocity field along itself,

$$\lim_{\Delta t\to 0}\frac{\langle\Delta V^A\rangle}{\Delta t} = (\partial_t + v^B\partial_B)v^A\,. \tag{30}$$

One can also show that

$$\langle\Delta U^A\rangle_{x''x'} = 0\,,\quad\text{and}\quad \langle\Delta U^A\Delta U^B\rangle_{x''x'} = 2\eta m^{AB}\Delta t\,, \tag{31}$$

which means that ΔU is a Wiener process and we deal with a Brownian motion of the Oernstein–Uhlenbeck type.

We conclude this section with some general remarks.

On the nature of clocks. In Newtonian mechanics time is defined to simplify the dynamics. The prototype of a clock is a free particle which moves equal distances in equal times. In ED time is also defined to simplify the dynamics of free particles (for sufficiently short times all particles are free) and the prototype of a clock is a free particle too: as we see in (23) *the particle's mean displacement increases by equal amounts in equal times.*

On the nature of mass. In standard quantum mechanics, "what is mass?" and "why quantum fluctuations?" are two independent mysteries. In ED the mystery is somewhat alleviated: as we see in Equation (25) mass and fluctuations are two sides of the same coin. *Mass is an inverse measure of the velocity fluctuations.*

The information metric of configuration space. In addition to defining the dynamics the transition probability Equation (18) serves to define the geometry of the N-particle configuration space, $\mathbf{X}_N$. Since the physical single particle space $\mathbf{X}$ is described by the Euclidean metric δ_{ab} we can expect that the N-particle configuration space, $\mathbf{X}_N = \mathbf{X}\times\ldots\times\mathbf{X}$, will also be flat, but for non-identical particles a question might be raised about the relative scales or weights associated with each $\mathbf{X}$ factor. Information geometry provides the answer.

The fact that to each point $x\in\mathbf{X}_N$ there corresponds a probability distribution $P(x'|x)$ means that to the space $\mathbf{X}_N$ we can associate a statistical manifold the geometry of which (up to an overall scale factor) is uniquely determined by the information metric [17,65],

$$\gamma_{AB} = \int dx'\,P(x'|x)\frac{\partial\log P(x'|x)}{\partial x^A}\frac{\partial\log P(x'|x)}{\partial x^B}\,. \tag{32}$$

Substituting Equations (18) into (32) yields

$$\gamma_{AB} = \frac{1}{\eta\Delta t^3}m_{AB}\,. \tag{33}$$

The divergence as $\Delta t\to 0$ arises because the information metric measures statistical distinguishability. As $\Delta t\to 0$ the distributions $P(x'|x)$ and $P(x'|x+\Delta x)$ become more sharply peaked and increasingly easier to distinguish so that $\gamma_{AB}\to\infty$. Thus, up to a scale factor the metric of configuration space is basically the mass tensor.

The practice of describing a many-particle system as a single point in an abstract configuration space goes back to the work of H. Hertz in 1894 [88]. Historically the choice of the mass tensor as the metric of configuration space has been regarded as being convenient but of no particular significance. We can now see that the choice is not just a merely useful convention: up to an overall scale the metric follows uniquely from information geometry. Furthermore, it suggests the intriguing possibility of a deeper connection between kinetic energy and information geometry.

Invariance under gauge transformations. The fact that constraints (5) and (6) are not independent—they are both linear in the same displacements $\langle \Delta x_n^a \rangle$—leads to a gauge symmetry. This is evident in Equation (7) where ϕ and A_a appear in the combination $\partial_{na}\phi - \beta_n A_a$ which is invariant under the gauge transformations,

$$A_a(x_n) \quad \rightarrow \quad A'_a(x_n) = A_a(x_n) + \partial_a \chi(x_n)\,, \tag{34}$$

$$\phi(x) \quad \rightarrow \quad \phi'(x) = \phi(x) + \sum_n \beta_n \chi(x_n)\,. \tag{35}$$

These transformations are local in 3d-space. Introducing

$$\bar{\chi}(x) = \sum_n \hbar \beta_n \chi(x_n)\,, \tag{36}$$

they can be written in the N-particle configuration space,

$$\bar{A}_A(x) \quad \rightarrow \quad \bar{A}'_A(x) = \bar{A}_A(x) + \partial_A \bar{\chi}(x)\,, \tag{37}$$

$$\Phi(x) \quad \rightarrow \quad \Phi'(x) = \Phi(x) + \bar{\chi}(x)\,. \tag{38}$$

Interpretation: The drift potential $\phi(x) = \phi(\vec{x}_1, \vec{x}_2, \ldots)$ is assumed to be an "angle"–$\phi(x)$ and $\phi(x) + 2\pi$ are meant to describe the same angle. The angle at $\vec{x}_1$ depends on the values of all the other positions $\vec{x}_2, \vec{x}_3, \ldots$, and the angle at $\vec{x}_2$ depends on the values of all the other positions $\vec{x}_1, \vec{x}_3, \ldots$, and so on. The fact that the origins from which these angles are measured can be redefined by different amounts at different places gives rise to a local gauge symmetry. To compare angles at different locations one introduces a *connection* field, the vector potential $A_a(\vec{x})$. It defines which origin at $\vec{x} + \Delta\vec{x}$ is the "same" as the origin at $\vec{x}$. This is implemented by imposing that as we change origins and $\Phi(x)$ changes to $\Phi + \bar{\chi}$ then the connection transforms as $A_a \rightarrow A_a + \partial_a \chi$ so that the quantity $\partial_A \Phi - \bar{A}_A$ remains invariant.

A fractional Brownian motion? The choices $\alpha_n \propto 1/\Delta t$ and $\alpha_n \propto 1/\Delta t^3$ lead to Einstein–Smoluchowski and Oernstein–Uhlenbeck processes, respectively. For definiteness throughout the rest of this paper we will assume that the sub-quantum motion is an OU process but more general fractional Brownian motions [89] are possible. Consider

$$\alpha_n = \frac{m_n}{\eta \Delta t^\gamma}\,, \tag{39}$$

where γ is some positive parameter. The corresponding transition probability (8),

$$P(x'|x) = \frac{1}{Z} \exp\left[-\frac{1}{2\eta\Delta t^\gamma} m_{AB} \left(\Delta x^A - v^A \Delta t\right)\left(\Delta x^B - v^B \Delta t\right)\right]\,, \tag{40}$$

leads to fluctuations such that

$$\langle \Delta w^A \rangle = 0 \quad \text{and} \quad \langle \Delta w^A \Delta w^B \rangle = \eta m^{AB} \Delta t^\gamma\,, \tag{41}$$

or

$$\left\langle \left(\frac{\Delta x^A}{\Delta t} - v^A\right)\left(\frac{\Delta x^B}{\Delta t} - v^B\right)\right\rangle = \eta m^{AB} \Delta t^{\gamma-2}. \tag{42}$$

We will not pursue this topic further except to note that since $\langle \Delta x^A \rangle \sim O(\Delta t)$ and $\Delta w^A \sim O(\Delta t^{\gamma/2})$ for $\gamma < 2$ the sub-quantum motion is dominated by fluctuations and the trajectories are non-differentiable, while for $\gamma > 2$ the drift dominates and velocities are well defined.

4. The Evolution Equation in Differential Form

Entropic dynamics is generated by iterating Equation (13): given the information that defines one instant, the integral Equation (13) is used to construct the next instant. As so often in physics it is more convenient to rewrite the equation of evolution in differential form. The result is

$$\partial_t \rho = -\partial_A \left(v^A \rho \right) , \tag{43}$$

where v^A is given by (19). Before we proceed to its derivation we note that Equation (43) is a consequence of the fact that the particles follow continuous paths. Accordingly, we will follow standard practice and call it *the continuity equation*. Also note that in the OU process considered here ($\gamma = 3$) the current velocity—the velocity with which the probability flows in configuration space—coincides with the drift velocity (19) and with the actual velocities of the particles (26) [90].

Next we derive (43) using a technique that is well known in diffusion theory [91]. (For an alternative derivation see [92].) The result of building up a *finite* change from an initial time t_0 to a later time t leads to the distribution

$$\rho(x,t) = \int dx_0 \, P(x,t|x_0,t_0)\rho(x_0,t_0) , \tag{44}$$

where the finite-time transition probability, $P(x,t|x_0,t_0)$, is constructed by iterating the infinitesimal changes described in Equation (13),

$$P(x,t+\Delta t|x_0,t_0) = \int dz \, P(x,t+\Delta t|z,t)P(z,t|x_0,t_0) . \tag{45}$$

For small times Δt the distribution $P(x,t+\Delta t|z,t)$, given in Equation (18), is very sharply peaked at $x = z$. In fact, as $\Delta t \to 0$ we have $P(x,t+\Delta t|z,t) \to \delta(x-z)$. Such singular behavior cannot be handled directly by Taylor expanding in z about the point x. Instead one follows an indirect procedure. Multiply by a smooth test function $f(x)$ and integrate over x,

$$\int dx \, P(x,t+\Delta t|x_0,t_0)f(x) = \int dz \left[\int dx \, P(x,t+\Delta t|z,t)f(x) \right] P(z,t|x_0,t_0) . \tag{46}$$

The test function $f(x)$ is assumed sufficiently smooth precisely so that it can be expanded about z. Then as $\Delta t \to 0$ the integral in the brackets, dropping all terms of order higher than Δt, is

$$\begin{aligned} [\cdots] &= \int dx \, P(x,t+\Delta t|z,t) \left(f(z) + \frac{\partial f}{\partial z^A}(x^A - z^A) + \ldots \right) \\ &= f(z) + v^A(z)\Delta t \frac{\partial f}{\partial z^A} + \ldots \end{aligned} \tag{47}$$

where we used Equation (23). Next substitute (47) into the right hand side of (46), divide by Δt, and let $\Delta t \to 0$. Since $f(x)$ is arbitrary the result is

$$\partial_t P(x,t|x_0,t_0) = -\partial_A [v^A(x)P(x,t|x_0,t_0)] , \tag{48}$$

which is the continuity equation for the finite-time transition probability. Differentiating Equation (44) with respect to t, and substituting (48) completes the derivation of the continuity Equation (43).

The continuity Equation (43) can be written in another equivalent but very suggestive form involving functional derivatives. For some suitably chosen functional $\tilde{H}[\rho,\Phi]$ we have

$$\partial_t \rho(x) = -\partial_A \left[\rho m^{AB} (\partial_B \Phi - \bar{A}_B) \right] = \frac{\delta \tilde{H}}{\delta \Phi(x)} . \tag{49}$$

It is easy to check that the appropriate functional $\tilde{H}$ is

$$\tilde{H}[\rho,\Phi]=\int dx\,\frac{1}{2}\rho m^{AB}\left(\partial_A\Phi-\bar{A}_A\right)\left(\partial_B\Phi-\bar{A}_B\right)+F[\rho]\,, \tag{50}$$

where the unspecified functional $F[\rho]$ is an integration constant [93].

The continuity Equation (49) describes a somewhat peculiar OU Brownian motion in which the probability density $\rho(x)$ is driven by the non-dynamical fields Φ, and $\bar{A}$. This is an interesting ED in its own right but it is not QM. Indeed, a *quantum* dynamics consists in the coupled evolution of two dynamical fields: the density $\rho(x)$ and the phase of the wave function. This second field can be naturally introduced into ED by allowing the phase field Φ in (19) to become dynamical which amounts to an ED in which the constraint (5) is continuously updated at each instant in time. Our next topic is to propose the appropriate updating criterion. It yields an ED in which the phase field Φ guides the evolution of ρ, and in return, the evolving ρ reacts back and induces a change in Φ.

5. The Epistemic Phase Space

In ED we deal with two configuration spaces. One is the *ontic configuration space* $\mathbf{X}_N=\mathbf{X}\times\mathbf{X}\times\ldots$ of all particle positions, $x=(x_1\ldots x_N)\in\mathbf{X}_N$. The other is the *epistemic configuration space* or *e-configuration space* $\mathbf{P}$ of all normalized probabilities,

$$\mathbf{P}=\left\{\rho\,\middle|\,\rho(x)\geq 0;\int dx\rho(x)=1\right\}. \tag{51}$$

To formulate the coupled dynamics of ρ and Φ we need a framework to study paths in the larger space $\{\rho,\Phi\}$ that we will call the *epistemic phase space* or *e-phase space*.

Given any manifold such as $\mathbf{P}$ the associated tangent and cotangent bundles, respectively $T\mathbf{P}$ and $T^*\mathbf{P}$, are geometric objects that are always available to us independently of any physical considerations. Both are manifolds in their own right but the cotangent bundle $T^*\mathbf{P}$—the space of all probabilities and all covectors—is of particular interest because it comes automatically endowed with a rich geometrical structure [56–62]. The point is that cotangent bundles are symplectic manifolds and this singles out as "natural" those dynamical laws that happen to preserve some privileged symplectic form. This observation will lead us to identify e-phase space $\{\rho,\Phi\}$ with the cotangent bundle $T^*\mathbf{P}$ and provides the natural criterion for updating constraints, that is, for updating the phase Φ [94].

5.1. Notation: Vectors, Covectors, Etc.

A point $X\in T^*\mathbf{P}$ will be represented as

$$X=(\rho(x),\pi(x))=(\rho^x,\pi_x)\,, \tag{52}$$

where ρ^x represents coordinates on the base manifold $\mathbf{P}$ and π_x represents some generic coordinates on the space $T^*\mathbf{P}_\rho$ that is cotangent to $\mathbf{P}$ at the point ρ. Curves in $T^*\mathbf{P}$ allow us to define vectors. Let $X=X(\lambda)$ be a curve parametrized by λ, then the vector $\bar{V}$ tangent to the curve at $X=(\rho,\pi)$ has components $d\rho^x/d\lambda$ and $d\pi_x/d\lambda$, and is written

$$\bar{V}=\frac{d}{d\lambda}=\int dx\left[\frac{d\rho^x}{d\lambda}\frac{\delta}{\delta\rho^x}+\frac{d\pi_x}{d\lambda}\frac{\delta}{\delta\pi_x}\right], \tag{53}$$

where $\delta/\delta\rho^x$ and $\delta/\delta\pi_x$ are the basis vectors. The directional derivative of a functional $F[X]$ along the curve $X(\lambda)$ is

$$\frac{dF}{d\lambda}=\tilde{\nabla}F[\bar{V}]=\int dx\left[\frac{\delta F}{\delta\rho^x}\frac{d\rho^x}{d\lambda}+\frac{\delta F}{\delta\pi_x}\frac{d\pi_x}{d\lambda}\right], \tag{54}$$

where $\tilde{\nabla}$ is the functional gradient in $T^*\mathbf{P}$, i.e., the gradient of a generic functional $F[X] = F[\rho, \pi]$ is

$$\tilde{\nabla}F = \int dx \left[\frac{\delta F}{\delta \rho^x} \tilde{\nabla}\rho^x + \frac{\delta F}{\delta \pi_x} \tilde{\nabla}\pi_x \right] . \tag{55}$$

The tilde '~' serves to distinguish the functional gradient $\tilde{\nabla}$ from the spatial gradient $\nabla f = \partial_a f \nabla x^a$ on $\mathbf{R}^3$.

The fact that the space $\mathbf{P}$ is constrained to normalized probabilities means that the coordinates ρ^x are not independent. This technical difficulty is handled by embedding the ∞-dimensional manifold $\mathbf{P}$ in a $(\infty + 1)$-dimensional manifold $\mathbf{P}^{+1}$ where the coordinates ρ^x are unconstrained [95]. Thus, strictly, $\tilde{\nabla}F$ is a covector on $T^*\mathbf{P}^{+1}$, i.e., $\tilde{\nabla}F \in T^*\left(T^*\mathbf{P}^{+1}\right)_X$ and $\tilde{\nabla}\rho^x$ and $\tilde{\nabla}\pi_x$ are the corresponding basis covectors. Nevertheless, the gradient $\tilde{\nabla}F$ will yield the desired directional derivatives (54) on $T^*\mathbf{P}$ provided its action is restricted to vectors $\bar{V}$ that are tangent to the manifold $\mathbf{P}$. Such tangent vectors are constrained to obey

$$\frac{d}{d\lambda} \int dx \rho(x) = \int dx \frac{d\rho^x}{d\lambda} = 0 . \tag{56}$$

Instead of keeping separate track of the ρ^x and π_x coordinates it is more convenient to combine them into a single index. A point $X = (\rho, \pi)$ will then be labelled by its coordinates

$$X^I = (X^{1x}, X^{2x}) = (\rho^x, \pi_x) . \tag{57}$$

We will use capital letters from the middle of the Latin alphabet $(I, J, K \ldots)$; $I = (\alpha, x)$ is a composite index where $\alpha = 1, 2$ keeps track of whether x is an upper index ($\alpha = 1$) or a lower index ($\alpha = 2$) [96]. Then Equations (53)–(55) are written as

$$\bar{V} = V^I \frac{\delta}{\delta X^I} , \quad \text{where} \quad V^I = \frac{dX^I}{d\lambda} = \begin{bmatrix} d\rho^x / d\lambda \\ d\pi_x / d\lambda \end{bmatrix} , \tag{58}$$

$$\frac{dF}{d\lambda} = \tilde{\nabla}F[\bar{V}] = \frac{\delta F}{\delta X^I} V^I \quad \text{and} \quad \tilde{\nabla}F = \frac{\delta F}{\delta X^I} \tilde{\nabla}X^I , \tag{59}$$

where the repeated indices indicate a summation over α and an integration over x.

5.2. The Symplectic Form in ED

In classical mechanics with configuration space $\{q^i\}$ the Lagrangian $L(q, \dot{q})$ is a function on the tangent bundle while the Hamiltonian $H(q, p)$ is a function on the cotangent bundle [97,98]. A symplectic form provides a mapping from the tangent to the cotangent bundles. Given a Lagrangian the map is defined by $p_i = \partial L / \partial \dot{q}^i$ and this automatically defines the corresponding symplectic form. In ED there is no Lagrangian so to define the symplectic map we must look elsewhere. We propose that the role played by the Lagrangian in classical mechanics will in ED be played by the continuity Equation (49).

The fact that the preservation of a symplectic structure must reproduce the continuity equation leads us to identify the phase Φ_x as the momentum canonically conjugate to ρ^x. This identification of the e-phase space $\{\rho, \Phi\}$ with $T^*\mathbf{P}$ is highly non-trivial. It amounts to asserting that the phase Φ_x transforms as the components of a Poincare 1-form

$$\theta = \int dx \, \Phi_x \mathrm{d}\rho^x , \tag{60}$$

where d is the exterior derivative and the corresponding symplectic 2-form $\Omega = -\mathrm{d}\theta$ is

$$\Omega = \int dx \, \mathrm{d}\rho^x \wedge \mathrm{d}\Phi_x = \int dx \, \left[\tilde{\nabla}\rho^x \otimes \tilde{\nabla}\Phi_x - \tilde{\nabla}\Phi_x \otimes \tilde{\nabla}\rho^x \right] . \tag{61}$$

By construction Ω is exact ($\Omega = -d\theta$) and closed ($d\Omega = 0$). The action of $\Omega[\cdot,\cdot]$ on two vectors $\bar{V} = d/d\lambda$ and $\bar{U} = d/d\mu$ is given by

$$\Omega[\bar{V},\bar{U}] = \int dx \left[V^{1x}U^{2x} - V^{2x}U^{1x} \right] = \Omega_{IJ}V^I U^J \,, \tag{62}$$

so that the components of Ω are

$$\Omega_{IJ} = \Omega_{\alpha x,\beta x'} = \begin{bmatrix} 0 & 1 \\ -1 & 0 \end{bmatrix} \delta(x,x') \,. \tag{63}$$

5.3. Hamiltonian Flows and Poisson Brackets

Next we reproduce the ∞-dimensional $T^*\mathbf{P}$ analogues of results that are standard in finite-dimensional classical mechanics [97,98]. Given a vector field $\bar{V}[X]$ in e-phase space we can integrate $V^I[X] = dX^I/d\lambda$ to find its integral curves $X^I = X^I(\lambda)$. We are particularly interested in those vector fields that generate flows that preserve the symplectic structure,

$$\pounds_V \Omega = 0 \,, \tag{64}$$

where the Lie derivative is given by

$$(\pounds_V\Omega)_{IJ} = V^K \tilde{\nabla}_K \Omega_{IJ} + \Omega_{KJ}\tilde{\nabla}_I V^K + \Omega_{IK}\tilde{\nabla}_J V^K \,. \tag{65}$$

Since by Equation (63) the components Ω_{IJ} are constant, $\tilde{\nabla}_K\Omega_{IJ} = 0$, we can rewrite $\pounds_V\Omega$ as

$$(\pounds_V\Omega)_{IJ} = \tilde{\nabla}_I(\Omega_{KJ}V^K) - \tilde{\nabla}_J(\Omega_{KI}V^K) \,, \tag{66}$$

which is the exterior derivative (basically, the curl) of the covector $\Omega_{KI}V^K$. By Poincare's lemma, requiring $\pounds_V\Omega = 0$ (a vanishing curl) implies that $\Omega_{KI}V^K$ is the gradient of a scalar function, which we will denote $\tilde{V}[X]$,

$$\Omega_{KI}V^K = \tilde{\nabla}_I \tilde{V} \,. \tag{67}$$

Using (63) this is more explicitly written as

$$\int dx \left[\frac{d\rho^x}{d\lambda}\tilde{\nabla}\Phi_x - \frac{d\Phi_x}{d\lambda}\tilde{\nabla}\rho^x \right] = \int dx \left[\frac{\delta\tilde{V}}{\delta\rho^x}\tilde{\nabla}\rho^x + \frac{\delta\tilde{V}}{\delta\Phi_x}\tilde{\nabla}\Phi_x \right] , \tag{68}$$

or

$$\frac{d\rho^x}{d\lambda} = \frac{\delta\tilde{V}}{\delta\Phi_x} \quad \text{and} \quad \frac{d\Phi_x}{d\lambda} = -\frac{\delta\tilde{V}}{\delta\rho^x} \,, \tag{69}$$

which we recognize as Hamilton's equations for a Hamiltonian function $\tilde{V}$. This justifies calling $\bar{V}$ the Hamiltonian vector field associated with the Hamiltonian function $\tilde{V}$.

From (62), the action of the symplectic form Ω on two Hamiltonian vector fields $\bar{V} = d/d\lambda$ and $\bar{U} = d/d\mu$ generated respectively by $\tilde{V}$ and $\tilde{U}$ is

$$\Omega[\bar{V},\bar{U}] = \int dx \left[\frac{d\rho^x}{d\lambda}\frac{d\Phi_x}{d\mu} - \frac{d\Phi_x}{d\lambda}\frac{d\rho^x}{d\mu} \right] , \tag{70}$$

which, using (69), gives

$$\Omega[\bar{V},\bar{U}] = \int dx \left[\frac{\delta\tilde{V}}{\delta\rho^x}\frac{\delta\tilde{U}}{\delta\Phi_x} - \frac{\delta\tilde{V}}{\delta\Phi_x}\frac{\delta\tilde{U}}{\delta\rho^x} \right] \overset{\text{def}}{=} \{\tilde{V},\tilde{U}\} \,, \tag{71}$$

where on the right we introduced the Poisson bracket notation.

To summarize these results: (1) The condition for a flow generated by the vector field V^I to preserve the symplectic structure, $\pounds_V\Omega = 0$, is that V^I be the Hamiltonian vector field associated to a Hamiltonian function $\tilde{V}$, Equation (69),

$$V^I = \frac{dX^I}{d\lambda} = \{X^I, \tilde{V}\} . \tag{72}$$

(2) The action of Ω on two Hamiltonian vector fields (71) is the Poisson bracket of the associated Hamiltonian functions,

$$\Omega[\bar{V}, \bar{U}] = \Omega_{IJ} V^I U^J = \{\tilde{V}, \tilde{U}\} . \tag{73}$$

We conclude that the ED that preserves the symplectic structure Ω and reproduces the continuity Equation (49) is described by the Hamiltonian flow of the scalar functional $\tilde{H}$ in (50). However, the full dynamics, which will obey the Hamiltonian evolution equations

$$\partial_t \rho^x = \frac{\delta \tilde{H}}{\delta \Phi_x} \quad \text{and} \quad \partial_t \Phi_x = -\frac{\delta \tilde{H}}{\delta \rho^x} , \tag{74}$$

is not yet fully determined because the integration constant $F[\rho]$ in (50) remains to be specified.

5.4. The Normalization Constraint

Since the particular flow that we will associate with time evolution is required to reproduce the continuity equation it will also preserve the normalization constraint,

$$\tilde{N} = 0 \quad \text{where} \quad \tilde{N} = 1 - |\rho| \quad \text{and} \quad |\rho| \stackrel{\text{def}}{=} \int dx\, \rho(x) . \tag{75}$$

Indeed, one can check that

$$\partial_t \tilde{N} = \{\tilde{N}, \tilde{H}\} = 0 . \tag{76}$$

The Hamiltonian flow (72) generated by $\tilde{N}$ and parametrized by α is given by the vector field

$$\bar{N} = N^I \frac{\delta}{\delta X^I} \quad \text{with} \quad N^I = \frac{dX^I}{d\alpha} = \{X^I, \tilde{N}\} , \tag{77}$$

or, more explicitly,

$$N^{1x} = \frac{d\rho^x}{d\alpha} = 0 \quad \text{and} \quad N^{2x} = \frac{d\Phi_x}{d\alpha} = 1 . \tag{78}$$

The conservation of $\tilde{N}$, Equation (76), implies that $\tilde{N}$ is the generator of a symmetry, namely

$$\frac{d\tilde{H}}{d\alpha} = \{\tilde{H}, \tilde{N}\} = 0 . \tag{79}$$

Integrating (78) one finds the integral curves generated by $\tilde{N}$,

$$\rho^x(\alpha) = \rho^x(0) \quad \text{and} \quad \Phi_x(\alpha) = \Phi_x(0) + \alpha . \tag{80}$$

This shows that the symmetry generated by $\tilde{N}$ is to shift the phase Φ by a constant α without otherwise changing the dynamics. This was, of course, already evident in the continuity Equation (43) with (19) but the implications are very significant. Not only does the constraint $\tilde{N} = 0$ reduce by one the (infinite) number of independent ρ^x degrees of freedom but the actual number of Φ_xs is also reduced by one because for any value of α the phases $\Phi_x + \alpha$ and Φ_x correspond to the same state. (This is the ED analogue of the fact that in QM states are represented by rays rather than vectors in a Hilbert space.) An immediate consequence is that two vectors $\bar{U}$ and $\bar{V}$ at X that differ by a vector proportional to $\bar{N}$,

$$\bar{U} = \bar{V} + k\bar{N}\,, \tag{81}$$

are "physically" equivalent. In particular the vector $\bar{N}$ is equivalent to zero.

The phase space of interest is $T^*\mathbf{P}$ but to handle the constraint $|\rho| = 1$ we have been led to using coordinates that are more appropriate to the larger embedding space $T^*\mathbf{P}^{+1}$. The price we pay for introducing one superfluous coordinate is to also introduce a superfluous momentum. We eliminate the extra coordinate by imposing the constraint $\tilde{N} = 0$. We eliminate the extra momentum by declaring it unphysical. All vectors that differ by a vector along the gauge direction $\bar{N}$ are declared equivalent; they belong to the same equivalence class. The result is a global gauge symmetry.

An equivalence class can be represented by any one of its members and choosing a convenient representative amounts to fixing the gauge. As we shall see below a convenient gauge condition is to impose

$$\int dx\, \rho^x V^{2x} = 0 \quad \text{or} \quad \langle V^2 \rangle = 0\,, \tag{82}$$

so that the representative "Tangent Gauge-Fixed" vectors (which we shall refer to as TGF vectors) will satisfy two conditions, Equations (56) and (82),

$$|V^1| = \int dx\, V^{1x} = 0 \quad \text{and} \quad \langle V^2 \rangle = \int dx\, \rho^x V^{2x} = 0\,. \tag{83}$$

The first condition enforces a flow tangent to the $|\rho| = 1$ surface; the second eliminates a superfluous vector component along the gauge direction $\bar{N}$.

We end this section with a comment on the symplectic form Ω which is non-degenerate on $T^*\mathbf{P}^{+1}$ but at first sight appears to be degenerate on $T^*\mathbf{P}$. Indeed, we have $\Omega(\bar{N}, \bar{V}) = 0$ for any tangent vector $\bar{V}$. However, we must recall that $\bar{N}$ is equivalent to 0. In fact, since the TGF equivalent of $\bar{N}$ is 0, Ω is not degenerate on $T^*\mathbf{P}$.

6. The Information Geometry of E-Phase Space

The construction of the ensemble Hamiltonian $\tilde{H}$—or *e-Hamiltonian*—is motivated as follows. The goal of dynamics is to determine the evolution of the state (ρ_t, Φ_t). From a given initial state (ρ_0, Φ_0) two slightly different Hamiltonians will lead to slightly different final states, say (ρ_t, Φ_t) or $(\rho_t + \delta\rho_t, \Phi_t + \delta\Phi_t)$. Will these small changes make any difference? Can we quantify the extent to which we can *distinguish* between two neighboring states? This is precisely the kind of question that metrics are designed to address. It is then natural that $\tilde{H}$ be in some way related to some choice of metric. But although $\mathbf{P}$ is naturally endowed with a unique information metric the space $T^*\mathbf{P}$ has none. Thus, our next goal is to construct a metric for $T^*\mathbf{P}$.

Once a metric structure is in place we can ask: does the distance between two neighboring states—the extent to which we can *distinguish* them—grow, stay the same, or diminish over time? There are many possibilities here but for pragmatic (and esthetic) reasons we are led to consider the simplest form of dynamics—one that preserves the metric. This leads us to study the Hamilton flows (those that preserve the symplectic structure) that are also Killing flows (those flows that preserve the metric structure).

In ED entropic time is constructed so that time (duration) is defined by a clock provided by the system itself. This leads to require that the generator $\tilde{H}$ of time translations be defined in terms of the very same clock that provides the measure of time. Thus, the third and final ingredient in the construction of $\tilde{H}$ is the requirement is that the e-Hamiltonian agree with (50) to reproduce the evolution of ρ given by the continuity Equation (49).

In this section, our goal is to transform e-phase space $T^*\mathbf{P}$ from a manifold that is merely symplectic to a manifold that is both symplectic and Riemannian. The implementation of the other two requirements on $\tilde{H}$—that it generates a Hamilton–Killing flow and that it agrees with the ED continuity equation—will be tackled in Sections 7 and 8.

6.1. The Metric on the Embedding Space $T^*\mathbf{P}^{+1}$

The configuration space $\mathbf{P}$ is a metric space. Our goal here is to extend its metric—given by information geometry—to the full cotangent bundle, $T^*\mathbf{P}$. It is convenient to first recall one derivation of the information metric. In the discrete case the statistical manifold is the k-simplex $\Sigma = \{p = (p^0 \ldots p^k) : \sum_{i=0}^k p^i = 1\}$. The basic idea is to find the most general metric consistent with a certain symmetry requirement. To suggest what that symmetry might be we change to new coordinates $\xi^i = (p^i)^{1/2}$. In these new coordinates the equation for the k-simplex Σ—the normalization condition—reads $\sum_{i=0}^k (\xi^i)^2 = 1$ which suggests the equation of a sphere.

We take this hint seriously and *declare* that the k-simplex is a k-sphere embedded in a generic $(k+1)$-dimensional spherically symmetric space Σ^{+1} [99]. In the ξ^i coordinates the metric of Σ^{+1} is of the form

$$d\ell^2 = [a(|p|) - b(|p|)] \left(\sum_{i=0}^{k} \xi^i d\xi^i \right)^2 + |p| b(|p|) \sum_{i=0}^{k} (d\xi^i)^2, \tag{84}$$

where $a(|p|)$ and $b(|p|)$ are two arbitrary smooth and positive functions of $|p| = \sum_{i=0}^k p^i$. Expressed in terms of the original p^i coordinates the metric of Σ^{+1} is

$$d\ell^2 = [a(|p|) - b(|p|)] \left(\sum_{i=0}^{k} dp^i \right)^2 + |p| b(|p|) \sum_{i=0}^{k} \frac{1}{p^i} (dp^i)^2 . \tag{85}$$

The restriction to normalized states, $|p| = 1$ with displacements tangent to the simplex, $\sum_{i=0}^k dp^i = 0$, gives the information metric induced on the k-simplex Σ,

$$d\ell^2 = b(1) \sum_{i=0}^{k} \frac{1}{p^i} (dp^i)^2 . \tag{86}$$

The overall constant $b(1)$ is not important; it amounts to a choice of the units of distance.

To extend the information metric from the k-simplex Σ to its cotangent bundle $T^*\Sigma$ we focus on the embedding spaces Σ^{+1} and $T^*\Sigma^{+1}$ and require that

(a) the metric on $T^*\Sigma^{+1}$ be compatible with the metric on Σ^{+1}; and
(b) that the spherical symmetry of the $(k+1)$-dimensional space Σ^{+1} be enlarged to full spherical symmetry for the $2(k+1)$-dimensional space $T^*\Sigma^{+1}$.

The simplest way to implement (a) is to follow as closely as possible the derivation that led to (85). The fact that Φ inherits from the drift potential ϕ the topological structure of an angle suggests introducing new coordinates,

$$\xi^i = (p^i)^{1/2} \cos \Phi_i/\hbar \quad \text{and} \quad \eta^i = (p^i)^{1/2} \sin \Phi_i/\hbar . \tag{87}$$

Then the normalization condition reads

$$|p| = \sum_{i=0}^{k} p^i = \sum_{i=0}^{k} \left[(\xi^i)^2 + (\eta^i)^2 \right] = 1 \tag{88}$$

which suggests the equation of a $(2k+1)$-sphere embedded in $2(k+1)$ dimensions. To implement (b) we take this spherical symmetry seriously. The most general metric in the embedding space that is invariant under rotations is

$$\begin{aligned} d\ell^2 &= [a(|p|)-b(|p|)]\left[\sum_{i=0}^{k}\left(\xi^i d\xi^i+\eta^i d\eta^i\right)\right]^2 \\ &\quad +|p|b(|p|)\sum_{i=0}^{k}\left[(d\xi^i)^2+(d\eta^i)^2\right], \end{aligned} \tag{89}$$

where the two functions $a(|p|)$ and $b(|p|)$ are smooth and positive but otherwise arbitrary. Therefore, changing back to the (p^i,Φ_i) coordinates, the most general rotationally invariant metric for the embedding space $T^*\Sigma^{+1}$ is

$$\begin{aligned} d\ell^2 &= \frac{1}{4}[a(|p|)-b(|p|)]\left[\sum_{i=0}^{k}dp^i\right]^2 \\ &\quad +|p|b(|p|)\frac{1}{2\hbar}\sum_{i=0}^{k}\left[\frac{\hbar}{2p^i}(dp^i)^2+\frac{2p^i}{\hbar}(d\Phi_i)^2\right]. \end{aligned} \tag{90}$$

Generalizing from the finite-dimensional case to the ∞-dimensional case yields the metric on the spherically symmetric space $T^*\mathbf{P}^{+1}$,

$$\delta\tilde{\ell}^2 = A\left[\int dx\,\delta\rho_x\right]^2 + B\int dx\left[\frac{\hbar}{2\rho_x}(\delta\rho_x)^2+\frac{2\rho_x}{\hbar}(\delta\Phi_x)^2\right]. \tag{91}$$

where we set

$$A(|\rho|)=\frac{1}{4}[a(|\rho|)-b(|\rho|)] \quad \text{and} \quad B(|\rho|)=\frac{1}{2\hbar}|\rho|b(|\rho|). \tag{92}$$

6.2. The Metric Induced on $T^*\mathbf{P}$

As we saw in Section 5.4 the normalization constraint $|\rho|=1$ induces a symmetry—points with phases differing by a constant are identified. Therefore, the e-phase space $T^*\mathbf{P}$ can be obtained from the spherically symmetric space $T^*\mathbf{P}^{+1}$ by the restriction $|\rho|=1$ and by identifying points (ρ_x,Φ_x) and $(\rho_x,\Phi_x+\alpha)$ that lie on the same gauge orbit, or on the same ray.

Consider two neighboring points (ρ_x,Φ_x) and (ρ'_x,Φ'_x). The metric induced on $T^*\mathbf{P}$ is defined as the shortest $T^*\mathbf{P}^{+1}$ distance between (ρ_x,Φ_x) and points on the ray defined by (ρ'_x,Φ'_x). Setting $|\delta\rho|=0$ the $T^*\mathbf{P}^{+1}$ distance between (ρ_x,Φ_x) and $(\rho_x+\delta\rho_x,\Phi_x+\delta\Phi_x+\delta\alpha)$ is given by

$$\delta\tilde{\ell}^2 = B(1)\int dx\left[\frac{\hbar}{2\rho_x}(\delta\rho_x)^2+\frac{2\rho_x}{\hbar}(\delta\Phi_x+\delta\alpha)^2\right]. \tag{93}$$

Let

$$\delta\tilde{s}^2 = \min_{\delta\alpha}\delta\tilde{\ell}^2. \tag{94}$$

Minimizing over $\delta\alpha$ gives the metric on $T^*\mathbf{P}$,

$$\delta\tilde{s}^2 = \int dx\left[\frac{\hbar}{2\rho_x}(\delta\rho_x)^2+\frac{2\rho_x}{\hbar}(\delta\Phi_x-\langle\delta\Phi\rangle)^2\right], \tag{95}$$

where we set $B(1)=1$ which amounts to a choice of units of length. This metric is known as the Fubini–Study metric.

The scalar product between two vectors $\bar{V}$ and $\bar{U}$ is

$$G(\bar{V},\bar{U}) = \int dx \left[\frac{\hbar}{2\rho_x} V^{1x}U^{1x} + \frac{2\rho_x}{\hbar}(V^{2x} - \langle V^2 \rangle)(U^{2x} - \langle U^2 \rangle) \right] . \tag{96}$$

It is at this point that we recognize the convenience of imposing the TGF gauge condition (83): the scalar product simplifies to

$$G(\bar{V},\bar{U}) = \int dx \left[\frac{\hbar}{2\rho_x} V^{1x}U^{1x} + \frac{2\rho_x}{\hbar} V^{2x}U^{2x} \right] . \tag{97}$$

An analogous expression can be written for the length $\delta\tilde{s}$ of a displacement $(\delta\rho_x, \delta\Phi_x)$,

$$\delta\tilde{s}^2 = \int dx \left[\frac{\hbar}{2\rho_x} (\delta\rho_x)^2 + \frac{2\rho_x}{\hbar} (\delta\Phi_x)^2 \right] , \tag{98}$$

where it is understood that $(\delta\rho_x, \delta\Phi_x)$ satisfies the TGF condition

$$|\delta\rho| = 0 \quad \text{and} \quad \langle \delta\Phi \rangle = 0 \, . \tag{99}$$

In index notation the metric (98) of $T^*\mathbf{P}$ is written as

$$\delta\tilde{s}^2 = G_{IJ}\delta X^I \delta X^J = \int dx dx' G_{\alpha x,\beta x'} \delta X^{x\alpha} \delta X^{x'\beta} \tag{100}$$

where the metric tensor G_{IJ} is

$$G_{IJ} = G_{\alpha x,\beta x'} = \begin{bmatrix} \frac{\hbar}{2\rho_x}\delta_{xx'} & 0 \\ 0 & \frac{2}{\hbar}\rho_x\delta_{xx'} \end{bmatrix} . \tag{101}$$

The tensor G_{IJ} in Equation (101) can act on arbitrary vectors whether they satisfy the TGF condition or not. It is only when G_{IJ} acts on TGF vectors that it is interpreted as a metric tensor on $T^*\mathbf{P}$.

6.3. A Complex Structure

Next we contract the symplectic form Ω_{IJ}, Equation (63), with the inverse of the metric tensor,

$$G^{IJ} = G^{\alpha x,\beta x'} = \begin{bmatrix} \frac{2}{\hbar}\rho_x\delta_{xx'} & 0 \\ 0 & \frac{\hbar}{2\rho_x}\delta_{xx'} \end{bmatrix} . \tag{102}$$

The result is a mixed tensor J with components

$$J^I{}_J = -G^{IK}\Omega_{KJ} = \begin{bmatrix} 0 & -\frac{2}{\hbar}\rho_x\delta_{xx'} \\ \frac{\hbar}{2\rho_x}\delta_{xx'} & 0 \end{bmatrix} . \tag{103}$$

(The reason for introducing an additional negative sign will become clear below.) The tensor $J^I{}_J$ maps vectors to vectors—as any mixed $(1,1)$ tensor should. What makes the tensor J special is that—as one can easily check— its action on a TGF vector $\bar{V}$ yields another vector $J\bar{V}$ that is also TGF and, furthermore, its square is

$$J^I{}_K J^K{}_J = -\delta_{xx'} \begin{bmatrix} 1 & 0 \\ 0 & 1 \end{bmatrix} = -\delta^I{}_J \, . \tag{104}$$

In words, when acting on vectors tangent to $T^*\mathbf{P}$ the action of J^2 (or Ω^2) is equivalent to multiplying by -1. This means that J plays the role of a complex structure.

We conclude that the cotangent bundle $T^*\mathbf{P}$ has a symplectic structure Ω, as all cotangent bundles do; that it can be given a Riemannian structure G_{IJ}; and that the mixed tensor J provides it with a complex structure.

6.4. Complex Coordinates

The fact that $T^*\mathbf{P}$ is endowed with a complex structure suggests introducing complex coordinates,

$$\Psi_x = \rho_x^{1/2} \exp i\Phi_x/\hbar\,, \tag{105}$$

so that a point $\Psi \in T^*\mathbf{P}^{+1}$ has coordinates

$$\Psi^{\mu x} = \begin{pmatrix} \Psi^{1x} \\ \Psi^{2x} \end{pmatrix} = \begin{pmatrix} \Psi_x \\ i\hbar\Psi_x^* \end{pmatrix}, \tag{106}$$

where the index μ takes two values, $\mu = 1, 2$.

We can check that the transformation from real coordinates (ρ, Φ) to complex coordinates $(\Psi, i\hbar\Psi^*)$ is canonical. Indeed, the action of Ω on two infinitesimal vectors δX^I and $\delta' X^J$ is

$$\Omega_{IJ}\delta X^I \delta' X^J = \int dx\ (\delta\rho_x \delta'\Phi_x - \delta\Phi_x \delta'\rho_x)\ ,$$

which, when expressed in Ψ coordinates, becomes

$$\Omega_{IJ}\delta X^I \delta' X^J = \int dx\ (\delta\Psi\delta' i\hbar\Psi^* - \delta i\hbar\Psi^* \delta'\Psi) = \Omega_{\mu x, \nu x'}\delta\Psi^{\mu x}\delta\Psi^{\nu x'} \tag{107}$$

where

$$\Omega_{\mu x, \nu x'} = \begin{bmatrix} 0 & 1 \\ -1 & 0 \end{bmatrix} \delta_{xx'}\,, \tag{108}$$

retains the same form as (63).

Expressed in Ψ coordinates the Hamiltonian flow generated by the normalization constraint (75),

$$\tilde{N} = 0 \quad \text{with} \quad \tilde{N} = 1 - \int dx\, \Psi_x^* \Psi_x\,, \tag{109}$$

and parametrized by α is given by the vector field

$$\bar{N} = -\begin{pmatrix} \Psi_x/i\hbar \\ i\hbar(\Psi_x/i\hbar)^* \end{pmatrix}. \tag{110}$$

Its integral curves are

$$\Psi_x(\alpha) = \Psi_x(0)e^{i\alpha/\hbar}\,. \tag{111}$$

The constraint $\tilde{N} = 0$ induces a gauge symmetry which leads us to restrict our attention to vectors $\bar{V} = d/d\lambda$ satisfying two real TGF conditions (83). In Ψ coordinates this is replaced by the single complex TGF condition,

$$\int dx\, \Psi_x^* \frac{d\Psi_x}{d\lambda} = 0\,. \tag{112}$$

In Ψ coordinates the metric on $T^*\mathbf{P}$, Equation (98), becomes

$$\delta\tilde{s}^2 = -2i \int dx\, \delta\Psi_x \delta i\hbar\Psi_x^* = \int dx dx' G_{\mu x, \nu x'}\, \delta\Psi^{\mu x} \delta\Psi^{\nu x'}\,, \tag{113}$$

where the metric tensor and its inverse are

$$G_{\mu x,\nu x'} = -i\delta_{xx'}\begin{bmatrix}0 & 1\\ 1 & 0\end{bmatrix} \quad \text{and} \quad G^{\mu x,\nu x'} = i\delta_{xx'}\begin{bmatrix}0 & 1\\ 1 & 0\end{bmatrix}. \tag{114}$$

Finally, using $G^{\mu x,\nu x'}$ to raise the first index of $\Omega_{\nu x',\gamma x''}$ gives the Ψ components of the tensor J

$$J^{\mu x}{}_{\gamma x''} \overset{\text{def}}{=} -G^{\mu x,\nu x'}\Omega_{\nu x',\gamma x''} = \begin{bmatrix}i & 0\\ 0 & -i\end{bmatrix}\delta_{xx'}. \tag{115}$$

7. Hamilton-Killing Flows

Our next goal will be to find those Hamiltonian flows Q^I that also happen to preserve the metric tensor, i.e., we want Q^I to be a Killing vector. The condition for Q^I is

$$(\pounds_Q G)_{IJ} = Q^K\tilde{\nabla}_K G_{IJ} + G_{KJ}\tilde{\nabla}_I Q^K + G_{IK}\tilde{\nabla}_J Q^K = 0. \tag{116}$$

In complex coordinates Equation (114) gives $\tilde{\nabla}_K G_{IJ} = 0$, and the Killing equation simplifies to

$$(\pounds_Q G)_{IJ} = G_{KJ}\tilde{\nabla}_I Q^K + G_{IK}\tilde{\nabla}_J Q^K = 0, \tag{117}$$

or

$$(\pounds_Q G)_{\mu x,\nu x'} = -i\begin{bmatrix}\frac{\delta Q^{2x'}}{\delta\Psi_x} + \frac{\delta Q^{2x}}{\delta\Psi_{x'}}\,; & \frac{\delta Q^{1x'}}{\delta\Psi_x} + \frac{\delta Q^{2x}}{\delta i\hbar\Psi^*_{x'}}\\ \frac{\delta Q^{2x'}}{\delta i\hbar\Psi^*_x} + \frac{\delta Q^{1x}}{\delta\Psi_{x'}}\,; & \frac{\delta Q^{1x'}}{\delta i\hbar\Psi^*_x} + \frac{\delta Q^{1x}}{\delta i\hbar\Psi^*_{x'}}\end{bmatrix} = 0. \tag{118}$$

If we further require that Q^I be a Hamiltonian flow, $\pounds_Q\Omega = 0$, then we substitute

$$Q^{1x} = \frac{\delta\tilde{Q}}{\delta i\hbar\Psi^*_x} \quad \text{and} \quad Q^{2x} = -\frac{\delta\tilde{Q}}{\delta\Psi_x} \tag{119}$$

into (118) to get

$$\frac{\delta^2\tilde{Q}}{\delta\Psi_x\delta\Psi_{x'}} = 0 \quad \text{and} \quad \frac{\delta^2\tilde{Q}}{\delta\Psi^*_x\delta\Psi^*_{x'}} = 0. \tag{120}$$

Therefore, to generate a flow that preserves both G and Ω the functional $\tilde{Q}[\Psi,\Psi^*]$ must be *linear* in both Ψ and Ψ^*,

$$\tilde{Q}[\Psi,\Psi^*] = \int dx dx'\,\Psi^*_x\hat{Q}_{xx'}\Psi_{x'}, \tag{121}$$

where $\hat{Q}_{xx'}$ is a possibly non-local kernel. The actual Hamilton–Killing flow is

$$\frac{d\Psi_x}{d\lambda} = Q^{1x} = \frac{\delta\tilde{Q}}{\delta i\hbar\Psi^*_x} = \frac{1}{i\hbar}\int dx'\,\hat{Q}_{xx'}\Psi_{x'}, \tag{122}$$

$$\frac{di\hbar\Psi^*_x}{d\lambda} = Q^{2x} = -\frac{\delta\tilde{Q}}{\delta\Psi_x} = -\int dx'\,\Psi^*_{x'}\hat{Q}_{xx'}. \tag{123}$$

Taking the complex conjugate of (122) and compared to (123), shows that the kernel $\hat{Q}_{xx'}$ is Hermitian,

$$\hat{Q}^*_{xx'} = \hat{Q}_{x'x}, \tag{124}$$

and we can check that the corresponding Hamiltonian functionals $\tilde{Q}$ are real,

$$\tilde{Q}[\Psi,\Psi^*]^* = \tilde{Q}[\Psi,\Psi^*].$$

The Hamiltonian flows that might potentially be of interest are those that generate symmetry transformations. For example, the generator of translations is total momentum. Under a spatial displacement by ε^a, $g(x) \to g_\varepsilon(x) = g(x-\varepsilon)$, the change in $f[\rho,\Phi]$ is

$$\delta_\varepsilon f[\rho,\Phi] = \int dx \left(\frac{\delta f}{\delta \rho_x}\delta_\varepsilon \rho_x + \frac{\delta f}{\delta \Phi_x}\delta_\varepsilon \Phi_x \right) = \{f, \tilde{P}_a \varepsilon^a\} \tag{125}$$

where

$$\tilde{P}_a = \int dx\, \rho \sum_n \frac{\partial \Phi}{\partial x_n^a} = \int dx\, \rho \frac{\partial \Phi}{\partial X^a} \tag{126}$$

is interpreted as the expectation of the total momentum, and X^a are the coordinates of the center of mass,

$$X^a = \frac{1}{M}\sum_n m_n x_n^a \,. \tag{127}$$

In complex coordinates,

$$\tilde{P}_a = \int dx\, \Psi^* \left(\sum_n \frac{\hbar}{i}\frac{\partial}{\partial x_n^a} \right) \Psi = \int dx\, \Psi^* \left(\frac{\hbar}{i}\frac{\partial}{\partial X^a} \right) \Psi \,, \tag{128}$$

and the corresponding kernel $\hat{P}_{axx'}$ is

$$\hat{P}_{axx'} = \delta_{xx'} \sum_n \frac{\hbar}{i}\frac{\partial}{\partial x_n^a} = \delta_{xx'}\frac{\hbar}{i}\frac{\partial}{\partial X^a} \,. \tag{129}$$

8. The E-Hamiltonian

In the previous sections we supplied the symplectic e-phase space $T^*\mathbf{P}$ with a Riemannian metric and, as a welcome by-product, also with a complex structure. Then we showed that the condition for the simplest form of dynamics—one that preserves all the metric, symplectic, and complex structures—is a Hamilton–Killing flow generated by a Hamiltonian $\tilde{H}$ that is linear in both Ψ and Ψ^*,

$$\tilde{H}[\Psi,\Psi^*] = \int dx dx'\, \Psi_x^* \hat{H}_{xx'} \Psi_{x'} \,. \tag{130}$$

The last ingredient in the construction of $\tilde{H}$ is that the e-Hamiltonian must agree with (50) to reproduce the entropic evolution of ρ given by the continuity Equation (49).

To proceed we use the identity

$$\frac{1}{2}\rho m^{AB}(\partial_A \Phi - \bar{A}_A)(\partial_B \Phi - \bar{A}_B) = \frac{\hbar^2}{2} m^{AB} (D_A \Psi)^* D_B \Psi - \frac{\hbar^2}{8\rho^2} m^{AB} \partial_A \rho \partial_B \rho \tag{131}$$

where

$$D_A = \partial_A - \frac{i}{\hbar}\bar{A}_A \quad \text{and} \quad \bar{A}_A(x) = \hbar \beta_n A_a(x_n) \,. \tag{132}$$

Rewriting $\tilde{H}[\rho,\Phi]$ in (50) in terms of Ψ and Ψ^* we get

$$\tilde{H}[\Psi,\Psi^*] = \int dx \left(\frac{-\hbar^2}{2} m^{AB} \Psi^* D_A D_B \Psi \right) + F'[\rho] \,. \tag{133}$$

where

$$F'[\rho] = F[\rho] - \frac{\hbar^2}{8\rho^2} m^{AB} \partial_A \rho \partial_B \rho \,. \tag{134}$$

According to (121) for $\tilde{H}[\Psi,\Psi^*]$ to generate an HK flow we must impose that $F'[\rho]$ be linear in both Ψ and Ψ^*,

$$F'[\rho] = \int dxdx' \, \Psi_x^* \hat{V}_{xx'} \Psi_{x'} \tag{135}$$

for some Hermitian kernel $\hat{V}_{xx'}$, but $F'[\rho]$ must remain independent of Φ,

$$\frac{\delta F'[\rho]}{\delta \Phi_x} = 0 \,. \tag{136}$$

Substituting $\Psi = \rho^{1/2} e^{i\Phi/\hbar}$ into (135) and using $\hat{V}^*_{x'x} = \hat{V}_{xx'}$ leads to

$$\frac{\delta F'}{\delta \Phi_x} = \frac{2}{\hbar} \rho_x^{1/2} \int dx' \rho_{x'}^{1/2} \mathrm{Im} \left(\hat{V}_{xx'} e^{-i(\Phi_x - \Phi_{x'})/\hbar} \right) = 0 \tag{137}$$

This equation must be satisfied for all choices of $\rho_{x'}$, which implies

$$\mathrm{Im} \left(\hat{V}_{xx'} e^{-i(\Phi_x - \Phi_{x'})/\hbar} \right) = 0 \,, \tag{138}$$

and also for all choices of Φ_x and $\Phi_{x'}$. Therefore, the kernel $\hat{V}_{xx'}$ must be local in x,

$$\hat{V}_{xx'} = \delta_{xx'} V_x \,, \tag{139}$$

where $V_x = V(x)$ is some real function.

We conclude that the Hamiltonian that generates a Hamilton–Killing flow and agrees with the ED continuity equation must be of the form

$$\tilde{H}[\Psi,\Psi^*] = \int dx \Psi^* \left(-\frac{\hbar^2}{2} m^{AB} D_A D_B + V(x) \right) \Psi \,. \tag{140}$$

The evolution of Ψ is given by the Hamilton equation,

$$\partial_t \Psi_x = \{\Psi_x, \tilde{H}\} = \frac{\delta \tilde{H}}{\delta(i\hbar \Psi^*(x))} \,, \tag{141}$$

which is the Schrödinger equation,

$$i\hbar \partial_t \Psi = -\frac{\hbar^2}{2} m^{AB} D_A D_B \Psi + V\Psi \,. \tag{142}$$

In more standard notation it reads

$$i\hbar \partial_t \Psi = \sum_n \frac{-\hbar^2}{2m_n} \delta^{ab} \left(\frac{\partial}{\partial x_n^a} - i\beta_n A_a(x_n) \right) \left(\frac{\partial}{\partial x_n^b} - i\beta_n A_b(x_n) \right) \Psi + V\Psi \,. \tag{143}$$

At this point we can finally provide the physical interpretation of the various constants introduced along the way. Since the Schrödinger Equation (143) is the tool we use to analyze experimental data we can identify $\hbar$ with Planck's constant, m_n will be interpreted as the particles' masses, and the β_n are related to the particles' electric charges q_n by

$$\beta_n = \frac{q_n}{\hbar c} \,. \tag{144}$$

For completeness we write the Hamiltonian in the (ρ, Φ) variables,

$$\begin{aligned}\tilde{H}[\rho,\Phi] &= \int d^{3N}x\,\rho\left[\sum_n \frac{\delta^{ab}}{2m_n}\left(\frac{\partial\Phi}{\partial x_n^a}-\frac{q_n}{c}A_a(x_n)\right)\left(\frac{\partial\Phi}{\partial x_n^b}-\frac{q_n}{c}A_b(x_n)\right)\right.\\ &\left.+\sum_n \frac{\hbar^2}{8m_n}\frac{\delta^{ab}}{\rho^2}\frac{\partial\rho}{\partial x_n^a}\frac{\partial\rho}{\partial x_n^b}+V(x_1\ldots x_n)\right]\,. \end{aligned} \tag{145}$$

The Hamilton equations for ρ and Φ are the continuity equation (49),

$$\partial_t\rho = \frac{\delta\tilde{H}}{\delta\Phi} = -\sum_n \frac{\partial}{\partial x_n^a}\left[\rho\frac{\delta^{ab}}{m_n}\left(\frac{\partial\Phi}{\partial x_n^b}-\frac{q_n}{c}A_b(x_n)\right)\right], \tag{146}$$

and the quantum analogue of the Hamilton-Jacobi equation,

$$\begin{aligned}\partial_t\Phi &= -\frac{\delta\tilde{H}}{\delta\rho} = \sum_n \frac{-\delta^{ab}}{2m_n}\left(\frac{\partial\Phi}{\partial x_n^a}-\frac{q_n}{c}A_a(x_n)\right)\left(\frac{\partial\Phi}{\partial x_n^b}-\frac{q_n}{c}A_b(x_n)\right)\\ &+\sum_n \frac{\hbar^2}{2m_n}\frac{\delta^{ab}}{\rho^{1/2}}\frac{\partial^2\rho^{1/2}}{\partial x_n^a\partial x_n^b}-V(x_1\ldots x_n)\Bigg]\,. \end{aligned} \tag{147}$$

To summarize: we have just shown that an ED that preserves both the symplectic and metric structures of the e-phase space $T^*\mathbf{P}$ leads to a linear Schrödinger equation. In particular, such an ED reproduces the quantum potential in (147) with the correct coefficients $\hbar^2/2m_n$.

9. Entropic Time, Physical Time, and Time Reversal

Now that the dynamics has been fully developed we revisit the question of time. The derivation of laws of physics as examples of inference led us to introduce the notion of entropic time which includes assumptions about the concept of instant, of simultaneity, of ordering, and of duration. It is clear that entropic time is useful but is this the actual, real, "physical" time? The answer is yes. By deriving the Schrödinger equation (from which we can obtain the classical limit) we have shown that the t that appears in the laws of physics is entropic time. Since these are the equations that we routinely use to design and calibrate our clocks we conclude that *what clocks measure is entropic time*. No notion of time that is in any way deeper or more "physical" is needed. Most interestingly, the entropic model automatically includes an arrow of time.

The statement that the laws of physics are invariant under time reversal has nothing to do with particles travelling backwards in time. It is instead the assertion that the laws of physics exhibit a certain symmetry. For a classical system described by coordinates q and momenta p the symmetry is the statement that if $\{q_t, p_t\}$ happens to be one solution of Hamilton's equations then we can construct another solution $\{q_t^T, p_t^T\}$ where

$$q_t^T = q_{-t} \quad \text{and} \quad p_t^T = -p_{-t}\,, \tag{148}$$

but both solutions $\{q_t, p_t\}$ and $\{q_t^T, p_t^T\}$ describe evolution forward in time. An alternative statement of time reversibility is the following: if there is one trajectory of the system that takes it from state $\{q_0, p_0\}$ at time t_0 to state $\{q_1, p_1\}$ at the later time t_1, then there is another possible trajectory that takes the system from state $\{q_1, -p_1\}$ at time t_0 to state $\{q_0, -p_0\}$ at the later time t_1. The merit of this re-statement is that it makes clear that nothing needs to travel back in time. Indeed, rather than time reversal the symmetry might be more appropriately described as momentum or motion reversal.

Since ED is a Hamiltonian dynamics one can expect that similar considerations will apply to QM and indeed they do. It is straightforward to check that given one solution $\{\rho_t(x), \Phi_t(x)\}$ that evolves forward in time, we can construct another solution $\{\rho_t^T(x), \Phi_t^T(x)\}$ that is also evolving forward in time. The reversed solution is

$$\rho_t^T(x) = \rho_{-t}(x) \quad \text{and} \quad \Phi_t^T(x) = -\Phi_{-t}(x) \,. \tag{149}$$

These transformations constitute a symmetry—e.g., the transformed $\Psi_t^T(x)$ is a solution of the Schrödinger equation— provided the motion of the sources of the external potentials is also reversed, i.e., the potentials $A_a(\vec{x}, t)$ and $V(x, t)$ are transformed according to

$$A_a^T(\vec{x}, t) = -A_a(\vec{x}, -t) \quad \text{and} \quad V^T(x, t) = V(x, -t) \,. \tag{150}$$

Expressed in terms of wave functions the time reversal transformation is

$$\Psi_t^T(x) = \Psi_{-t}^*(x) \,. \tag{151}$$

The proof that this is a symmetry is straightforward; just take the complex conjugate of (143), and let $t \to -t$.

10. Linearity and the Superposition Principle

The Schrödinger equation is linear, i.e., a linear combination of solutions is a solution too. However, this *mathematical* linearity does not guarantee the *physical* linearity that is usually referred to as the superposition principle. The latter is the physical assumption that if there is one experimental setup that prepares a system in the (epistemic) state Ψ_1 and there is another setup that prepares the system in the state Ψ_2 then, at least in principle, it is possible to construct yet a third setup that can prepare the system in the superposition

$$\Psi_3 = \alpha_1 \Psi_1 + \alpha_2 \Psi_2 \,, \tag{152}$$

where α_1 and α_2 are arbitrary complex numbers. Mathematical linearity refers to the fact that solutions can be expressed as sums of solutions. There is no implication that any of these solutions will necessarily describe physical situations. Physical linearity on the other hand—the Superposition Principle—refers to the fact that the superposition of physical solutions is also a physical solution. The point to be emphasized is the Superposition Principle is not a principle; it is a physical hypothesis that need not be universally true.

10.1. The Single-Valuedness of Ψ

The question "Why should wave functions be single-valued?" has been around for a long time. In this section we build on and extend recent work [100] to argue that the single- or multi-valuedness of the wave functions is closely related to the question of linearity and the superposition principle. Our discussion parallels that by Schrödinger [101,102]. (See also [103–110].)

To show that the mathematical linearity of (143) is not sufficient to imply the superposition principle, we argue that even when $|\Psi_1|^2 = \rho_1$ and $|\Psi_2|^2 = \rho_2$ are probabilities it is not generally true that $|\Psi_3|^2$, Equation (152), will also be a probability. Consider moving around a closed loop Γ in configuration space. Since phases $\Phi(x)$ can be multi-valued the corresponding wave functions could in principle be multi-valued too. Let a generic Ψ change by a phase factor,

$$\Psi \to \Psi' = e^{i\delta} \Psi \,, \tag{153}$$

then the superposition Ψ_3 of two wave functions Ψ_1 and Ψ_2 changes into

$$\Psi_3 \to \Psi_3' = \alpha_1 e^{i\delta_1} \Psi_1 + \alpha_2 e^{i\delta_2} \Psi_2 \,. \tag{154}$$

The problem is that even if $|\Psi_1|^2 = \rho_1$ and $|\Psi_2|^2 = \rho_2$ are single-valued (because they are probability densities), the quantity $|\Psi_3|^2$ need not in general be single-valued. Indeed,

$$|\Psi_3|^2 = |\alpha_1|^2\rho_1 + |\alpha_2|^2\rho_2 + 2\mathrm{Re}[\alpha_1\alpha_2^*\Psi_1\Psi_2^*]\,, \tag{155}$$

changes into

$$|\Psi_3'|^2 = |\alpha_1|^2\rho_1 + |\alpha_2|^2\rho_2 + 2\mathrm{Re}[\alpha_1\alpha_2^* e^{i(\delta_1-\delta_2)}\Psi_1\Psi_2^*]\,, \tag{156}$$

so that in general

$$|\Psi_3'|^2 \neq |\Psi_3|^2\,, \tag{157}$$

which precludes the interpretation of $|\Psi_3|^2$ as a probability. That is, even when the epistemic states Ψ_1 and Ψ_2 describe actual physical situations, their superpositions need not.

The problem does not arise when

$$e^{i(\delta_1-\delta_2)} = 1\,. \tag{158}$$

If we were to group the wave functions into classes each characterized by its own δ then we could have a limited version of the superposition principle that applies within each class. We conclude that beyond the linearity of the Schrödinger equation we have a superselection rule that restricts the validity of the superposition principle to wave functions belong to the same δ-class.

To find the allowed values of δ we argue as follows. It is natural to assume that if $\{\rho, \Phi\}$ (at some given time t_0) is a physical state then the state with reversed momentum $\{\rho, -\Phi\}$ (at the same time t_0) is an equally reasonable physical state. Basically, the idea is that if particles can be prepared to move in one direction, then they can also be prepared to move in the opposite direction. In terms of wave functions the statement is that if Ψ_{t_0} is a physically allowed initial state, then so is $\Psi_{t_0}^*$ [111]. Next we consider a generic superposition

$$\Psi_3 = \alpha_1\Psi + \alpha_2\Psi^*\,. \tag{159}$$

Is it physically possible to construct superpositions such as (159)? The answer is that while constructing Ψ_3 for an arbitrary Ψ might not be feasible in practice there is strong empirical evidence that there exist no superselection rules to prevent us from doing so in principle. Indeed, it is easy to construct superpositions of wavepackets with momentum $\vec{p}$ and $-\vec{p}$, or superpositions of states with opposite angular momenta, $Y_{\ell m}$ and $Y_{\ell,-m}$. *We shall assume that in principle the superpositions (159) are physically possible.*

According to Equation (153) as one moves in a closed loop Γ the wave function Ψ_3 will transform into

$$\Psi_3' = \alpha_1 e^{i\delta}\Psi + \alpha_2 e^{-i\delta}\Psi^*\,, \tag{160}$$

and the condition (158) for $|\Psi_3|^2$ to be single-valued is

$$e^{2i\delta} = 1 \quad \text{or} \quad e^{i\delta} = \pm 1\,. \tag{161}$$

Thus, we are restricted to two discrete possibilities ± 1. Since the wave functions are assumed sufficiently well behaved (continuous, differentiable, etc.) we conclude that they must be either single-valued, $e^{i\delta} = 1$, or double-valued, $e^{i\delta} = -1$.

We conclude that the Superposition Principle appears to be valid in a sufficiently large number of cases to be a useful rule of thumb but it is restricted to single-valued (or double-valued) wave functions. The argument above does not exclude the possibility that a multi-valued wave function might describe an actual physical situation. What the argument implies is that the Superposition Principle would not extend to such states.

10.2. Charge Quantization

Next we analyze the conditions for the electromagnetic gauge symmetry to be compatible with the superposition principle. We shall confine our attention to systems that are described by single-valued wave functions ($e^{i\delta} = +1$) [112]. The condition for the wave function to be single-valued is

$$\Delta \frac{\Phi}{\hbar} = \oint_\Gamma d\ell^A \partial_A \frac{\Phi}{\hbar} = 2\pi k_\Gamma \,, \tag{162}$$

where k_Γ is an integer that depends on the loop Γ. Under a local gauge transformation

$$A_a(\vec{x}) \to A_a(\vec{x}) + \partial_a \chi(\vec{x}) \tag{163}$$

the phase Φ transforms according to (38),

$$\Phi(x) \to \Phi'(x) = \Phi(x) + \sum_n \frac{q_n}{c} \chi(\vec{x}_n) \,. \tag{164}$$

The requirement that the gauge symmetry and the superposition principle be compatible amounts to requiring that the gauge transformed states also be single-valued,

$$\Delta \frac{\Phi'}{\hbar} = \oint_\Gamma d\ell^A \partial_A \frac{\Phi'}{\hbar} = 2\pi k'_\Gamma \,. \tag{165}$$

Thus, the allowed gauge transformations are restricted to functions $\chi(\vec{x})$ such that

$$\sum_n \frac{q_n}{\hbar c} \oint_\Gamma d\ell_n^a \partial_{na} \chi(\vec{x}_n) = 2\pi \Delta k_\Gamma \tag{166}$$

where $\Delta k_\Gamma = k'_\Gamma - k_\Gamma$ is an integer. Consider now a loop γ in which we follow the coordinates of the nth particle around some closed path in 3-dimensional space while all the other particles are kept fixed. Then

$$\frac{q_n}{\hbar c} \oint_\gamma d\ell_n^a \partial_{an} \chi(\vec{x}_n) = 2\pi \Delta k_{n\gamma} \tag{167}$$

where $\Delta k_{n\gamma}$ is an integer. Since the gauge function $\chi(\vec{x})$ is just a function in 3-dimensional space it is the same for all particles and the integral on the left is independent of n. This implies that the charge q_n divided by an integer $\Delta k_{n\gamma}$ must be independent of n which means that q_n must be an integer multiple of some basic charge q_0. We conclude that the charges q_n are quantized.

The issue of charge quantization is ultimately the issue of deciding which is the gauge group that generates electromagnetic interactions. We could for example decide to restrict the gauge transformations to single-valued gauge functions $\chi(\vec{x})$ so that (167) is trivially satisfied irrespective of the charges being quantized or not. Under such a restricted symmetry group the single-valued (or double-valued) nature of the wave function is unaffected by gauge transformations. If, on the other hand, the gauge functions $\chi(\vec{x})$ are allowed to be multi-valued, then the compatibility of the gauge transformation (163)–(164) with the superposition principle demands that charges be quantized.

The argument above cannot fix the value of the basic charge q_0 because it depends on the units chosen for the vector potential A_a. Indeed since the dynamical equations show q_n and A_a appearing only in the combination $q_n A_a$ we can change units by rescaling charges and potentials according to $Cq_n = q'_n$ and $A_a/C = A'_a$ so that $q_n A_a = q'_n A'_a$. For conventional units such that the basic charge is $q_0 = e/3$ with $\alpha = e^2/\hbar c = 1/137$ the scaling factor is $C = (\alpha\hbar c)^{1/2}/3q_0$. A more natural set of units might be to set $q_0 = \hbar c$ so that all β_ns are integers and the gauge functions $\chi(\vec{x})$ are angles.

A similar conclusion—that charge quantization is a reflection of the compactness of the gauge group—can be reached following an argument due to C. N. Yang [113]. Yang's argument assumes that a Hilbert space has been established and one has access to the unitary representations of symmetry groups. Yang considers a gauge transformation

$$\Psi(x) \to \Psi(x) \exp i \sum_n \frac{q_n}{c} \chi(\vec{x}_n) \,, \tag{168}$$

with $\chi(\vec{x})$ independent of $\vec{x}$. If the q_ns are not commensurate there is no value of χ (except 0) that makes (168) be the identity transformation. The gauge group—translations on the real line—would not be compact. If, on the other hand, the charges are integer multiples of a basic charge q_0, then two values of χ that differ by an integer multiple of $2\pi c/q_0$ give identical transformations and the gauge group is compact. In the present ED derivation, however, we deal with the space $T^*\mathbf{P}$ which is a complex projective space. We cannot adopt Yang's argument because a gauge transformation χ independent of $\vec{x}$ is already an identity transformation—it leads to an equivalent state in the same ray—and cannot therefore lead to any constraints on the allowed charges.

11. The Classical Limits and the Bohmian Limit

11.1. Classical Limits

There are two classical limits that one might wish to consider. One is the mathematical limit $\hbar \to 0$. Taking $\hbar \to 0$ leaves unchanged both the velocities v_n^a of the particles, Equation (19), and the probability flow, Equation (146). The main effect is to suppress the quantum potential so that Equation (147) becomes the classical Hamilton-Jacobi equation. The symplectic form, Equation (63), survives unscathed but the metric and the complex structures, Equations (101) and (103), do not. However, this is not quite classical mechanics. Since the velocity fluctuations, Equation (25), remain unaffected the resulting dynamics is a non-dissipative version of the classical Oernstein–Uhlenbeck Brownian motion. To recover a deterministic classical mechanics one must also take the limit $\eta \to 0$.

The other classical limit arises in the more physically relevant situation where one deals with a system with a large number N of particles—for example, a speck of dust—and one wishes to study the motion of an effective macrovariable such as the center of mass (CM), Equation (127). The large N limit of ED with particles undergoing an ES Brownian motion was studied in [77]. The same argument goes through essentially unchanged for the OU Brownian motion discussed here. Skipping all details we find that because of the central limit theorem the continuity equation for $\rho_{\text{cm}}(X^a)$ and the velocity fluctuations are given by the analogues of (43) and (25) for a single particle of mass $M = \sum_{n=1}^{N} m_n$,

$$\partial_t \rho_{\text{cm}} = \frac{\partial}{\partial X^a} \left(\rho_{\text{cm}} V^a \right) \quad \text{with} \quad V^a = \frac{\langle \Delta X^a \rangle}{\Delta t} = \frac{1}{M} \frac{\partial \Phi_{\text{cm}}}{\partial X^a} \,, \tag{169}$$

$$\left\langle \left(\frac{\Delta X^a}{\Delta t} - V^a \right) \left(\frac{\Delta X^b}{\Delta t} - V^b \right) \right\rangle = \frac{\eta \Delta t}{M} . \tag{170}$$

We also find that under rather general conditions the CM motion decouples from the motion of the component particles and obeys the single particle HJ equation

$$-\partial_t \Phi_{\text{cm}} = \frac{1}{2M} \left(\frac{\partial \Phi_{\text{cm}}}{\partial X^a} \right)^2 - \frac{\hbar^2}{2M} \frac{\nabla^2 \rho_{\text{cm}}^{1/2}}{\rho_{\text{cm}}^{1/2}} + V_{\text{ext}}(X) \,. \tag{171}$$

In the large N limit $M \sim O(N)$ and we obtain a finite velocity V^a in (169) provided $\Phi_{\text{cm}} \sim O(N)$. In Equation (171) we see that for a sufficiently large system the quantum potential for the CM motion vanishes. Therefore, for $N \to \infty$, the CM follows smooth trajectories described by a classical Hamilton-Jacobi equation. Furthermore, Equation (170) shows that as $N \to \infty$ the velocity fluctuations vanish irrespective of the value of η. This is a truly deterministic classical mechanics.

An important feature of this derivation is that $\hbar$ and η remain finite which means that a mesoscopic or macroscopic object will behave classically while all its component particles remain fully quantum mechanical.

11.2. The Bohmian Limit

ED models with different values of η lead to the same Schrödinger equation. In other words, different sub-quantum models lead to the same emergent quantum behavior. The limit of vanishing η deserves particular attention because the velocity fluctuations, Equation (25), are suppressed and the motion becomes deterministic. This means that ED includes the Bohmian form of quantum mechanics [51–53] as a special limiting case—but with the important caveat that the difference in physical interpretation remains enormous. It is only with respect to the mathematical formalism that ED includes Bohmian mechanics as a special case.

Bohmian mechanics attempts to provide an actual description of reality. In the Bohmian view the universe consists of real particles that have definite positions and their trajectories are guided by a real field, the wave function Ψ. Not only does this pilot wave live in $3N$-dimensional configuration space but it manages to act on the particles without the particles reacting back upon it. These are peculiarities that have stood in the way of a wider acceptance of the Bohmian interpretation. In contrast, ED's pragmatic goal is much less ambitious: to make the best possible predictions based on very incomplete information. As in Bohmian mechanics, in ED the particles also have definite positions and its formalism includes a function Φ that plays the role of a pilot wave. However, Φ is an epistemic tool for reasoning; it is not meant to represent anything real. There is no implication that the particles move the way they do because they are pushed around by a pilot wave or by some stochastic force. In fact, ED is silent on the issue of what if anything is pushing the particles. What the probability ρ and the phase Φ are designed to do is not to guide the particles but to guide our inferences. They guide our expectations of where and when to find the particles but they do not exert any causal influence on the particles themselves.

12. Hilbert Space

The formulation of the ED of spinless particles is now complete. We note, in particular, that the notion of Hilbert spaces turned out to be unnecessary to the formulation of quantum mechanics. As we shall see next, while strictly unnecessary in principle, the introduction of Hilbert spaces is nevertheless very convenient for calculational purposes.

A vector space. As we saw above the infinite-dimensional e-phase space—the cotangent bundle $T^*\mathbf{P}$—is difficult to handle. The problem is that the natural coordinates are probabilities ρ_x which, due to the normalization constraint, are not independent. In a discrete space one could single out one of the coordinates and its conjugate momentum and then proceed to remove them. Unfortunately, with a continuum of coordinates and momenta the removal is not feasible. The solution is to embed $T^*\mathbf{P}$ in a larger space $T^*\mathbf{P}^{+1}$. This move allows us to keep the natural coordinates ρ_x but there is a price: we are forced to deal with a constrained system and its attendant gauge symmetry.

We also saw that the geometry of the embedding space was not fully determined: any spherically symmetric space would serve our purposes. This is a freedom we can further exploit. For calculational purposes the linearity of the Schrödinger Equation (143) is very convenient but its usefulness is severely limited by the normalization constraint. If Ψ_1 and Ψ_2 are flows in $T^*\mathbf{P}$ then the superposition Ψ_3 in (152) will also be a flow in $T^*\mathbf{P}$ but only if the coefficients α_1 and α_2 are such that Ψ_3 is properly normalized. This restriction can be removed by choosing the extended embedding space $T^*\mathbf{P}^{+1}$ to be flat—just set $A = 0$ and $B = 1$ in Equation (91). (The fact that this space is flat is evident in the metric (89) for the discrete case.) We emphasize that this choice is not at all obligatory; it is optional.

The fact that in the flat space $T^*\mathbf{P}^{+1}$ superpositions are allowed for arbitrary constants α_1 and α_2 means that $T^*\mathbf{P}^{+1}$ is not just a manifold; it is also a vector space. Each point Ψ in $T^*\mathbf{P}^{+1}$ is itself a vector. Furthermore, since the vector tangent to a curve is just a difference of two vectors Ψ we see that that points on the manifold and vectors tangent to the manifold are objects of the same kind. In other words, the tangent spaces $T[T^*\mathbf{P}^{+1}]_\Psi$ are identical to the space $T^*\mathbf{P}^{+1}$ itself.

The symplectic form Ω and the metric tensor G on the extended space $T^*\mathbf{P}^{+1}$ are given by Equations (108) and (114). Since they are tensors Ω and G are meant to act on vectors but now they can also act on all points $\Psi \in T^*\mathbf{P}^{+1}$ and not just on those that happen to be normalized and gauge fixed according to (83). For example, the action of the mixed tensor J, Equation (115), on a wave function Ψ is

$$J^{\mu x}{}_{\nu x'}\Psi^{\nu x'} = \begin{bmatrix} i & 0 \\ 0 & -i \end{bmatrix} \begin{pmatrix} \Psi_x \\ i\hbar\Psi_x^* \end{pmatrix} = \begin{pmatrix} i\Psi_x \\ i\hbar(i\Psi_x)^* \end{pmatrix}, \tag{172}$$

which indicates that J plays the role of multiplication by i, i.e., when acting on a point Ψ the action of J is $\Psi \xrightarrow{J} i\Psi$.

Dirac notation. We can at this point introduce the Dirac notation to represent the wave functions Ψ_x as vectors $|\Psi\rangle$ in a Hilbert space. The scalar product $\langle\Psi_1|\Psi_2\rangle$ is defined using the metric G and the symplectic form Ω,

$$\langle\Psi_1|\Psi_2\rangle \overset{\text{def}}{=} \frac{1}{2\hbar}\int dx\,dx'\,(\Psi_{1x}, i\hbar\Psi_{1x}^*)\,(G+i\Omega)\begin{pmatrix} \Psi_{2x'} \\ i\hbar\Psi_{2x'}^* \end{pmatrix}. \tag{173}$$

A straightforward calculation gives

$$\langle\Psi_1|\Psi_2\rangle = \int dx\,\Psi_1^*\Psi_2\,. \tag{174}$$

The map $\Psi_x \leftrightarrow |\Psi\rangle$ is defined by

$$|\Psi\rangle = \int dx|x\rangle\Psi_x \quad \text{where} \quad \Psi_x = \langle x|\Psi\rangle\,, \tag{175}$$

where, in this "position" representation, the vectors $\{|x\rangle\}$ form a basis that is orthogonal and complete,

$$\langle x|x'\rangle = \delta_{xx'} \quad \text{and} \quad \int dx\,|x\rangle\langle x| = \hat{1}\,. \tag{176}$$

Hermitian and unitary operators. The bilinear Hamilton functionals $\tilde{Q}[\Psi,\Psi^*]$ with kernel $\hat{Q}(x,x')$ in Equation (121) can now be written in terms of a Hermitian operator $\hat{Q}$ and its matrix elements,

$$\tilde{Q}[\Psi,\Psi^*] = \langle\Psi|\hat{Q}|\Psi\rangle \quad \text{and} \quad \hat{Q}(x,x') = \langle x|\hat{Q}|x'\rangle\,. \tag{177}$$

The corresponding Hamilton–Killing flows are given by

$$i\hbar\frac{d}{d\lambda}\langle x|\Psi\rangle = \langle x|\hat{Q}|\Psi\rangle \quad \text{or} \quad i\hbar\frac{d}{d\lambda}|\Psi\rangle = \hat{Q}|\Psi\rangle\,. \tag{178}$$

These flows are described by unitary transformations

$$|\Psi(\lambda)\rangle = \hat{U}_Q(\lambda)|\Psi(0)\rangle \quad \text{where} \quad \hat{U}_Q(\lambda) = \exp\left(-\frac{i}{\hbar}\hat{Q}\lambda\right). \tag{179}$$

Commutators. The Poisson bracket of two Hamiltonian functionals $\tilde{U}[\Psi,\Psi^*]$ and $\tilde{V}[\Psi,\Psi^*]$,

$$\{\tilde{U},\tilde{V}\} = \int dx\left(\frac{\delta\tilde{U}}{\delta\Psi_x}\frac{\delta\tilde{V}}{\delta i\hbar\Psi_x^*} - \frac{\delta\tilde{U}}{\delta i\hbar\Psi_x^*}\frac{\delta\tilde{V}}{\delta\Psi_x}\right),$$

can be written in terms of the commutator of the associated operators, then

$$\{\tilde{U},\tilde{V}\} = \frac{1}{i\hbar}\langle\Psi|[\hat{U},\hat{V}]|\Psi\rangle\,. \tag{180}$$

Thus, the Poisson bracket is the expectation of the commutator. This *identity* is much sharper than Dirac's pioneering discovery that the quantum commutator of two q-variables is *analogous* to the Poisson bracket of the corresponding classical variables. Further parallels between the geometric and the Hilbert space formulation of QM can be found in [56–64].

13. Remarks on ED and Quantum Bayesianism

Having discussed the ED approach in some detail it is now appropriate to comment on how ED differs from the interpretations known as Quantum Bayesianism [20–22] and its closely related descendant QBism [23,24]; for simplicity, I shall refer to both as QB. Both ED and QB adopt an epistemic degree-of-belief concept of probability but there are important differences:

(a) QB adopts a personalistic de Finetti type of Bayesian interpretation while ED adopts an impersonal entropic Bayesian interpretation somewhat closer but not identical to Jaynes' [15–18]. In ED, the probabilities do not reflect the subjective beliefs of any particular person. They are tools designed to assist us in those all too common situations in which are confused and due to insufficient information we do not know what to believe. The probabilities will then provide guidance as to what agents ought to believe if only they were ideally rational. More explicitly, probabilities in ED describe the objective degrees of belief of ideally rational agents who have been supplied with the maximal allowed information about a particular quantum system.

(b) ED derives or reconstructs the mathematical framework of QM—it explains where the symplectic, metric, and complex structures, including Hilbert spaces and time evolution come from. In contrast, at its current stage of development QB consists of appending a Bayesian interpretation to an already existing mathematical framework. Indeed, assumptions and concepts from quantum information are central to QB and are implicitly adopted from the start. For example, a major QB concern is the justification of the Born rule starting from the Hilbert space framework while ED starts from probabilities and its goal is to justify the construction of wave functions; the Born rule follows as a trivial consequence.

(c) ED is an application of entropic/Bayesian inference. Of course, the choices of variables and of the constraints that happen to be physically relevant are specific to our subject matter—quantum mechanics—but the inference method itself is of universal applicability. It applies to electrons just as well as to the stock market or to medical trials. In contrast, in QB the personalistic Bayesian framework is not of universal validity. For those special systems that we call 'quantum' the inference framework is itself modified into a new "Quantum-Bayesian coherence" in which the standard Bayesian inference must be supplemented with concepts from quantum information theory. The additional technical ingredient is a hypothetical structure called a "symmetric informationally complete positive-operator-valued measure". In short, in QB Born's Rule is not derived but constitutes an addition beyond the raw probability theory.

(d) QB is an anti-realist neo-Copenhagen interpretation; it accepts complementarity. (Here complementarity is taken to be the common thread that runs through all Copenhagen interpretations.) Probabilities in QB refer to the outcomes of experiments and not to ontic pre-existing values. In contrast, in ED probabilities refer to ontic positions—including the ontic positions of pointer variables. In the end, this is what solves the problem of quantum measurement (see [70,71]).

14. Some Final Remarks

We conclude with a summary of the main assumptions:

- Particles have definite but unknown positions and follow continuous trajectories.
- The probability of a short step is given by the method of maximum entropy subject to a drift potential constraint that introduces directionality and correlations, plus gauge constraints that account for external electromagnetic fields.

- The accumulation of short steps requires a notion of time as a book-keeping device. This involves the introduction of the concept of an instant and a convenient definition of the duration between successive instants.
- The e-phase space $\{\rho, \Phi\}$ has a natural symplectic geometry that results from treating the pair (ρ_x, Φ_x) as canonically conjugate variables.
- The information geometry of the space of probabilities is extended to the full e-phase space by imposing the latter be spherically symmetric.
- The drift potential constraint is updated instant by instant in such a way as to preserve both the symplectic and metric geometries of the e-phase space.

The resulting entropic dynamics is described by the Schrödinger equation. Different sub-quantum Brownian motions all lead to the same emergent quantum mechanics. In previous work we dealt with an Einstein–Smoluchowski process; here we have explored an Oernstein–Uhlenbeck process. Other "fractional" Brownian motions might be possible but have not yet been studied.

A natural question is whether these different sub-quantum Brownian motions might have observable consequences. At this point our answer can only be tentative. To the extent that we have succeeded in deriving QM and not some other theory one should not expect deviations in the predictions for the standard experiments that are the subject of the standard quantum theory— at least not in the non-relativistic regime. As the ED program is extended to other regimes involving higher energies and/or gravity it is quite possible that those different sub-quantum motions might not be empirically equivalent.

ED achieves ontological clarity by sharply separating the ontic elements from the epistemic elements — positions of particles on one side and probabilities ρ and phases Φ on the other. ED is a dynamics of probabilities and not a dynamics of particles. Of course, if probabilities at one instant are large in one place and at a later time they are large in some other place one infers that the particles must have moved—but nothing in ED describes what it is that has pushed the particles around. ED is a mechanics without a mechanism.

We can elaborate on this point from a different direction. The empirical success of ED suggests that its epistemic probabilities agree with ontic features of the physical world. It is highly desirable to clarify the precise nature of this agreement. Consider, for example, a fair die. Its property of being a perfect cube is an ontic property of the die which is reflected at the epistemic level in the equal assignment of probabilities to each face of the die. In this example we see that the epistemic probabilities achieve objectivity, and therefore usefulness, by corresponding to something ontic. The situation in ED is similar except for one crucial aspect. The ED probabilities are objective, and they are empirically successful. They must therefore reflect something real. However, it is not yet known what those underlying ontic properties might possibly be. Fortunately, for the purposes of making predictions knowing those epistemic probabilities is all we need.

The trick of embedding the e-phase space $T^*\mathbf{P}$ in a *flat* vector space $T^*\mathbf{P}^{+1}$ is clever but optional. It allows one to make use of the calculational advantages of linearity. This recognition that Hilbert spaces are not fundamental is one of the significant contributions of the entropic approach to our understanding of QM. The distinction—whether Hilbert spaces are necessary in principle as opposed to merely convenient in practice—is not of purely academic interest. It can be important in the search for a quantum theory that includes gravity: Shall we follow the usual approaches to quantization that proceed by replacing classical dynamical variables by an algebra of linear operators acting on some abstract space? Or, in the spirit of an entropic dynamics, shall we search for an appropriately constrained dynamics of probabilities and information geometries? First steps towards formulating a first-principles theory along these lines are given in [114,115].

Acknowledgments: I would like to thank M. Abedi, D. Bartolomeo, C. Cafaro, N. Carrara, N. Caticha, F. Costa, S. DiFranzo, K. Earle, A. Giffin, S. Ipek, D.T. Johnson, K. Knuth, O. Lunin, S. Nawaz, P. Pessoa, M. Reginatto,

C. Rodríguez, and K. Vanslette, for valuable discussions on entropic inference and entropic dynamics and for their many insights and contributions at various stages of this program.

References and Notes

1. Dirac, P.A.M. *Quantum Mechanics*, 3rd ed; Oxford University Press: Oxford, UK, 1930.
2. Von Neumann, J. *Mathematical Foundations of Quantum Mechanics*; Princeton University Press: Princeton, NJ, USA, 1955.
3. Bell, J. Against 'measurement'. *Phys. World* **1990**, *8*, 33. [CrossRef]
4. Excellent reviews with extended references to the literature are given in e.g., [5–8].
5. Stapp, H.P. The Copenhagen Interpretation. *Am. J. Phys.* **1972**, *40*, 1098. [CrossRef]
6. Schlösshauer, M. Decoherence, the measurement problem, and interpretations of quantum mechanics. *Rev. Mod. Phys.* **2004**, *76*, 1267. [CrossRef]
7. Jaeger, G. *Entanglement, Information, and the Interpretation of Quantum Mechanics*; Springer: Berlin/Heidelberger, Germany, 2009.
8. Leifer, M.S. Is the Quantum State Real? An Extended Review of Ψ-ontology Theorems. *Quanta* **2014**, *3*, 67, arXiv.org:1409.1570. [CrossRef]
9. Since the terms 'ontic' and 'epistemic' are not yet of widespread use outside the community of Foundations of QM, a clarification might be useful. A concept is referred as 'ontic' when it describes something that is supposed to be real, to exist out there independently of any observer. A concept is referred as 'epistemic' when it is related to the state of knowledge, opinion, or belief of an agent, albeit an ideally rational agent. Examples of epistemic quantities are probabilities and entropies. An important point is that the distinction ontic/epistemic is not the same as the distinction objective/subjective. For example, probabilities are fully epistemic—they are tools for reasoning with incomplete information— but they can lie anywhere in the spectrum from being completely subjective (two different agents can have different beliefs) to being completely objective. In QM, for example, probabilities are epistemic and objective. Indeed, at the non-relativistic level anyone who computes probabilities that disagree with QM will be led to experimental predictions that are demonstrably wrong. We will say that the wave function Ψ, which is fully epistemic and objective, represents a "physical" state when it represents information about an actual "physical" situation.
10. Caticha, A. Entropic Dynamics, Time, and Quantum Theory. *J. Phys. A Math. Theor.* **2011**, *44*, 225303, arXiv.org:1005.2357.
11. Caticha, A. Entropic Dynamics. *Entropy* **2015**, *17*, 6110–6128, arXiv.org:1509.03222. [CrossRef]
12. Caticha, A. Entropic Dynamics: Quantum Mechanics from Entropy and Information Geometry. *Ann. Physik* **2018**, 1700408, arXiv.org:1711.02538.
13. The principle of maximum entropy as a method for inference can be traced to the pioneering work of E. T. Jaynes [14–16]. For a pedagogical overview including more modern developments see [17,18]
14. Jaynes, E.T. Information Theory and Statistical Mechanics I and II. *Phys. Rev.* **1957**, *106*, 108, and *171*, 620. [CrossRef]
15. Jaynes, E.T. *Papers on Probability, Statistics and Statistical Physics*; Rosenkrantz, R.D., Ed.; D. Reidel: Dordrecht, The Netherlands, 1983.
16. Jaynes, E.T. *Probability Theory: The Logic of Science*; Bretthorst, G.L., Ed.; Cambridge University Press: Cambridge, UK, 2003.
17. Caticha, A. Entropic Inference and the Foundations of Physics. Available online: http://www.albany.edu/physics/ACaticha-EIFP-book.pdf (accessed on 20 September 2019).
18. Caticha, A. Towards an Informational Pragmatic Realism. *Mind Mach.* **2014**, *24*, 37, arXiv.org:1412.5644. [CrossRef]
19. There exist many different Bayesian interpretations of probability. In Section 13 we comment on how ED differs from the frameworks known as Quantum Bayesianism [20–22] and its closely related descendant QBism [23,24].
20. Brun, T.A.; Caves, C.M.; Schack, R. Quantum Bayes rule. *Phys. Rev. A* **2001**, *63*, 042309. [CrossRef]
21. Caves, C.M.; Fuchs, C.A.; Schack, R. Unknown quantum states: The quantum de Finetti representation. *J. Math. Phys.* **2002**, *43*, 4547. [CrossRef]

22. Caves, C.M.; Fuchs, C.A.; Schack, R. Quantum Probabilities as Bayesian Probabilities. *Phys. Rev. A* **2002**, *65*, 022305. [CrossRef]
23. Fuchs, C.A.; Schack, R. Quantum-Bayesian Coherence. *Rev. Mod. Phys.* **2013**, *85*, 1693. [CrossRef]
24. Fuchs, C.A.; Mermin, N.D.; Schack, R. An introduction to QBism with an application to the locality of quantum mechanics. *Am. J. Phys.* **2014**, *82*, 749. [CrossRef]
25. Nelson, E. *Quantum Fluctuations*; Princeton University Press: Princeton, NJ, USA, 1985.
26. Adler, S. *Quantum Theory as an Emergent Phenomenon*; Cambridge University Press: Cambridge, UK, 2004.
27. Smolin, L. Could quantum mechanics be an approximation to another theory? *arXiv* **2006**, arXiv.org/abs/quant-ph/0609109.
28. de la Peña, L.; Cetto, A.M. *The Emerging Quantum: The Physics Behind Quantum Mechanics*; Springer: Berlin/Heidelberg, Germany, 2014.
29. Grössing, G. The Vacuum Fluctuation Theorem: Exact Schrödinger Equation via Nonequilibrium Thermodynamics. *Phys. Lett. A* **2008**, *372*, 4556, arXiv:0711.4954.
30. Grössing, G.; Fussy, S.; Mesa Pascasio, J.; Schwabl, H. The Quantum as an Emergent System. *J. Phys.Conf. Ser.* **2012**, *361*, 012008. [CrossRef]
31. Hooft, G.T. *The Cellular Automaton Interpretation of Quantum Mechanics*; Springer: Berlin/Heidelberg, Germany, 2016.
32. Wootters, W.K. Statistical distance and Hilbert space. *Phys. Rev. D* **1981**, *23*, 357. [CrossRef]
33. Caticha, A. Consistency and Linearity in Quantum Theory. *Phys. Lett. A* **1998**, *244*, 13. [CrossRef]
34. Caticha, A. Consistency, Amplitudes, and Probabilities in Quantum Theory. *Phys. Rev. A* **1998**, *57*, 1572. [CrossRef]
35. Caticha, A. Insufficient Reason and Entropy in Quantum Theory. *Found. Phys.* **2000**, *30*, 227. [CrossRef]
36. Brukner, C.; Zeilinger, A. Information and Fundamental Elements of the Structure of Quantum Theory. In *Time, Quantum, Information*; Castell, L., Ischebeck, O., Eds.; Springer: Berlin/Heidelberg, Germany, 2003.
37. Mehrafarin, M. Quantum mechanics from two physical postulates. *Int. J. Theor. Phys.* **2005**, *44*, 429, arXiv:quant-ph/0402153. [CrossRef]
38. Spekkens, R. Evidence for the epistemic view of quantum states: A toy theory. *Phys. Rev. A* **2007**, *75*, 032110. [CrossRef]
39. Goyal, P. From Information Geometry to Quantum Theory. *New J. Phys.* **2010**, *12*, 023012. [CrossRef]
40. Goyal, P.; Knuth, K.; Skilling, J. Origin of complex quantum amplitudes and Feynman's rules. *Phys. Rev. A* **2010**, *81*, 022109. [CrossRef]
41. Chiribella, G.; D'Ariano, G.M.; Perinotti, P. Informational derivation of quantum theory. *Phys. Rev.* **2011**, *84*, 012311. [CrossRef]
42. Hardy, L. Reformulating and Reconstructing Quantum Theory. *arXiv* **2011**, arXiv:1104.2066.
43. Reginatto, M.; Hall, M.J.W. Quantum theory from the geometry of evolving probabilities. *AIP Conf. Proc.* **2012**, *1443*, 96, arXiv:1108.5601.
44. Reginatto, M.; Hall, M.J.W. Information geometry, dynamics and discrete quantum mechanics. *AIP Conf. Proc.* **2013**, *1553*, 246, arXiv:1207.6718.
45. Hardy, L. Reconstructing Quantum Theory. In *Quantum Theory: Informational Foundations and Foils*; Chiribella, G., Spekkens, R., Eds.; Springer: Berlin/Heidelberg, Germany, 2015.
46. D'Ariano, G.M. Physics without physics: The power of information-theoretical principles. *Int. J. Theor. Phys.* **2017**, *56*, 97. [CrossRef]
47. Nelson, E. *Dynamical Theories of Brownian Motion*, 2nd ed.; Princeton University Press: Princeton, NJ, USA, 1967. Available online: http://www.math.princeton.edu/nelson/books.html (accessed on 20 September 2019).
48. In both the ES and the OU processes, which were originally meant to model the actual physical Brownian motion, friction and dissipation play essential roles. In contrast, ED is non-dissipative. ED formally resembles Nelson's stochastic mechanics [25] but the conceptual differences are significant. Nelson's mechanics attempted an ontic interpretation of QM as an ES process driven by real stochastic classical forces while ED is a purely epistemic model that does not appeal to an underlying classical mechanics.
49. Bartolomeo, D.; Caticha, A. Entropic Dynamics: The Schrödinger equation and its Bohmian limit. *AIP Conf. Proc.* **2016**, *1757*, 030002, arXiv.org:1512.09084.
50. Bartolomeo, D.; Caticha, A. Trading drift and fluctuations in entropic dynamics: Quantum dynamics as an emergent universality class. *J. Phys. Conf. Ser.* **2016**, *701*, 012009, arXiv.org:1603.08469. [CrossRef]

51. Bohm, D. A suggested interpretation of the quantum theory in terms of "hidden" variables, I and II. *Phys. Rev.* **1952**, *85*, 166, 180. [CrossRef]
52. Bohm, D.; Hiley, B. J. *The Undivided Universe—An Ontological Interpretation of Quantum Theory*; Routlege: New York, NY, USA, 1993.
53. Holland, P. R. *The Quantum Theory of Motion*; Cambridge University Press: Cambridge, UK, 1993.
54. Caticha, A.; Bartolomeo, D.; Reginatto, M. Entropic Dynamics: From entropy and information geometry to Hamiltonians and quantum mechanics. *AIP Conf. Proc.* **2015**, *1641*, 155, arXiv.org:1412.5629.
55. Nelson, E. Connection between Brownian motion and quantum mechanics. *Lect. Notes Phys.* **1979**, *100*, 168.
56. Kibble, T.W.B. Geometrization of Quantum Mechanics. *Commun. Math. Phys.* **1979**, *65*, 189. [CrossRef]
57. Heslot, A. Quantum mechanics as a classical theory. *Phys. Rev.* **1985**, *31*, 1341. [CrossRef]
58. Anandan, J.; Aharonov, Y. Geometry of Quantum Evolution. *Phys. Rev. Lett.* **1990**, *65*, 1697. [CrossRef]
59. Cirelli, R.; Manià, A.; Pizzochero, L. Quantum mechanics as an infinite-dimensional Hamiltonian system with uncertainty structure: Part I and II. *J. Math. Phys.* **1990**, *31*, 2891 and 2898. [CrossRef]
60. Abe, S. Quantum-state space metric and correlations. *Phys. Rev. A* **1992**, *46*, 1667. [CrossRef]
61. Hughston, L.P. Geometric aspects of quantum mechanics. In *Twistor Theory*; Huggett, S.A., Ed.; Marcel Dekker: New York, NY, USA, 1995.
62. Ashtekar, A.; Schilling, T.A. Geometrical Formulation of Quantum Mechanicss. In *On Einstein's Path*; Harvey, A., Ed.; Springer: New York, NY, USA, 1998.
63. de Gosson, M.A.; Hiley, B.J. Imprints of the Quantum World in Classical Mechanics. *Found. Phys.* **2011**, *41*, 1415. [CrossRef]
64. Elze, H.T. Linear dynamics of quantum-classical hybrids. *Phys. Rev. A* **2012**, *85*, 052109. [CrossRef]
65. Amari, S. *Differential-Geometrical Methods in Statistics*; Springer: Berlin/Heidelberger, Germany, 1985.
66. Campbell, L.L. An extended Čencov characterization of the information metric. *Proc. Am. Math. Soc.* **1986**, *98*, 135.
67. Rodríguez, C.C. The metrics generated by the Kullback number. In *Maximum Entropy and Bayesian Methods*; Skilling, J., Ed.; Kluwer: Dordrecht, The Netherlands, 1989.
68. Ay, N.; Jost, J.; Vân Lê, H.; Schwachhöfer, L. *Information Geometry*; Springer: Berlin, Germany, 2017.
69. Brodie, D.J.; Hughston, L.P. Statistical Geometry in Quantum Mechanics. *Philos. Trans. R. Soc. Lond. A* **1998**, *454*, 2445, arXiv:gr-qc/9701051.
70. Johnson, D.T.; Caticha, A. Entropic dynamics and the quantum measurement problem. *AIP Conf. Proc.* **2012**, *1443*, 104, arXiv:1108.2550
71. Vanslette, K.; Caticha, A. Quantum measurement and weak values in entropic quantum dynamics. *AIP Conf. Proc.* **2017**, *1853*, 090003, arXiv:1701.00781.
72. Nawaz, S.; Caticha, A. Momentum and uncertainty relations in the entropic approach to quantum theory. *AIP Conf. Proc.* **2012**, *1443*, 112, arXiv:1108.2629.
73. These are the well-known uncertainty relations due to Heisenberg and to Schrödinger. The entropic uncertainty relations proposed by Deutsch [74–76] have not yet been explored within the context of ED.
74. Deutsch, D. Uncertainty in Quantum Measurements. *Phys. Rev. Lett.* **1983**, *50*, 631. [CrossRef]
75. Partovi, M.H. Entropic Formulation of Uncertainty for Quantum Measurements. *Phys. Rev. Lett.* **1983**, *50*, 1883. [CrossRef]
76. Maassen, H.; Uffink, J. Generalized Entropic Uncertainty Relations. *Phys. Rev. Lett.* **1988**, *60*, 1103. [CrossRef]
77. Demme, A.; Caticha, A. The Classical Limit of Entropic Quantum Dynamics. *AIP Conf. Proc.* **2017**, *1853*, 090001, arXiv.org:1612.01905.
78. Nawaz, S.; Abedi, M.; Caticha, A. Entropic Dynamics on Curved Spaces. *AIP Conf. Proc.* **2016**, *1757*, 030004, arXiv.org:1601.01708.
79. Ipek, S.; Caticha, A. Entropic quantization of scalar fields. *AIP Conf. Proc.* **2015**, *1641*, 345, arXiv.org:1412.5637.
80. Ipek, S.; Abedi, M.; Caticha, A. Entropic Dynamics: Reconstructing Quantum Field Theory in Curved Spacetime. *Class. Quantum Grav.* **2019**, in press, arXiv:1803.07493.
81. Caticha, A.; Carrara, N. The Entropic Dynamics of Spin. In preparation.
82. In this work ED is a model for the quantum mechanics of particles. The same framework can be deployed to construct models for the quantum mechanics of fields, in which case it is the fields that are ontic and have well-defined albeit unknown values [79,80].
83. In Cartesian coordinates $\mu = const$ and may be ignored.

84. The angular nature of the drift potential is explained when the ED framework is extended to particles with spin [81].
85. In a relativistic theory there is more freedom in the choice of instants and this translates into a greater flexibility with the notion of simultaneity. Conversely, the requirement of consistency among the different notions of simultaneity severely limits the allowed forms of relativistic ED [80].
86. Price, H. *Time's Arrow and Archimedes' Point*; Oxford University Press: Oxford, UK, 1996.
87. Zeh, H.D. *The Physical Basis of the Direction of Time*; Springer: Berlin, Germany, 2007.
88. Lanczos, C. *The Variational Principles of Mechanics*, 4th ed.; Dover: New York, NY, USA, 1986.
89. Mandelbrot, B.B.; Van Ness, J.W. Fractional Brownian motions, fractional noises, and applications. *SIAM Rev.* **1968**, *10*, 422. [CrossRef]
90. In the ES type of ED considered in previous papers ($\gamma = 1$) [10–12] the probability also satisfies a continuity equation—a Fokker-Planck equation—and the current velocity is the sum of the drift velocity plus an osmotic component
$$u^A = -\hbar m^{AB} \partial_B \log \rho^{1/2}$$
due to diffusion.
91. Chandrasekhar, S. Stochastic Problems in Physics and Astronomy. *Rev. Mod. Phys.* **1943**, *15*, 1. [CrossRef]
92. DiFranzo, S. The Entropic Dynamics Approach to the Paradigmatic Quantum Mechanical Phenomena. Ph.D. Thesis, University at Albany, Albany, NY, USA, 2018.
93. Equations (49) and (50) show the reason to have introduced the new variable $\Phi = \hbar\phi$. With this choice Φ will eventually be recognized as the momentum that is canonically conjugate to the generalized coordinate ρ with Hamiltonian $\tilde{H}$.
94. We deal with ∞-dimensional spaces. The level of mathematical rigor in what follows is typical of theoretical physics—which is a euphemism for "from very low to none at all." For a more sophisticated treatment, see [59,62].
95. At this point the act of embedding $\mathbf{P}$ into $\mathbf{P}^{+1}$ represents no loss of generality because the embedding space $\mathbf{P}^{+1}$ remains unspecified.
96. This allows us, among other things, the freedom to switch from ρ^x to ρ_x as convenience dictates; from now on $\rho_x = \rho^x = \rho(x)$.
97. Arnold, V. I. *Mathematical Methods of Classical Mechanics*; Springer: Berlin/Heidelberger, Germany, 1997.
98. Schutz, B. *Geometrical Methods of Mathematical Physics*; Cambridge University Press: Cambridge, UK, 1980.
99. We are effectively determining the metric by imposing a symmetry, namely rotational invariance. One might be concerned that choosing this symmetry is an ad hoc assumption but the result proves to be very robust. It turns out that exactly the same metric is obtained by several other criteria that may appear more natural in the context of inference and probability. Such criteria include invariance under Markovian embeddings, the geometry of asymptotic inference, and the metrics induced by relative entropy [66,67] (see also [17]).
100. Carrara, N.; Caticha, A. Quantum phases in entropic Dynamics. *Springer Proc. Math. Stat.* **2018**, *239*, 1.
101. Schrödinger, E. The multi-valuedness of the wave function. *Ann. Phys.* **1938**, *32*, 49. [CrossRef]
102. Schrödinger invoked time reversal invariance which was a very legitimate move back in 1938 but today it is preferable to develop an argument which does not invoke symmetries that are already known to be violated.
103. The answer proposed by Pauli is also worthy of note [104–106]. He proposed that admissible wave functions must form a basis for representations of the transformation group that happens to be pertinent to the problem at hand. Pauli's argument serves to discard double-valued wave functions for describing the orbital angular momentum of scalar particles. The question of single-valuedness was revived by Takabayashi [107,108] in the context of the hydrodynamical interpretation of QM, and later rephrased by Wallstrom [109,110] as an objection to Nelson's stochastic mechanics: Are these theories equivalent to QM or do they merely reproduce a subset of its solutions? Wallstrom's objection is that Nelson's stochastic mechanics leads to phases and wave functions that are either both multi-valued or both single-valued. Both alternatives are unsatisfactory because on one hand QM requires single-valued wave functions, while on the other hand single-valued phases exclude states that are physically relevant (e.g., states with non-zero angular momentum).
104. Pauli, W. Über ein Kriterium für Ein-oder Zweiwertigkeit der Eigenfunktionen in der Wellenmechanik. *Helv. Phys. Acta* **1939**, 12, 147.
105. Pauli, W. *General Principles of Quantum Mechanics*; Springer: Berlin, Germany, 1980.

106. Merzbacher, E. Single Valuedness of Wave Functions. *Am. J. Phys.* **1962**, *30*, 237. [CrossRef]
107. Takabayasi, T. On the Formulation of Quantum Mechanics associated with Classical Pictures. *Prog. Theor. Phys.* **1952**, *8*, 143. [CrossRef]
108. Takabayasi, T. Vortex, Spin and Triad for Quantum Mechanics of Spinning Particle. *Prog. Theor. Phys.* **1983**, *70*, 1. [CrossRef]
109. Wallstrom, T.C. On the derivation of the Schrödinger equation from stochastic mechanics. *Found. Phys. Lett.* **1989**, *2*, 113. [CrossRef]
110. Wallstrom, T.C. The inequivalence between the Schrödinger equation and the Madelung hydrodynamic equations. *Phys. Rev. A* **1994**, *49*, 1613. [CrossRef] [PubMed]
111. We make no symmetry assumptions such as parity or time reversibility. It need not be the case that there is any symmetry that relates the time evolution of $\Psi^*_{t_0}$ to that of Ψ_{t_0}.
112. Double-valued wave functions with $e^{i\delta} = -1$ will, of course, find use in the description of spin-1/2 particles [81].
113. Yang, C. N. Charge Quantization, Compactness of the Gauge Group, and Flux Quantization. *Phys. Rev. D* **1979**, *1*, 2360. [CrossRef]
114. Caticha, A. The information geometry of space-time. In Proceedings of the 39th International Workshop on Bayesian Inference and Maximum Entropy Methods in Science and Engineering, Garching, Germany, 30 June–5 July 2019.
115. Ipek, S.; Caticha, A. An entropic approach to geometrodynamics. In Proceedings of the 39th International Workshop on Bayesian Inference and Maximum Entropy Methods in Science and Engineering, Garching, Germany, 30 June–5 July 2019.

8

Quantum Discord, Thermal Discord, and Entropy Generation in the Minimum Error Discrimination Strategy

Omar Jiménez [1,*], Miguel Angel Solís-Prosser [2,†], Leonardo Neves [3] and Aldo Delgado [2,†]

1 Centro de Óptica e Información Cuántica, Facultad de Ciencias, Universidad Mayor, Camino La Pirámide N°5750, Huechuraba, Santiago, Chile

2 Departamento de Física, Universidad de Concepción, 160-C Concepción, Chile; msolisp@udec.cl (M.A.S.-P.); aldelgado@udec.cl (A.D.)

3 Departamento de Física, Universidade Federal de Minas Gerais, Belo Horizonte-MG 31270-901, Brazil; lneves@fisica.ufmg.br

* Correspondence: omar.jimenez@umayor.cl

† Current address: Millennium Institute for Research in Optics, Universidad de Concepción, 160-C Concepción, Chile.

Abstract: We study the classical and quantum correlations in the minimum error discrimination (ME) of two non-orthogonal pure quantum states. In particular, we consider quantum discord, thermal discord and entropy generation. We show that ME allows one to reach the accessible information between the two involved parties, Alice and Bob, in the discrimination process. We determine the amount of quantum discord that is consumed in the ME and show that the entropy generation is, in general, higher than the thermal discord. However, in certain cases the entropy generation is very close to thermal discord, which indicates that, in these cases, the process generates the least possible entropy. Moreover, we also study the ME process as a thermodynamic cycle and we show that it is in agreement with the second law of thermodynamics. Finally, we study the relation between the accessible information and the optimum success probability in ME.

Keywords: minimum error discrimination; accessible information; discord; second law of thermodynamics

---

1. Introduction

Recently, quantum communication protocols have been studied from the point of view of the quantum correlations between the involved parties [1–6]. This allows one to quantify the resources that are required to carry out quantum communication protocols. Quantum correlations also allow us to differentiate between quantum and classical properties of a quantum state. The implementation of protocols for quantum communication requires that at least one of the parties implements a quantum measurement. This is, in general, an irreversible process [7] that changes the quantum state, produces decoherence and, also, entropy. Studied quantum correlations in bipartite scenarios are entanglement [8], quantum discord [9–11], thermal discord [12] and global discord [13,14].

The total amount of correlations contained in a bipartite state is quantified by the quantum mutual information, which represents the minimal amount of noise that is required to erase or destroy the total correlations in a many-copy scenario [15]. Quantum mutual information is also directly connected with Landauer's original idea [16], which states that any logical irreversible process must dissipate entropy into the environment [17]. It can be cast as the sum of two terms: classical correlations and quantum discord. The latter, quantum discord, is defined in order to minimize the loss of quantum correlations due to quantum measurements [9]. Another measure of quantum correlation is the thermal discord.

This takes into account the entropic cost of measurements and minimizes the entropy generation produced by quantum measurements [7,12]. Quantum discord and thermal discord are defined when one of the parties implements measurements in specific bases. These bases, in general, are different from each other. On the other hand, in some particular cases, such as in minimum error discrimination, we need to implement a measurement in a determined basis. For those cases, the basis-dependent versions of the aforementioned quantum correlations are considered [11]. In general, the discrimination of non-orthogonal states must satisfy the second law of thermodynamics [18,19] and, to demonstrate this, a central argument is the Landauer's cost of erasure [12].

One of the most elemental quantum communication protocols is the discrimination among states belonging to a predefined set of known quantum states. If the set contains two or more non-orthogonal states, the discrimination of the quantum states cannot be carried out with certainty and deterministically [20]. In this case, we can resort to several discrimination strategies that optimize a predefined figure of merit. Minimum error discrimination (ME) is one of these strategies, where the discrimination of the non-orthogonal states is carried out in such a way that the probability of mistaking a retrodiction is minimized [21,22]. ME finds application in several quantum information processes such as quantum teleportation [23,24], entanglement swapping [25,26], quantum cryptography [27] and dense coding [28], among many others. The discrimination of non-orthogonal states by means of ME has been experimentally implemented for two states [29,30], for four states [31] in a two-dimensional Hilbert space, and for several sets of symmetric pure states in dimensions as high as 29 [32].

Here, we study the minimum error discrimination of two non-orthogonal states generated with arbitrary probabilities, in terms of the quantum correlations involved in the process. For this purpose, we consider the cases when Alice and Bob share a separable quantum channel. We determine the quantum discord and the thermal discord and compare these with the loss of correlations and the entropy generation in the case of ME. Moreover, we show that ME, when considered as a thermodynamic cycle, is in agreement with the second law of thermodynamics. Finally, we study the relation between the accessible information and the optimal probability of success in the ME protocol.

This article is organized as follows: In Section 2 we briefly review minimum error discrimination for two non-orthogonal states at the Helstrom limit. In Section 3 we describe the initial and final states of Alice and Bob after the measurement implemented by Bob according to ME. In Section 4 we study quantum and the thermal discord together with their relationship to the second law of thermodynamics. Moreover, the relation between the information gained by Bob and the optimum success probability in the ME, is studied. Finally, in Section 5 we summarize our results and conclude.

2. Minimum Error Discrimination

In the minimum error discrimination protocol, one of the communicating parties, let us say Bob, receives a single copy of a quantum system. This can be described by one of two possible non-orthogonal states in the set $\Omega = \{|\phi_0\rangle_B, |\phi_1\rangle_B\}$. These states are generated with a priori probabilities η_0 and $\eta_1 = 1 - \eta_0$, respectively. The set of states and a priori probabilities are known by Bob beforehand. Bob's task is to identify, with the lowest average error, the probability of the state in Ω that describes the quantum system.

The states in Ω can be written as

$$|\phi_0\rangle_B = \cos\frac{\beta}{2}|0\rangle_B + \sin\frac{\beta}{2}|1\rangle_B, \tag{1a}$$

$$|\phi_1\rangle_B = \sin\frac{\beta}{2}|0\rangle_B + \cos\frac{\beta}{2}|1\rangle_B, \tag{1b}$$

where the two orthonormal states $\{|0\rangle_B, |1\rangle_B\}$ form a base of the two-dimensional Hilbert space of Bob's quantum system. The inner product between the non-orthogonal states is denoted by $\alpha = \langle\phi_0|\phi_1\rangle = \sin\beta$, where $\beta \in [0, \pi/2]$, so that $\alpha \in [0, 1]$.

In order to discriminate with ME between the non-orthogonal states $\{|\phi_0\rangle_B, |\phi_1\rangle_B\}$, Bob first applies a unitary transformation U_B. In this case, U_B can be written in the following form

$$U_B|\phi_0\rangle_B = \sqrt{p_0}|0\rangle_B + \sqrt{r_0}|1\rangle_B, \tag{2a}$$
$$U_B|\phi_1\rangle_B = \sqrt{r_1}|0\rangle_B + \sqrt{p_1}|1\rangle_B, \tag{2b}$$

where, now, the orthonormal states $\{|0\rangle_B, |1\rangle_B\}$ represent the base in which Bob must implement his measurement in order to discriminate with ME probability. Here, r_i (p_i) represents the probability of failure (success) in the identification of the state $|\phi_i\rangle_B$, where $p_i + r_i = 1$. The unitarity of U_B implies that the following constraint must be satisfied

$$\alpha = \sqrt{r_0 p_1} + \sqrt{r_1 p_0}. \tag{3}$$

The average probability of error in the discrimination between the non-orthogonal states $\{|\phi_0\rangle_B, |\phi_1\rangle_B\}$ is

$$P_e = \eta_0 r_0 + \eta_1 r_1, \tag{4}$$

where, $r_i = |\langle j|U_B|\phi_i\rangle|^2$ for $i \neq j$. We reach the minimum average error probability in the discrimination process when the probabilities r_i are given by [33]

$$r_i = \frac{1}{2}\left(1 - \frac{1 - 2\eta_j\alpha^2}{\sqrt{1 - 4\eta_i\eta_j\alpha^2}}\right), \tag{5}$$

for $i \neq j$. Therefore, the minimal average error probability attained by the minimum error discrimination strategy is given by

$$P_e^{min} = \frac{1}{2}(1 - \sqrt{1 - 4\eta_0\eta_1\alpha^2}), \tag{6}$$

which is equal to the Helstrom limit [20,22]. The optimal average success probability in the discrimination is equal to $P_s^{opt} = 1 - P_e^{min}$. In what follows, given the symmetry with respect to the a priori probabilities, we consider the case $\eta_1 \geq \eta_0$ for $0 \leq \eta_0 \leq 1/2$.

3. Channel without Entanglement

Let us consider initially that the communicating parties, Alice and Bob, share a separable quantum state of the form

$$\rho_{AB} = \sum_{i=0}^{1} \eta_i |i\rangle_A\langle i| \otimes |\phi_i\rangle_B\langle\phi_i|, \tag{7}$$

where the states $\{|0\rangle_A, |1\rangle_A\}$ form an orthonormal base for Alice's two-dimensional quantum system, and $\{|\phi_0\rangle_B, |\phi_1\rangle_B\}$ are the two possible non-orthogonal states of Bob's quantum system given by Equations (1). Alice prepares a single copy of a quantum system in the state $|\phi_i\rangle_B$ and sends it to Bob with a priori probability η_i. Thereby, Alice and Bob share quantum and classical correlations encoded in the joint state ρ_{AB} of Equation (7). The initial state ρ_A of Alice's quantum system, that is, prior to the application of any transformation or measurement, is obtained by $\rho_A = \mathrm{tr_B}(\rho_{AB})$, where

$$\rho_A = \sum_{i=0}^{1} \eta_i |i\rangle_A\langle i|. \tag{8}$$

In a similar form, the initial state of Bob's quantum system can be obtained from $\rho_B = \mathrm{tr_A}(\rho_{AB})$, where

$$\rho_B = \sum_{i=0}^{1} \eta_i |\phi_i\rangle_B\langle\phi_i|. \tag{9}$$

Once Bob has received the single copy of the quantum system in the state $|\phi_i\rangle_B$, he implements the optimal strategy of ME. For that purpose, Bob first applies the unitary transformation U_B onto his quantum system. Thereby, the initial joint state ρ_{AB} of Equation (7) changes to $\hat{\rho}_{AB} = (\mathbf{1}_A \otimes U_B)\rho_{AB}(\mathbf{1}_A \otimes U_B^\dagger)$, where

$$\hat{\rho}_{AB} = \sum_{i=0}^{1} \eta_i |i\rangle_A\langle i| \otimes |\hat{\phi}_i\rangle_B\langle\hat{\phi}_i|, \tag{10}$$

with $|\hat{\phi}_i\rangle_B = U_B|\phi_i\rangle_B$. The unitary transformation U_B of Equation (2), applied by Bob onto his quantum system, is a reversible process [7]. Therefore, it does not change the quantum correlations between Alice and Bob and it does not produce entropy either.

We consider that Bob can implement his measurement in an arbitrary basis $\{|0'\rangle_B, |1'\rangle_B\}$, which is given by

$$|0'\rangle_B = x|0\rangle_B - y|1\rangle_B, \tag{11a}$$
$$|1'\rangle_B = y|0\rangle_B + x|1\rangle_B, \tag{11b}$$

where the coefficients x and y are real positive numbers that satisfy $x^2 + y^2 = 1$. The measurement carried out by Bob on his quantum system generates two conditional post-measurement states $\rho^i_{A|b}$ for Alice's quantum system. Provided that Bob's measurement projects his quantum system onto the state $|i'\rangle_B$, Alice's post-measurement states can be

$$\rho^0_{A|b} = (t_{00}^2|0\rangle_A\langle 0| + t_{10}^2|1\rangle_A\langle 1|)/p_0^b, \tag{12}$$
$$\rho^1_{A|b} = (t_{01}^2|0\rangle_A\langle 0| + t_{11}^2|1\rangle_A\langle 1|)/p_1^b, \tag{13}$$

respectively, where

$$t_{00} = xt_0 - ym_0, \tag{14}$$
$$t_{01} = xm_0 + yt_0, \tag{15}$$
$$t_{10} = xm_1 - yt_1, \tag{16}$$
$$t_{11} = xt_1 + ym_1, \tag{17}$$

with

$$m_i = \sqrt{\eta_i r_i}, \tag{18}$$
$$t_i = \sqrt{\eta_i(1 - r_i)}, \tag{19}$$
$$p_i^b = t_{0i}^2 + t_{1i}^2, \tag{20}$$

for $i = 0, 1$ and

$$\sum_{i,j=0}^{1} t_{ij}^2 = 1. \tag{21}$$

The final average joint state between Alice and Bob ρ'_{AB}, when Bob implements his measurement in the basis $\{|0'\rangle_B, |1'\rangle_B\}$, takes the following form:

$$\rho'_{AB} = \sum_{i=0}^{1} p_i^b \rho^i_{A|b} \otimes \Pi_b^{i'}, \tag{22}$$

where, $\Pi_b^{i'}$ are the projectors $|i'\rangle_B\langle i'|$ onto the Hilbert space of Bob's quantum system. The state ρ'_{AB} is a classical state because there are local measurements in Alice's and Bob's systems that do not perturb it [11]. The final reduced states for Alice's and Bob's quantum systems are given by

$$\rho'_A = p_0^b \rho^0_{A|b} + p_1^b \rho^1_{A|b}, \quad (23)$$
$$\rho'_B = p_0^b |0'\rangle_B\langle 0'| + p_1^b |1'\rangle_B\langle 1'|. \quad (24)$$

Thereby, the final reduced state for Alice's system does not change, that is, $\rho'_A = \eta_0|0\rangle_A\langle 0| + \eta_1|1\rangle_A\langle 1| = \rho_A$.

In the particular case of ME, that is, for $x = 1$ in Equations (11), the average joint state ρ'_{AB} of Equation (22) takes the following form

$$\rho'_{AB}(x=1) = t_0^2|00\rangle_{AB}\langle 00| + m_0^2|01\rangle_{AB}\langle 01| + m_1^2|10\rangle_{AB}\langle 10| + t_1^2|11\rangle_{AB}\langle 11|. \quad (25)$$

In this particular case, if Bob successfully discriminates the state $|\phi_i\rangle_B$, the final state that Alice and Bob share will be $|ii\rangle_{AB}$. Otherwise, when the discrimination attempt is unsuccessful, the final state is $|ij\rangle$ with $i \neq j$. The average error probability is equal to $p_e^{min} = m_0^2 + m_1^2$ and agrees with the Helstrom bound given by Equation (6).

4. Correlations between Alice and Bob

4.1. Classical Correlations and Quantum Discord

In a bipartite state ρ_{AB}, the total amount of correlation, in the many copy scenario [15], is given by the quantum mutual information. This is defined as [9,15]

$$I(\rho_{AB}) = S(\rho_A) + S(\rho_B) - S(\rho_{AB}), \quad (26)$$

where $S(\rho)$ is the von Neumann entropy of the state ρ, given by $S(\rho) = -\sum_i \lambda_i \log_2 \lambda_i$, where λ_i are the eigenvalues of ρ. Hence, in our scheme of ME discrimination, we consider that Alice emits many copies of independent identically distributed (i.i.d.) data, that is $\sigma = \rho_{AB}^n$ for some large n [34].

The quantum mutual information can be written as [9,11]

$$I(\rho_{AB}) = J(A|\{\Pi_b\}) + D(A|\{\Pi_b\}), \quad (27)$$

where, $J(A|\{\Pi_b\})$ are the classical correlations and $D(A|\{\Pi_b\})$ is the quantum discord. These two quantities depend on the measurement implemented by Bob, through the set of projectors $\{\Pi_b\}$, but their sum does not [2], i.e., they are complementary to each other. The classical correlations $J(A|\{\Pi_b\})$ between Alice and Bob are defined as [10,11]

$$J(A|\{\Pi_b\}) = S(\rho_A) - \sum_{i=0}^{1} p_i^b S(\rho^i_{A|b}), \quad (28)$$

which can be interpreted as the information about Alice's system gained by Bob by means of the measurement $\{\Pi_b\}$. Here, we are interested in maximizing the classical correlation $J(A|\{\Pi_b\})$ with respect to all possible measurements implemented by Bob, that is

$$J(A|B) = \max_{\{\Pi_b\}} J(A|\{\Pi_b\}) = S(\rho_A) - \min_{\{\Pi_b\}} \sum_{i=0}^{1} p_i^b S(\rho^i_{A|b}), \quad (29)$$

which is called the accessible information [35,36]. This represents the classical mutual information maximized with respect to the detection strategy [31].

On the other hand, the minimum quantum discord, which quantifies the quantum correlations that are consumed or lost in the process, is given by

$$D(A|B) = I(\rho_{AB}) - J(A|B). \tag{30}$$

To quantify the accessible information, we minimize the expression $f(x) = \sum_i p_i^b S(\rho^i_{A|b})$, with respect to the variable x that defines the measurement base in Equation (11). The function $f(x)$ can be cast as

$$f(x) = -\sum_{i,j=0}^{1} t_{ij}^2 \log_2\left(\frac{t_{ij}^2}{p_j^b}\right), \tag{31}$$

and its derivative with respect to x is

$$\frac{d}{dx}f(x) = -\frac{2}{\sqrt{1-x^2}}\sum_{i=0}^{1} t_{i0}t_{i1}\log_2\left(\frac{t_{i0}^2 p_1^b}{t_{i1}^2 p_0^b}\right). \tag{32}$$

This derivative vanishes at $x = 0$. In fact, this a minimum. However, the function $f(x)$ is such that $f(x=0) = f(x=1)$ and consequently $x = 1$ is also a minimum. In both cases the entropy of Alice's conditional states are equal, $S(\rho^0_{A|b}) = S(\rho^1_{A|b})$. In fact, the cases $x = 0$ or $x = 1$ are physically equivalent since they are connected by the transformations $|0'\rangle \to |1'\rangle$ and $|1'\rangle \to |0'\rangle$, which is equivalent to the exchange $|\phi_0\rangle_B \to |\phi_1\rangle_B$ and $|\phi_1\rangle_B \to |\phi_0\rangle_B$. On the other hand, to carry out ME discrimination, Bob must consider the basis of measurement with $x = 1$ (or $x = 0$) in Equations (11). Thereby, in these particular cases, the same measurement base reaches simultaneously ME discrimination and accessible information [37,38].

To determine the amount of quantum discord, we consider the expression

$$D(A|B) = S(\rho_B) + \min_{\{\Pi_b\}}\sum_{i=0}^{1} p_i^b S(\rho^i_{A|b}) - S(\rho_{AB}), \tag{33}$$

where the eigenvalues of Bob's state ρ_B, given by Equation (9), are $\lambda_0^b = (1-\sqrt{1-4\eta_0\eta_1\cos^2\beta})/2$ and $\lambda_1^b = 1-\lambda_0^b$. Moreover, the entropy of the joint initial state ρ_{AB}, given by Equation (7), is $S(\rho_{AB}) = S(\rho_A) + \sum\eta_i S(|\phi_i\rangle_B\langle\phi_i|) = S(\rho_A)$.

Figure 1a,b show the accessible information and the quantum discord, respectively, for $x = 1$, as a function of the inner product α for three decreasing values of η_0 (from top to bottom). The accessible information $J(A|B)$ takes its maximum value, for a fixed value of η_0, when the states $\{|\phi_0\rangle_B, |\phi_1\rangle_B\}$ are orthogonal ($\alpha = 0$) and it decreases with the inner product α, reaching its minimum value, $J(A|B) = 0$, when $\alpha = 1$. Moreover, in the particular case $\eta_0 = 1/2$, Alice and Bob share the maximal mutual information available from any ensemble of quantum states [37]. If we consider the case $\eta_0 = 0$, the ME has only one state in the discrimination and the ME strategy does not convey any information at all. In this case we have that $J(A|B) = 0$.

The quantum discord $D(A|B)$ takes its minimum value equal to zero, for any value of η_0, when the states $\{|\phi_0\rangle_B, |\phi_1\rangle_B\}$ are orthogonal ($\alpha = 0$) or their inner product is equal to one. In the aforementioned cases, Bob's measurement does not change the joint state $\rho'_{AB} = \hat{\rho}_{AB}$ and therefore, the states $\{|\phi_0\rangle_B, |\phi_1\rangle_B\}$ behave like classical states. On the other hand, for a fixed value of α the quantum discord takes its maximal value when $\eta_0 = 0.5$. Simultaneously, the accessible information for Bob is maximal in this case. As is apparent from Figure 1b, the maximum of the quantum discord occurs at $\alpha = 1/\sqrt{2}$ for any value of η_0. This happens because $D(A|B)$ is a concave function of α^2 and it is symmetric under interchange of α^2 and $1-\alpha^2$. Physically, we can say that when $\alpha^2 = 1/2$, the states $\{|\phi_0\rangle_B, |\phi_1\rangle_B\}$ are in an intermediate position between the two cases $\alpha^2 = 0$ and $\alpha^2 = 1$, where the quantum discord vanishes and, consequently, there are no quantum correlations. On the other hand, the case $\alpha^2 = 1/2$ is the least similar to the aforementioned ones and, therefore, a maximum value of the quantum discord is to be expected.

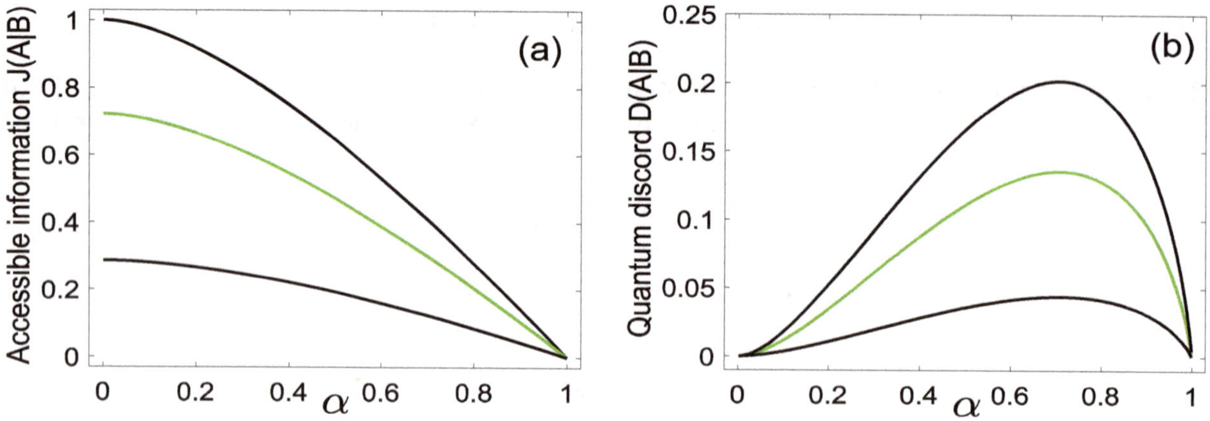

Figure 1. (**a**) Accessible information and (**b**) quantum discord, when $x = 1$, as a function of the inner product α for $\eta_0 = 0.5$ (red line), $\eta_0 = 0.2$ (green line), and $\eta_0 = 0.05$ (black line).

4.2. Thermal Discord

Another measure of quantum correlations that we consider in the analysis of ME discrimination, is the thermal discord. This takes into account the entropic cost of realizing local measurements. The thermal discord is defined by [11,12]

$$D_{th}(A|B) = \min_{\{\Pi_b\}}[S(\rho'_B) + \sum_{i=0}^{1} p_i^b S(\rho'^i_{A|b})] - S(\rho_{AB}). \tag{34}$$

In this case, the term $S(\rho'_B) + \sum_{i=0}^{1} p_i^b S(\rho'^i_{A|b})$ to be minimized in the Equation (34) turns out to be equal to the entropy $S(\rho'_{AB})$ of the state ρ'_{AB} of Equation (22), which corresponds to the joint state after Bob's measurement. Hence, the thermal discord is equal to the minimum entropy generation due to the measurement implemented by Bob, which is also equal to the one-way quantum deficit [7,11,39]. Then the measurement-dependent thermal discord (DTD) is

$$D_{th}(A|\{\Pi_b\}) = S(\rho'_{AB}) - S(\rho_{AB}), \tag{35}$$

and it corresponds to entropy generation due to Bob's measurement in a particular basis $\{\Pi_b\}$.

In general, the thermal discord is higher than or equal to the quantum discord [7,12,40], that is,

$$D_{th}(A|B) \geq D(A|B). \tag{36}$$

This indicates that the minimal generation of entropy must be higher or equal to the minimum of the quantum correlations that are destroyed due to Bob's measurement [16]. This inequality is related to the second law of thermodynamics [12], by means of the entropic cost to take Bob's system to the initial state. Given that a projective measurement does not decrease the entropy [41], we have $S(\rho'_B) \geq S(\rho_B)$ and this is valid for any base chosen by Bob for implementing his measurement. On the other hand, from the condition $I(\rho_{AB}) \geq J(A|B)$ we obtain that $S(\rho_B) \geq J(A|B)$. Thus, we have that

$$J(A|B) \leq S(\rho'_B). \tag{37}$$

Equation (37) can be understood if we consider the scheme of ME as a thermodynamic cycle [17,42]. The second law establishes that the net work after completing a cycle cannot be positive, that is $W_{out} - W_{in} \leq 0$. The work W_{out} that can be extracted is proportional to the classical mutual information $J(A|B)$ between Alice and Bob. The work invested W_{in} is proportional to the Holevo bound $S(\rho_B) - \sum \eta_i S(|\phi_i\rangle_B\langle\phi_i|) = S(\rho_B)$ plus the minimal erasure work to take the final state ρ'_B to the initial state ρ_B, which is equal to $S(\rho'_B) - S(\rho_B) \geq 0$. From this, it is clear that $W_{in} = k_B T S(\rho'_B)$ and $W_{out} = k_B T J(A|B)$, where k_B is the Boltzmann constant and T is the temperature of the thermal reservoir. Thereby, the second law of thermodynamics is satisfied [17–19].

In our case, to determine the thermal discord and the entropy generation in Bob's measurement, we consider that $S(\rho'_B) = -p_0^b \log_2 p_0^b - p_1^b \log_2 p_1^b$. For the entropy generation in ME, we evaluate the function of Equation (35) for $x = 1$. On the other hand, the DTD in Equation(35) is a function of the variable x that defines the measurement base in Equations (11) and it can be cast as

$$D_{th}(A|\{\Pi_b\})(x) = -\sum_{i,j=0}^{1} t_{ij}^2 \log_2(t_{ij}^2) - S(\rho_A). \tag{38}$$

In order to find the thermal discord, we need to minimize it with respect to the variable x, then we consider the restriction $\frac{d}{dx} D_{th}(A|\{\Pi_b\})(x) = 0$. The last condition takes the following form

$$\frac{1}{\sqrt{1-x^2}} \left(t_{00} t_{01} \log_2 \left(\frac{t_{00}^2}{t_{01}^2} \right) + t_{10} t_{11} \log_2 \left(\frac{t_{10}^2}{t_{11}^2} \right) \right) = 0. \tag{39}$$

We solve Equation (39) numerically in order to find the values of x where the function $D_{th}(A|\{\pi_b\})(x)$ takes its minimum values.

Figure 2a displays the values of x that allow us to attain the minimal value of the measurement-dependent thermal discord, that is, the thermal discord as a function of the inner product α for three values of η_0. For $\eta_0 = 0.05$ and $\eta_0 = 0.2$ the variable x is close to the unity in the complete interval of α values. For $\eta \approx 0.5$ the values of x depart from the unity when α is approximately equal to or larger than 0.5.

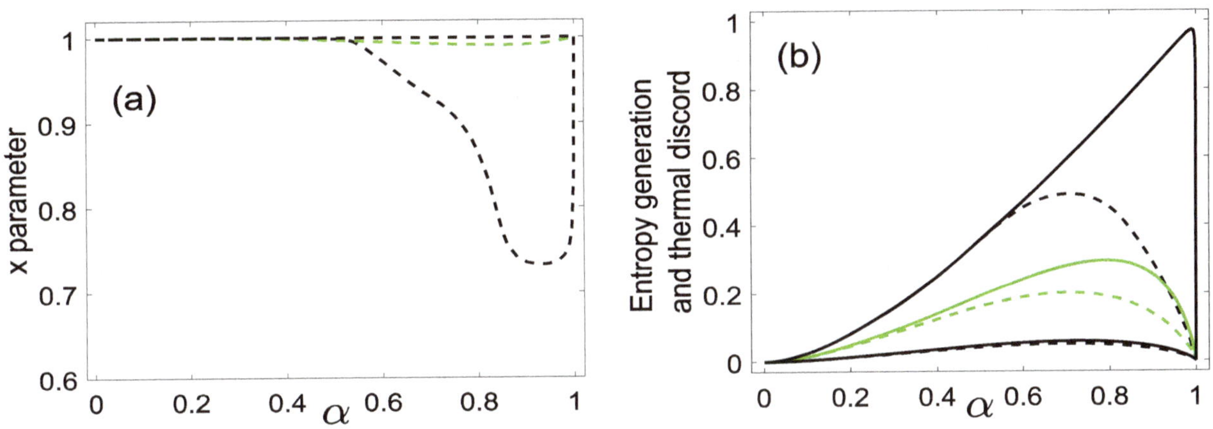

Figure 2. (**a**) Values of x that attain the thermal discord. (**b**) Entropy generation in minimum error (ME) (solid line) and thermal discord (dashed line) as a function of the inner product α for: $\eta_0 = 0.49$ (red line), $\eta_0 = 0.2$ (green line), and $\eta_0 = 0.05$ (black line).

In the interval $\alpha \in [0, 0.5]$ the minimum error probability increases very slowly and is upper bounded by the value 0.07 for any value of η_0. Therefore, the discrimination process achieves a high accuracy. In this regime, the coefficients m_i^2 are very small. As can be seen from Equation (5), the m_i^2 coefficients are upper bounded by the value 0.035. In this case, the post-measurement state for ME is approximately given by $\rho'_{AB} \approx t_0^2|00\rangle_{AB}\langle 00| + t_1^2|11\rangle_{AB}\langle 11|$. A similar state can also be obtained from Equation (22) considering a first order Taylor series expansion for m_i and $\sqrt{1-x^2}$. The latter is suggested because it is easy to show that the base-dependent thermal discord is optimized at $x = 1$ for $\alpha^2 = 0$. Thus, for small values of α^2 the Taylor series expansion is a good guess. The state obtained in this way has four eigenvalues that are functions of x. In order to minimize the entropy of this state, we can vanish two of the eigenvalues by choosing $x = 1$ independently of the value of η_0. This procedure leads to a state that agrees with the state ρ'_{AB} above. Therefore, when the minimum error probability P_e^{min} is very small, that is, for $\alpha \leq 0.5$, entropy generation in the ME process approximately agrees with the value of the thermal discord.

Figure 2b shows the generation of entropy in ME in solid lines and the thermal discord in dashed lines, as a function of the inner product α and for three values of η_0. We obtained that the entropy generation in ME is higher than or equal to the thermal discord. This means that, in general, the base for the thermal discord and for ME do not coincide. However, for small values of α the entropy generation is very close to the thermal discord and consequently, in these cases, ME is also almost optimal from the point of view of the entropy generation.

When the states $\{|\phi_0\rangle_B, |\phi_1\rangle_B\}$ are orthogonal ($\alpha = 0$) the entropy generation and thermal discord are zero because in this case, as already pointed out, the joint state does not change, that is, $\rho'_{AB} = \rho_{AB}$. On the other hand, in the case $\alpha = 1$ and $\eta_0 = 0$, the state ρ_{AB} is a product state and thus the entropy generation and thermal discord are also zero. The maximum value of entropy generation in ME occurs for $\eta_0 = 0.5$ as in the case of the maximum accessible information $J(A|B)$ between Alice and Bob. As in the case of quantum discord, thermal discord is a concave function of α^2 and also has its maximum when $\alpha^2 = 1/2$ for any value of η_0. This means that quantum correlations (quantum discord and thermal discord) indicate that the biggest disturbance, due to the quantum measurement, of the initial joint state ρ_{AB} will be produced if $\alpha^2 = 1/2$. However, in general, the process of ME generates more entropy than thermal discord as we see in Figure 2b. The maximum of the entropy generation now arises at values closer to $\alpha^2 = 1$, depending on the value of η_0. We see that for large overlaps (which means a more difficult discrimination process) the entropy generation departs from the thermal discord. In addition, as η_0 increases, this departure becomes more "dramatic". Large values of η_0 implies in less bias towards some state in the set. So, large overlaps and large η_0 bring more difficulty in the discrimination process and require a larger effort to perform ME. The extreme case would be $\eta_0 = 0.5$ and $\alpha = 1$. Finally, we would like to point out that the choice of $\eta_0 = 0.49$ was considered in order to show that when $\alpha = 1$ the entropy generation goes to zero.

Figure 3a compares the average error probability of ME (solid lines) to the average error probability obtained when employing the base that optimizes the thermal discord (dashed lines). As is apparent from the figure, the base associated with the thermal discord discriminates the states $\{|\phi_0\rangle_B, |\phi_1\rangle_B\}$ almost as good as ME in the interval $\alpha \in [0, 0.5]$ for the three inspected values of η_0. For α equal or larger than 0.5, ME delivers the smallest average error probability. Thus, the base that optimizes the measurement-dependent thermal discord provides a discrimination almost as good as optimal ME in the interval $\alpha \in [0, 0.5]$ for the inspected values of η_0.

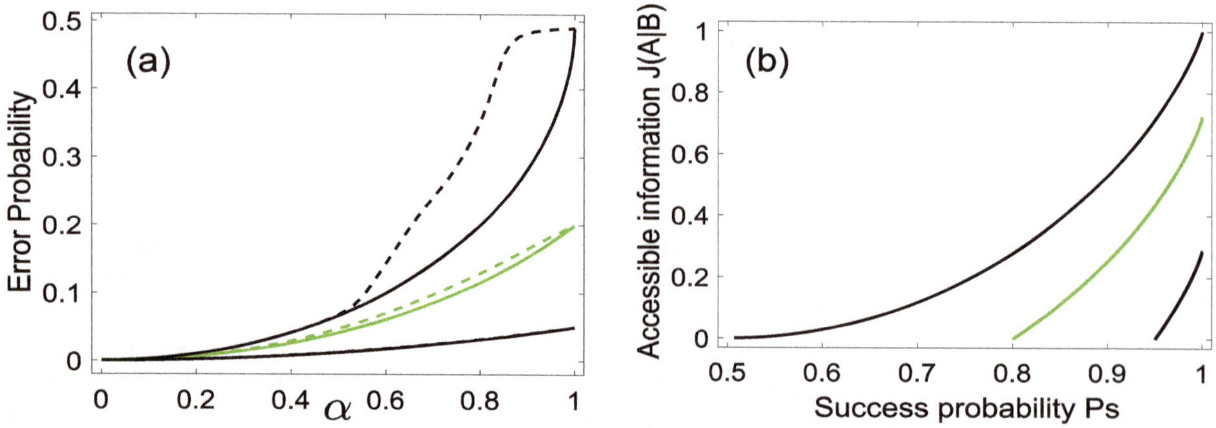

Figure 3. (**a**) Average error probability in ME (solid lines) and for the case when the measurement base leads to the thermal discord (dashed lines). (**b**) Accessible information versus average success probability in ME for $\eta_0 = 0.49$ (red line), $\eta_0 = 0.2$ (green line), and $\eta_0 = 0.05$ (black line).

4.3. Accessible Information and Optimum Success Probability

Finally, we study the relation between the accessible information $J(A|B)$ and the optimal average success probability in the minimum error discrimination. This can be done because the bases for accessible information and optimal ME coincide. Figure 3b shows that for a fixed value of η_0, the accessible information $J(A|B)$ increases as a function of the optimum probability of discrimination P_s^{opt}. When the states are orthogonal, i.e., $\alpha = 0$, then $P_s^{opt} = 1$ and the accessible information is maximal for a fixed value of η_0. On the other hand, if the states are equal, i.e., $\alpha = 1$, then $P_s^{opt} = \eta_1$ and the accessible information is equal to zero.

5. Conclusions

We studied ME discrimination for two non-orthogonal states generated with arbitrary a priori probabilities from the point of view of the quantum correlations involved in the process. We recovered a previously known result, namely, optimal ME allows us also to attain the accessible information between the communicating parties, Alice and Bob. Thereby, in the ME discrimination process it is possible to optimize simultaneously the average success probability as well as the information gained by Bob through the measurement.

In general, the base that optimizes the measurement-dependent thermal discord does not agree with the base that leads to ME. This implies that ME generates more entropy than the minimum possible given by the thermal discord. However, for values of the inner product α in the interval $[0, 0.5]$, the entropy generated in the ME process is very close to the one obtained in the thermal discord. This indicates that for these cases, the ME process is also efficient in terms of the generation of entropy. Furthermore, when the discrimination process is carried out by measuring onto the base that leads to the thermal discord, the average error probability becomes very close to the optimal value when $\alpha \in [0, 0.5]$.

Quantum discord and thermal discord are zero when the states are orthogonal ($\alpha = 0$) or when we have only one state $\alpha = 1$ or $\eta_0 = 0$. Hence, in these cases, the ME protocol presents only a classical behavior, i.e., the initial state does not change and there is no entropy generation with Bob's measurement. Otherwise, the scheme of ME presents quantum properties given that the quantum discord and the thermal discord are greater than zero. Moreover, the quantum discord maximum and the thermal discord maximum occur when $\alpha^2 = 1/2$, which is the intermediate point between the two classical cases $\alpha = 0$ and $\alpha = 1$.

We showed that the process of ME discrimination satisfies the second law of thermodynamics when it is considered as a thermodynamic cycle. Finally, we obtained that the amount of accessible information increases as a function of the optimal average success probability of discrimination in the minimum error discrimination strategy.

Here, we have studied the case of ME for two pure states. ME can also be formulated for an arbitrary finite number of states, which points out to the possibility of generalizing our results to this scenario. However, analytical solutions for ME are known in very few special cases such as, for instance, sets of equally likely pure symmetric states [43] or two states with arbitrary prior probabilities. Furthermore, in certain situations, when the number of states is larger than the dimension of the underlying Hilbert space, the solution of ME requires the use of a positive-operator valued measure, which is not analytically known. Therefore, the lack of analytical solutions prevents us from extrapolating our results to more complex scenarios for ME. This also applies to other optimal discrimination strategies, being these also complex optimization problems.

A feasible extension of our results might appear in the so-called sequential discrimination [44]. In this scenario, several parties attempt to discriminate among a set of states in such away that each one of them has access to the post-measurement states generated by the other parties. All parties cannot resort to classical communications. The strategy optimizes the joint discrimination probability. Here, we could study the change of the classical and quantum correlations as the quantum system encoding the unknown states passes from party to party.

Author Contributions: O.J. provided the idea of the study, did the calculations of the study and performed critical revision of the manuscript. M.A.S.-P., L.N. and A.D., provided critical advice and revision of the manuscript. All authors have read and approved the final manuscript.

Acknowledgments: This work was supported by FONDECYT grant 11121318.

References

1. Gu, M.; Chrzanowski, H.M.; Assad, S.M.; Symul, T.; Modi, K.; Ralph, T.C.; Vedral, V.; Lam, P.K. Observing the operational significance of discord consumption. *Nat. Phys.* **2012**, *8*, 671–675. [CrossRef]
2. Zwolak, M.; Zurek, W.H. Complementarity of quantum discord and classically accessible information. *Sci. Rep.* **2013**, *3*, 1729. [CrossRef]
3. Xu, J.-S.; Xu, X.-Y.; Li, C.-J.; Zhang, C.-J.; Zou, X.-B.; Guo, G.-C. Experimental investigation of classical and quantum correlations under decoherence. *Nat. Comms.* **2010**, *1*, 7. [CrossRef] [PubMed]
4. Roa, L.; Retamal, J.C.; Alid-Vaccarezza, M. Dissonance is Required for assited optimal state discrimination. *Phys. Rev. Lett.* **2011**, *107*, 080401. [CrossRef] [PubMed]
5. Li, B.; Fei, S.M.; Wang, Z.-X.; Fan H. Assisted state discrimination without entanglement. *Phys. Rev. A* **2012**, *85*, 022328. [CrossRef]
6. Zhang, F.-L.; Chen, J.-L.; Kwek, L.C.; Vedral, V. Requirement of Dissonance in Assisted Optimal State Discrimination. *Sci. Rep.* **2013**, *3*, 2134. [CrossRef] [PubMed]
7. Horodecki, M.; Horodecki, P.; Horodecki, R.; Oppenheim, J.; Sen, A.; Sen U.; Synak-Radtke, B. Local versus nonlocal information in quantum-information theory: Formalism and phenomena. *Phys. Rev. A* **2005**, *71*, 062307. [CrossRef]
8. Horodecki, R.; Horodecki, P.; Horodecki, M.; Horodecki K. Quantum entanglement. *Rev. Mod. Phys.* **2009**, *81*, 865. [CrossRef]
9. Ollivier, H.; Zurek, W.H. Quantum Discord: A Measure of the Quantumness of Correlations. *Phys. Rev. Lett.* **2001**, *88*, 017901. [CrossRef]
10. Herderson, L.; Vedral, V. Classical, quantum and total correlations. *J. Phys. A Math. Gen.* **2001**, *34*, 6899. [CrossRef]
11. Modi, K.; Brodutch, A.; Cable, H.; Paterek, T.; Vedral, V. The classical-quantum boundary for correlations: Discord and related measures. *Rev. Mod. Phys.* **2012**, *84*, 1655. [CrossRef]
12. Zurek, W.H. Quantum discord and Maxwell's demons. *Phys. Rev. A* **2003**, *67*, 012320. [CrossRef]
13. Rulli, C.C.; Sarandy, M.S. Global quantum discord in multipartite systems. *Phys. Rev. A* **2011**, *84*, 042109. [CrossRef]
14. Coto, R.; Orszag, M. Determination of the maximum global quantum discord via measurements of excitations in a cavity QED network. *J. Phys. B At. Mol. Opt. Phys.* **2014**, *47*, 095501. [CrossRef]
15. Groisman, B.; Popescu, S.; Winter, A. Quantum, classical, and total amount of correlations in a quantum state. *Phys. Rev. A* **2005**, *72*, 032317. [CrossRef]
16. Landauer, R. Irreversibility and Heat Generation in the Computing Process. *IBM J. Res. Dev.* **1961**, *5*, 183–191. [CrossRef]
17. Maruyama, K.; Nori, F.; Vedral, V. *Colloquium:* The physics of Maxwell's demon and information. *Rev. Mod. Phys.* **2009**, *81*, 1. [CrossRef]
18. Peres, A. *Complexity, Entropy and the Physics of Information*; Zurek, W.H., Ed.; Addison-Wesley: Boston, MA, USA, 1990; 345p.
19. Peres, A. *Quantum Theory: The Concepts and Methods*; Kluwer Academic: Boston, MA, USA, 1993.
20. Bergou, J.; Hillery, M. *Introduction to the Theory of Quantum Information Processing*; Springer: Berlin/Heidelberg, Germany, 2013.
21. Holevo, A.S. Statistical Decision Theory for Quantum Systems. *J. Multivar. Anal.* **1973**, *3*, 337–394. [CrossRef]
22. Helstrom, C.W. *Quantum Detection and Estimation Theory*; Academic Press: Cambridge, MA, USA, 1976.
23. Roa, L.; Delgado A.; Fuentes-Guridi, I. Optimal conclusive teleportation of quantum states. *Phys. Rev. A* **2003**, *68*, 022310. [CrossRef]
24. Neves, L.; Solís-Prosser, M.A.; Delgado, A.; Jiménez O. Quantum teleportation via maximum-confidence quantum measurements. *Phys. Rev. A* **2012**, *85*, 062322. [CrossRef]

25. Delgado, A.; Roa, L.; Retamal, J.C.; Saavedra, C. Entanglement swapping via quantum state discrimination. *Phys. Rev. A* **2005**, *71*, 012303. [CrossRef]
26. Solís-Prosser, M.A.; Delgado, A.; Jiménez, O.; Neves, L. Deterministic and probabilistic entanglement swapping of nonmaximally entangled states assisted by optimal quantum state discrimination. *Phys. Rev. A* **2014**, *89*, 012337. [CrossRef]
27. Phoenix, S.J.D.; Barnett, S.M.; Chefles, A. Three-state quantum cryptography. *J. Mod. Opt.* **2000**, *47*, 507–516. [CrossRef]
28. Pati, A.K.; Parashar, P.; Agrawal, P. Probabilistic superdense coding. *Phys. Rev. A* **2005**, *72*, 012329. [CrossRef]
29. Barnett, S.M.; Riis, E. Experimental demonstration of polarization discrimination at the Helstrom bound. *J. Mod. Opt.* **1997**, *44*, 1061–1064. [CrossRef]
30. Waldherr, G.; Dada, A.C.; Neumann, P.; Jelezko, F.; Andersson, E.; Wrachtrup, J. Distinguishing between Nonorthogonal Quantum States of a Single Nuclear Spin. *Phys. Rev. Lett.* **2012**, *109*, 180501. [CrossRef]
31. Clarke, R.B.M.; Kendon, V.M.; Chefles, A.; Barnett, S.M.; Riis, E.; Sasaki, M. Experimental realization of optimal detection strategies for overcomplete states. *Phys. Rev. A* **2001**, *64*, 012303. [CrossRef]
32. Solís-Prosser, M.A.; Fernandes, M.F.; Jiménez, O.; Delgado, A.; Neves, L. Experimental Minimum-Error Quantum-State Discrimination in High Dimensions. *Phys. Rev. Lett.* **2017**, *118*, 100501. [CrossRef]
33. Andi, S. Quantum State Discrimination and Quantum Cloning: Optimization and Implementation. Ph.D. Thesis, CUNY Academic Works, City University of New York, New York, NY, USA, 2015.
34. Dahlsten, O.; Renner, R.; Rieper, E.; Vedral, V. Inadequacy of von Neumann entropy for characterizing extractable work. *New J. Phys.* **2011**, *13*, 053015. [CrossRef]
35. Dall'Arno, M.; D'Ariano, G.M.; Sacchi, M.F. Informational power of quantum measurements. *Phys. Rev. A* **2011**, *83*, 062304. [CrossRef]
36. Sasaki, M.; Barnett, S.; Jozsa, R.; Osaki, M.; Hirota, O. Accessible information and optimal strategies for real symmetrical quantum sources. *Phys. Rev. A* **1999**, *59*, 3325. [CrossRef]
37. Levitin, L.B. Optimal quantum measurements for two pure and mixed states. In *Quantum Communications and Measurement*; Belavkin, V.P., Hirota, O., Hudson, R.L., Eds.; Plenum: New York, NY, USA, 1995; pp. 439–448.
38. Barnett, S.M.; Croke, S. Quantum State Discrimination. *Adv. Opt. Photon.* **2009**, *1*, 238–278. [CrossRef]
39. Streltsov, A.; Kampermann, H.; Bruss, D. Linking Quantum Discord to Entanglement in a Measurement. *Phys. Rev. Lett.* **2011**, *106*, 160401. [CrossRef] [PubMed]
40. Brodutch, A.; Terno, D. Quantum discord, local operations, and Maxwell's demons. *Phys. Rev. A* **2010**, *81*, 062103. [CrossRef]
41. Nielsen, M.A.; Chuang, I.L. *Quantum Computation and Quantum Information*; Cambridge University Press: Cambridge, UK, 2000.
42. Maruyama, K.; Brukner, C.; Vedral, V. Thermodynamical cost of accessing quantum information. *J. Phys. A: Math. Gen.* **2005**, *38*, 7175. [CrossRef]
43. Ban, M.; Kurokawa, K.; Momose, R.; Hirota, O. Optimum Measurements for Discrimination among Symmetric Quantum States and Parameter Estimation. *Int. J. Theor. Phys.* **1997**, *36*, 1269–1288. [CrossRef]
44. Bergou, J.; Feldman, E.; Hillery, M. Extracting Information from a Qubit by Multiple Observers: Toward a Theory of Sequential State Discrimination. *Phys. Rev. Lett.* **2013**, *111*, 100501. [CrossRef]

9

Security Analysis of Unidimensional Continuous-Variable Quantum Key Distribution using Uncertainty Relations

Pu Wang [1], **Xuyang Wang** [1,2,*] **and Yongmin Li** [1,2,*]

1 State Key Laboratory of Quantum Optics and Quantum Optics Devices, Institute of Opto-Electronics, Shanxi University, Taiyuan 030006, China; 201622607026@email.sxu.edu.cn
2 Collaborative Innovation Center of Extreme Optics, Shanxi University, Taiyuan 030006, China
* Correspondence: wangxuyang@sxu.edu.cn (X.W.); yongmin@sxu.edu.cn (Y.L.)

Abstract: We study the equivalence between the entanglement-based scheme and prepare-and-measure scheme of unidimensional (UD) continuous-variable quantum key distribution protocol. Based on this equivalence, the physicality and security of the UD coherent-state protocols in the ideal detection and realistic detection conditions are investigated using the Heisenberg uncertainty relation, respectively. We also present a method to increase both the secret key rates and maximal transmission distances of the UD coherent-state protocol by adding an optimal noise to the reconciliation side. It is expected that our analysis will aid in the practical applications of the UD protocol.

Keywords: continuous-variable quantum key distribution; unidimensional modulation; Heisenberg uncertainty relations

---

1. Introduction

Quantum key distribution (QKD), which is a prominent application of the quantum information, enables two remote parties, conventionally called Alice and Bob, to share a common secret key through an insecure quantum channel and an authenticated classical channel [1,2]. This unconditional security is guaranteed by the basic principles of quantum mechanics. Continuous-variable quantum key distribution (CV-QKD) has attracted considerable attention over the past years because of its good performances in the secret key rates and compatibility with the current optical networks [3–16]. A particular class of CV-QKD protocols that is based on the Gaussian modulation of coherent states has experienced a rapid development [17–27]. In a coherent-state protocol, Alice encodes her information in the amplitude and phase quadratures of the coherent light field by using amplitude and phase modulators, and Bob performs homodyne or heterodyne detection.

Recently, a further simplified unidimensional (UD) CV-QKD protocol has been proposed [28]. In such protocol, Alice, still using coherent states, encodes her information by using one modulator (e.g., amplitude modulator) instead of two, whereas Bob performs a homodyne detection, hence simplifying both the modulation scheme and the key extraction task. The security against collective attacks has been proved in asymptotic regime. However, this early work only considered the UD model under an idea homodyne detector. It does not refer to the realistic condition, such as the efficiency and electronic noise of the homodyne detector. Then, a model of the UD protocol under realistic condition was designed and realized in an experiment [29]. Furthermore, the finite size effect was analyzed in paper [30], and an optimum ratio in parameters estimation was proposed.

In the UD protocol, due to the fact that the phase quadrature is not modulated in Alice's side, we cannot estimate the correlation in the phase quadrature between Alice and Bob. However, this unknown parameter is bounded by the requirement of the physicality of the state. A Gaussian state can typically be characterized by a covariance matrix. However, not all covariance matrices correspond to physical states, as the covariance matrix must respect the Heisenberg uncertainty relation [31,32]. By using this uncertainty relation, we can calculate the physical region boundary of a covariance matrix, which is crucial for the security of the protocol. We can see that the UD CV-QKD protocol is very different from the previous symmetrical (SY) coherent-state protocol [18,21]. Due to the equivalence between the prepare-and-measure (PM) and entanglement-based (EB) scheme of UD protocol, the differences of the Heisenberg uncertainty relations under the idea and realistic condition, and the effect of noise from Bob's setup on secret key rate under realistic condition are not described or investigated in depth [28–30], a further study about above questions is required.

In this paper, we first consider the equivalence between the PM scheme and the EB scheme of the UD CV-QKD protocol. Then, we analyze the boundary of the physical region of the symmetrical coherent-state protocol based on the Heisenberg uncertainty relation. We also study the variances of the physical region of the UD coherent-state protocol under the conditions of different detection efficiency and electronic noise. Secure and unsecure regions of both the protocols are further analyzed under ideal and realistic detection conditions. It is found that adding an optimal noise to Bob's side can truly help the improvement of the secret key rate and increase the transmission distance of the UD coherent-state protocol under the assumption of reverse reconciliation.

The paper is organized as follows. In Section 2, we introduce the equivalence between the EB scheme and the PM scheme of the UD CV-QKD protocol. In Section 3, a comparison between the physical and secure regions of the UD protocol under ideal and realistic detection conditions is shown, and a method to improve the performance of the UD coherent-state protocol by adding an optimal noise to Bob's side is proposed. In Section 4, we give our conclusions and discussions.

2. Unidimensional Quantum Key Distribution

2.1. Equivalence between the EB Scheme and the PM Scheme

Generally, most of the experimental systems in CV-QKD are focused on PM schemes currently, given their ease of implementation in practice. However, it's hard to analyze the security in theory. On the contrary, the theoretical analysis based on EB scheme is maturity. The involved entangled states make the calculations feasible and simpler [33]. Especially in UD CV-QKD protocol, the security analysis based on EB scheme has more advantages. The covariance matrices achieved from the EB schemes contain the constraints of phase amplitude quadrature. However these constraints is difficult to achieve from the PM scheme. More details about the security analysis will be shown later. Now, it is necessary to study the equivalence of EB and PM schemes, firstly. This equivalence is based on the indistinguishability between these two protocols for Bob and Eve. The consequent advantage of this equivalence is that it is sufficient to implement the PM scheme and study the EB scheme.

In the PM scheme, as depicted in Figure 1a, the sender, Alice, prepares coherent states using a laser source. Then, she encodes the information in the amplitude or phase quadratures of coherent states by using either amplitude or phase modulators. Here, without losing generality, we assume that Alice uses an amplitude modulator with a modulation variance V_M, which is assumed to be expressed in shot-noise units, and that the coherent states follow the uncertainty principle of variance 1. Thus, the mixture of Gaussian-modulated coherent states gives rise to a unidimensional chain structure with a thickness of 1 and a length of $\sqrt{1+V_M}$ in the phase space. These quantum states are then sent to Bob through an untrusted quantum channel with transmittance T_x, T_y and excess noise ε_x, ε_y.

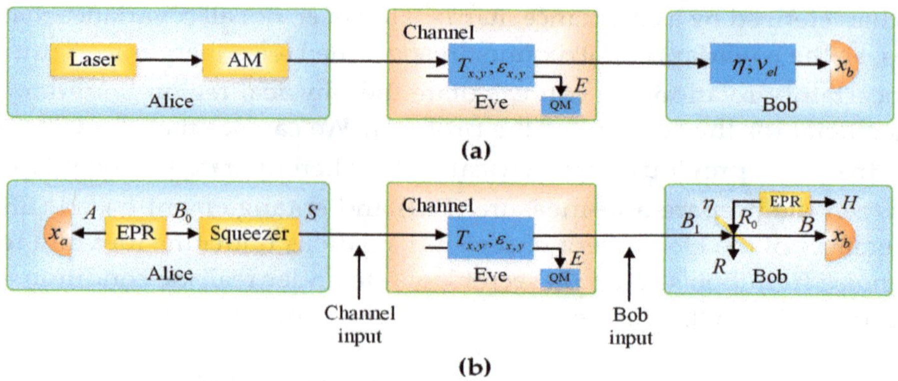

Figure 1. Unidimensional (UD) protocol schemes under realistic conditions. (**a**) Prepare-and-measure (PM) scheme of the UD protocol; (**b**) Entanglement-based (EB) scheme of the UD protocol.

In the EB scheme, as shown in Figure 1b, Alice starts with a two-mode squeezed vacuum state ρ_{AB_0} with variance V. Then, she performs homodyne detection on the first half of the state and squeezes the second half by $r = \ln\sqrt{V}$. The result is the covariance matrix

$$\gamma_{AS} = \begin{bmatrix} V & 0 & \sqrt{V(V^2-1)} & 0 \\ 0 & V & 0 & -\sqrt{(V^2-1)/V} \\ \sqrt{V(V^2-1)} & 0 & V^2 & 0 \\ 0 & -\sqrt{(V^2-1)/V} & 0 & 1 \end{bmatrix} \tag{1}$$

The covariance matrix of mode S, conditioned on Alice's measurement result (x_a), can be written as

$$\gamma_S^{x_a} = \gamma_S - \sigma_{AS}^T (X\gamma_A X)^{MP} \sigma_{AS}, \tag{2}$$

and the displacement vector can be expressed as

$$d_S^{x_a} = \sigma_{AS}^T (X\gamma_A X)^{MP} d_A, \tag{3}$$

where d_A is the result of the homodyne measurement, γ_A and γ_S are the covariance matrices of the modes A and S, respectively, σ_{AS} is the correlation matrix of the two modes, $X = diag(1,0)$, and MP denotes the Moore–Penrose inverse of the matrix [34].

Then, we obtain

$$\gamma_S^{x_a} = \begin{bmatrix} 1 & 0 \\ 0 & 1 \end{bmatrix} \text{ and } d_S^{x_a} = \sqrt{\frac{V^2-1}{V}}(x_a, 0), \tag{4}$$

which is a coherent state centered on $d_S^{x_a}$. Furthermore, the variance of $d_S^{x_a}$ is

$$\left\langle \Delta^2 d_S^{x_a} \right\rangle = \frac{V^2-1}{V}\left\langle x_a^2 \right\rangle = \frac{V^2-1}{V} \cdot V = V^2 - 1, \tag{5}$$

where $V^2 - 1$ is exactly the variance of the Alice's V_M. Then, we can establish a one-to-one correspondence between the EB scheme and the PM scheme by multiplying the outcome of Alice's measurements by the factor $\alpha = \sqrt{\frac{V^2-1}{V}}$.

2.2. Calculation of Secret Key Rate with Reverse Reconciliation

Thus far, we have established the equivalence between the EB scheme and the PM scheme of the UD CV-QKD protocol. In this subsection, we present a brief overview of the calculation of the secret key rates. In the EB protocol, the realistic Bob's detector can be modeled by an ideal balanced homodyne detector and a beam splitter, with transmission efficiency η and input noise $V_N = 1 + v_{el}/(1-\eta)$, as the one shown in Figure 1b. The secret key rate against collective attacks for reverse reconciliation in the asymptotic regime can be calculated as [29,30]

$$K_{RR}^{\infty} = \beta \cdot I_{AB} - \chi_{BE}, \tag{6}$$

where β is the reverse reconciliation efficiency and I_{AB} is the mutual information between Alice and Bob. I_{AB} can expressed as

$$I_{AB} = \frac{1}{2}\log_2(1 + \frac{V_M}{1+\chi_{\text{totx}}}), \tag{7}$$

where

$$\begin{aligned} \chi_{\text{hom}} &= (1+v_{el})/\eta - 1 \\ \chi_{\text{linex}} &= (1-T_x)/T_x + \varepsilon_x \ . \\ \chi_{\text{totx}} &= \chi_{\text{linex}} + \chi_{\text{hom}}/T_x \end{aligned} \tag{8}$$

Still from Equation (6), χ_{BE} is the Holevo bound, which represents an upper bound on the information acquired for reverse reconciliation by the potential eavesdropper Eve. The procedures to calculate χ_{BE} can be written as:

$$\begin{aligned} \chi_{BE} &= S(\rho_E) - S(\rho_E^{x_b}) \\ &= S(\rho_{AB_1}) - S(\rho_{ARH}^{x_b}) \\ &= \sum_{i=1}^{2} g(\tfrac{\lambda_i - 1}{2}) - \sum_{i=3}^{5} g(\tfrac{\lambda_i - 1}{2}) \end{aligned} , \tag{9}$$

where $S(\rho)$ is the von Neumann entropy of the quantum state ρ, $g(x) = (x+1)\log_2(x+1) - x\log_2 x$ and λ_i are the symplectic eigenvalues of the covariance matrix γ, with

$$\begin{aligned} \lambda_{1,2}^2 &= \tfrac{1}{2}(a \pm \sqrt{a^2 - 4b}) \\ \lambda_{3,4}^2 &= \tfrac{1}{2}(c \pm \sqrt{c^2 - 4d}) \ , \\ \lambda_5 &= 1 \end{aligned} \tag{10}$$

$$\begin{aligned} a &= 1 + V_M + V_y^{B_1}(1 + V_M + \chi_{\text{linex}})T_x + 2C_y^{B_1}(1+V_M)^{1/4}\sqrt{V_M T_x} \\ b &= (V_y^{B_1}(1+V_M) - (C_y^{B_1})^2\sqrt{1+V_M})(1+\varepsilon_x T_x) \\ c &= (a(\chi_{\text{hom}}+1) + ((1+\varepsilon_x T_x)(V_M+2) + V_M T_x - a))/e \\ d &= (b\chi_{\text{hom}} + (1+V_M)(1+\varepsilon_x T_x))/e \\ e &= T_x(1 + V_M + \chi_{\text{totx}}) \end{aligned} , \tag{11}$$

where $V_y^{B_1}$ is the variance of the mode B_1 in phase quadrature with $V_y^{B_1} = 1 + T_y\varepsilon_y$ and $C_y^{B_1}$ is the correlation between A and B_1 in phase quadrature with $C_y^{B_1} = -\sqrt{T_y V_M}(1+V_M)^{-1/4}$.

3. Security Analysis Using Uncertainty Relations

In this section, we provide a security analysis of continuous variable quantum key distribution with coherent states based on the Heisenberg uncertainty relation. Before describing the UD coherent-state protocol case, it is useful to first consider the SY coherent-state protocol case.

3.1. Uncertainty Relations for Symmetrical Coherent-State Protocol

Let us consider a n-mode quantum mechanical system that is described by the canonical conjugate operators $\hat{x}_j$ and $\hat{p}_j$, with $j = 1, 2, \cdots, n$. In terms of the annihilation and creation operators ($\hat{a}_j$ and $\hat{a}_j^{\dagger}$, respectively), one has

$$\hat{x}_j = \frac{1}{\sqrt{2}}(\hat{a}_j + \hat{a}_j^{\dagger}) \text{ and } \hat{p}_j = -\frac{i}{\sqrt{2}}(\hat{a}_j - \hat{a}_j^{\dagger}), \tag{12}$$

which are the dimensionless position and momentum operators. Such operators also satisfy the bosonic canonical commutation relations (CCR)

$$[\hat{x}_i, \hat{p}_j] = i\delta_{i,j}, \ [\hat{x}_i, \hat{x}_j] = [\hat{p}_i, \hat{p}_j] = 0, \tag{13}$$

Furthermore, if we group together the canonical conjugate operators in a vector $\hat{\gamma}$ as

$$\hat{\gamma} = (\hat{r}_1, \hat{r}_2, \cdots \hat{r}_{2n})^T = (\hat{x}_1, \hat{p}_1, \hat{x}_2, \hat{p}_2, \cdots, \hat{x}_n, \hat{p}_n)^T, \tag{14}$$

we can express the CCR in a compact form:

$$[\hat{\gamma}_j, \hat{\gamma}_k] = i\Omega_{jk}, \tag{15}$$

where Ω is defined as

$$\Omega = \bigoplus_{i=1}^{n} \begin{bmatrix} 0 & 1 \\ -1 & 0 \end{bmatrix}. \tag{16}$$

By combining this CCR relation and the positive semi-definiteness of the density operator ρ, we obtain the following uncertainty relation [35]

$$\gamma + i \cdot \Omega \geq 0, \tag{17}$$

which is a more precise and complete version of the Heisenberg uncertainty relation. This well-known inequality is the only constraint that γ has to respect to be a covariance matrix satisfying a physical state.

Let us consider the physicality of the SY coherent-state protocol by using the uncertainty relation in Equation (17). In the EB protocol, as shown in Figure 2, we have:

$$\gamma_{AB_1}^{sym} = \begin{bmatrix} V & 0 & \sqrt{T(V^2-1)} & 0 \\ 0 & V & 0 & -\sqrt{T(V^2-1)} \\ \sqrt{T(V^2-1)} & 0 & T(V+\chi_{\text{line}}) & 0 \\ 0 & -\sqrt{T(V^2-1)} & 0 & T(V+\chi_{\text{line}}) \end{bmatrix}, \tag{18}$$

$$\gamma_{AB}^{sym} = \begin{bmatrix} V & 0 & \sqrt{\eta T(V^2-1)} & 0 \\ 0 & V & 0 & -\sqrt{\eta T(V^2-1)} \\ \sqrt{\eta T(V^2-1)} & 0 & \eta T(V+\chi_{\text{tot}}) & 0 \\ 0 & -\sqrt{\eta T(V^2-1)} & 0 & \eta T(V+\chi_{\text{tot}}) \end{bmatrix}, \tag{19}$$

where $\chi_{\text{tot}} = \chi_{\text{line}} + \chi_{\text{hom}}/T$, $\chi_{\text{line}} = (1-T)/T + \varepsilon$, $\chi_{\text{hom}} = (1-\eta)/\eta + v_{el}/\eta$, and $V = V_A + 1$, V_A is the modulation variance of the Alice's side. According to the Heisenberg uncertainty relation, we have:

$$\begin{cases} \gamma_{AB_1}^{sym} + i \cdot \Omega \geq 0 \\ \gamma_{AB}^{sym} + i \cdot \Omega \geq 0 \end{cases}. \tag{20}$$

Then, we obtain

$$\begin{cases} \varepsilon T(2+(\varepsilon-2)T)(V^2-1) \geq 0 \\ (\varepsilon T\eta(2+(\varepsilon-2)T\eta) + 2v_{el}(1+(\varepsilon-1)T\eta) + v_{el}^2)(V^2-1) \geq 0 \end{cases}. \tag{21}$$

The two inequalities in Equation (21) are simultaneously satisfied if $\varepsilon, v_{el} \geq 0$ and $T, \eta \in [0,1]$. Here, we further consider the secure and unsecure regions of the protocol for both ideal and realistic Bob's detectors, which are shown in Figure 3a. In the secure region, the secret key rate is greater than zero; in the unsecure region, the secret key rate is less than zero. We observe that the realistic protocol can provide a bigger secure region. The secret key rate as a function of the excess noise, in correspondence of three values of channel losses, under ideal and realistic detection conditions, is shown in Figure 3b. We can see that the realistic Bob detection improves the resistance of the protocol to the excess noise, although the total noise is increased, which will lead to the appearance of a phenomenon called "fighting noise with noise" [36], and will be discussed in detail in the following. Here, we set the values of the actual parameters: the reconciliation efficiency is $\beta = 0.99$ [37] and the modulation variance is $V_A = 10$.

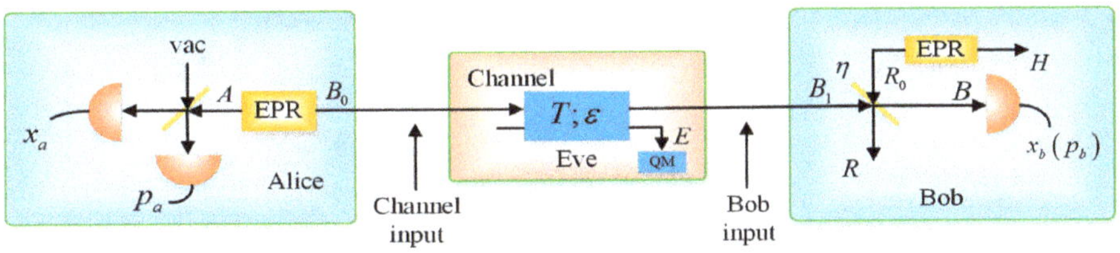

Figure 2. EB scheme of the SY protocol under realistic conditions.

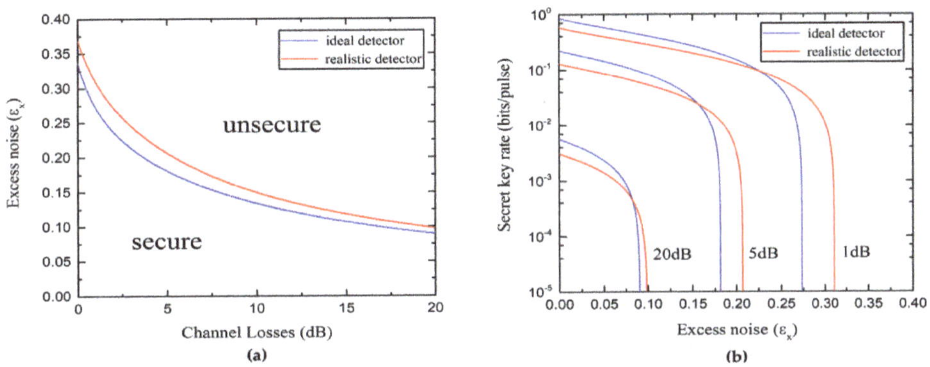

Figure 3. (**a**) Secure and unsecure regions of the SY protocol using ideal homodyne detector ($\eta = 1, v_{el} = 0$) and realistic homodyne detector ($\eta = 0.6, v_{el} = 0.1$); (**b**) Secret key rate versus the excess noise for different channel losses.

3.2. Uncertainty Relations for Unidimensional Coherent-State Protocol

In the above, we have discussed the physicality of the SY coherent-state protocol by using the Heisenberg uncertainty relation. The securities under ideal and realistic homodyne detectors have also been analyzed. Next, let us consider the UD coherent-state protocol. As shown in Figure 1b, in the EB scheme, we have

$$\gamma_{AB_1}^{uni} = \begin{bmatrix} \sqrt{1+V_M} & 0 & \sqrt{T_x V_M}(1+V_M)^{1/4} & 0 \\ 0 & \sqrt{1+V_M} & 0 & C_y^{B_1} \\ \sqrt{T_x V_M}(1+V_M)^{1/4} & 0 & T_x(V_M+1+\chi_{\text{linex}}) & 0 \\ 0 & C_y^{B_1} & 0 & V_y^{B_1} \end{bmatrix} \text{ and} \tag{22}$$

$$\gamma_{AB}^{uni} = \begin{bmatrix} \sqrt{1+V_M} & 0 & \sqrt{\eta T_x V_M}(1+V_M)^{1/4} & 0 \\ 0 & \sqrt{1+V_M} & 0 & C_y^{B_1}\sqrt{\eta} \\ \sqrt{\eta T_x V_M}(1+V_M)^{1/4} & 0 & \eta T_x(V_M+1+\chi_{\text{totx}}) & 0 \\ 0 & C_y^{B_1}\sqrt{\eta} & 0 & \eta(V_y^{B_1}+\chi_{\text{hom}}) \end{bmatrix}. \tag{23}$$

In the UD protocol, in order to estimate the information of the Eve eavesdropping, χ_{BE}, we have to know the parameters $C_y^{B_1}$ and $V_y^{B_1}$. Here, $V_y^{B_1}$ can be estimated by randomly measuring the phase quadrature in Bob's side, while $C_y^{B_1}$ is unknown due to the fact that the phase quadrature is not modulated in Alice's side. However, such an unknown parameter is constrained by the requirement of the physicality of the state. Differently from Ref. [30], under realistic condition, when the mode

B_1 is transformed into mode B after the beam splitter, there will have to be a new constraint on the covariance matrix γ_{AB}^{uni} in order to make it correspondent to a physical state. According to the Heisenberg uncertainty relation, we have

$$\begin{cases} \gamma_{AB_1}^{uni} + i \cdot \Omega \geq 0 \\ \gamma_{AB}^{uni} + i \cdot \Omega \geq 0 \end{cases}. \tag{24}$$

Then, we obtain the following two parabolic equations:

$$\begin{cases} \left(C_y^{B_1} - C_0\right)^2 \leq \frac{V_M}{\sqrt{(1+V_M)}} \frac{\chi_{\text{linex}}}{1+\chi_{\text{linex}}} \left(V_y^{B_1} - V_0\right) \\ \left(C_y^{B_1} - C'_0\right)^2 \leq \frac{V_M}{\sqrt{(1+V_M)}} \frac{\chi_{\text{totx}}}{1+\chi_{\text{totx}}} \left(V_y^{B_1} - V'_0\right) \end{cases}, \tag{25}$$

where $C_0 = -\frac{V_0\sqrt{T_x V_M}}{(1+V_M)^{1/4}}$, $V_0 = \frac{1}{T_x(1+\chi_{\text{linex}})}$, $C'_0 = -\frac{\sqrt{T_x V_M}}{(1+V_M)^{1/4}\eta T_x(1+\chi_{\text{totx}})}$ and $V'_0 = \frac{1}{\eta^2 T_x(1+\chi_{\text{totx}})} - \chi_{\text{hom}}$.

The parabolic curves between $C_y^{B_1}$ and $V_y^{B_1}$, under ideal and realistic detection conditions, are shown in Figure 4. The whole plane is divided into two regions: the unphysical and physical regions. In the unphysical region, the values of the parameters $C_y^{B_1}$ and $V_y^{B_1}$ cannot be satisfied simultaneously, otherwise, the Heisenberg uncertainty principle will be violated. In the physical region, the whole region is divided into two parts, R1 and R2. The R1 represents the real physical region, which is delimited by the ideal parabolic curve and ensures the attacks of Eve to the quantum channel complying with the physical principles. The red dashed line further divides the region R1 into unsecure and secure regions. The R2 represents the pseudo physical region, which is the overlapped part between the physical region contained by the realistic parabolic curve and the unphysical region, as defined by the ideal parabolic curve. The appearance of the pseudo physical region is due to the fact that, even if some attacks of Eve are unphysical, after the transform of the realistic homodyne detection of Bob, the final covariance matrix can satisfy a physical state. Hence, the physical region should be delimited at the input side of Bob, or equivalently, Bob performs an ideal detection. Furthermore, in Figure 5, we see how the physical region delimited by the realistic parabolic curve changes according to different conditions of detection efficiency and electronic noise. We also compare such regions with the one delimited by the ideal parabolic curve (black solid line in Figure 5). We find that the physical region defined by the realistic parabolic curve gradually decreases as the detection efficiency increases and the electronic noise decreases. Therefore, also in this case, in order to ensure the physicality of the UD protocol, we select the smaller region R1.

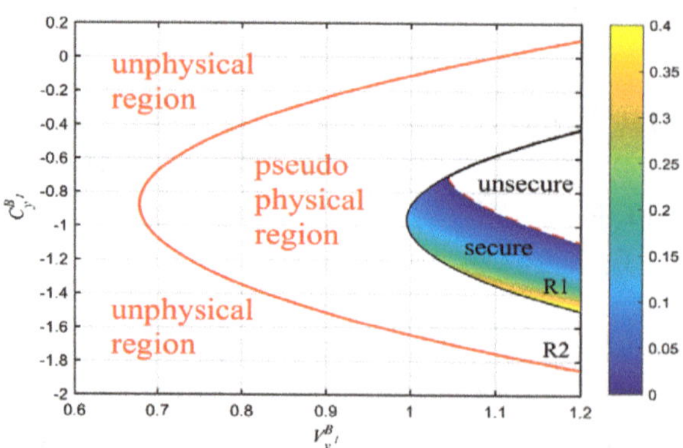

Figure 4. Comparison among physical regions of the UD protocol under both ideal and realistic detection conditions. The red solid line represents the realistic parabolic curve (equivalent to Bob using a realistic homodyne detector with $\eta = 0.6, v_{el} = 0.1$) and black solid line is the ideal parabolic curve (equivalent to Bob using an ideal homodyne detector with $\eta = 1, v_{el} = 0$). The red dashed line represents the part where the key rate is zero under realistic detection condition. Here, we set: $\beta = 0.99$, $T_x = 0.4$ (corresponding to a distance of 20 km fiber), $\varepsilon_x = 0.01$ and $V_M = 6.35$.

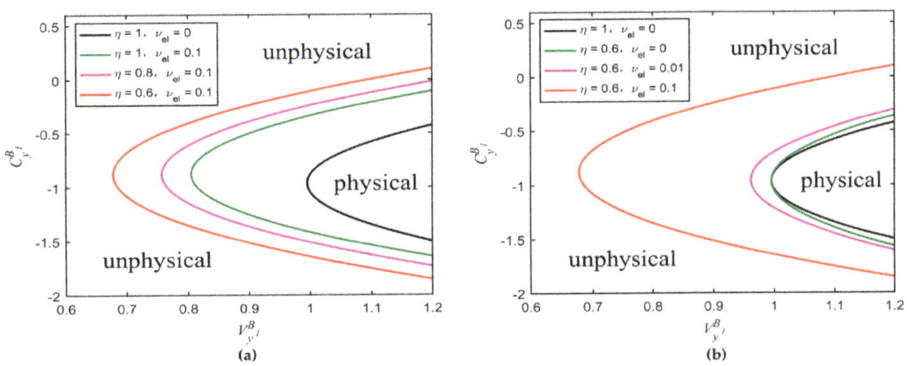

Figure 5. Comparison among physical regions delimited by the parabolic curves of the UD protocol. The black solid curve corresponds to the ideal parabolic curve, whereas the others to the realistic parabolic curves obtained for different parameter conditions. (**a**) Changes of the physical region extension according to different values of η (v_{el} remains constant); (**b**) Changes of the physical region extension according to different values of v_{el} (η remains constant). The values of the parameters T_x, ε_x, and V_M are the same as in Figure 4.

In Figure 6, we consider the dependence of the ideal parabolic curve (R1) on related parameters, including V_M, T_x, ε_x, and β. From Figure 6a, we can find that the parabolic curve moves down and gradually becomes broader as the modulation variance increases. In Figure 6b, the parabolic curve moves towards bottom-left corner and gradually becomes narrower as the transmission efficiency increases. In Figure 6c, as the excess noise increases, the parabolic curve moves towards left and gradually becomes larger. The reconciliation efficiency β does not change the shape of the parabola, but rather expands the secure region. In Figure 6d, the red solid line represents the minimum secret key rate, which was obtained by scanning the parameter $C_y^{B_1}$. The black solid line represents the ideal parabolic curve. It is interesting that a larger $C_y^{B_1}$ does not always give a higher secret key rate, more details about the red solid line can be seen in paper [30]. Later, we can see that the minimum secret key rate can also be achieved by scanning T_y and ε_y simultaneously.

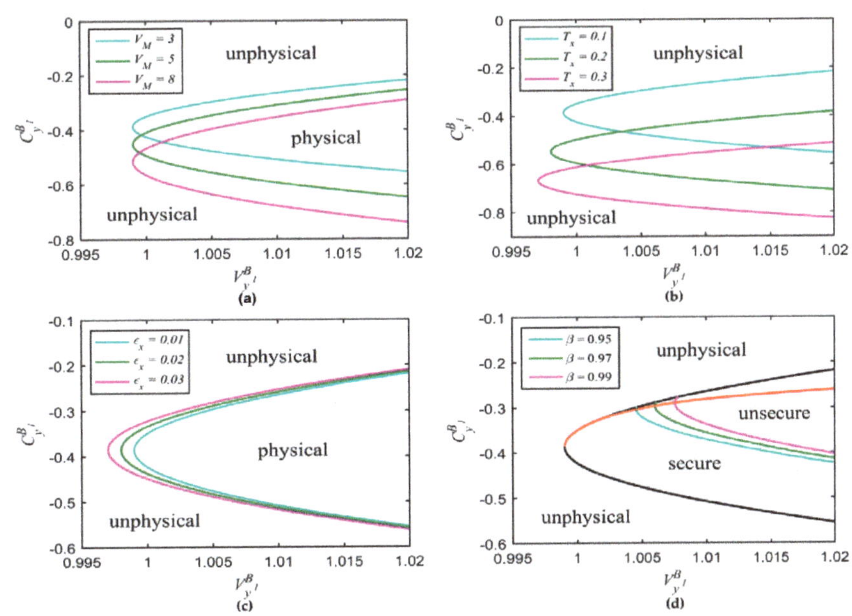

Figure 6. Ideal parabolic curve versus related parameters. (**a**) Different modulation variance values with $T_x = 0.1$ and $\varepsilon_x = 0.01$; (**b**) Different transmission efficiency values with $\varepsilon_x = 0.01$ and $V_M = 3$; (**c**) Different excess noise values with $T_x = 0.1$ and $V_M = 3$; (**d**) Different reconciliation efficiency values with $T_x = 0.1$, $\varepsilon_x = 0.01$, and $V_M = 3$.

Furthermore, if we assume $C_y^{B_1} = -\sqrt{T_y V_M}(1+V_M)^{-1/4}$ and $V_y^{B_1} = 1 + T_y\varepsilon_y$, the parabolic equations (Equation (25)), as determined by the Heisenberg uncertainty relation under ideal and realistic detection conditions transform into

$$\begin{cases} (k\sqrt{T_x} - \sqrt{T_y})^2 \leq (1 - kT_x)(1 + T_y\varepsilon_y - k) \\ (k'\sqrt{T_x} - \sqrt{T_y})^2 \leq (1 - k'T_x\eta)(1 + T_y\varepsilon_y - k'/\eta + \chi_{\text{hom}}) \end{cases}, \tag{26}$$

where $k = \frac{1}{T_x(1+\chi_{\text{linex}})}$ and $k' = \frac{1}{\eta T_x(1+\chi_{\text{totx}})}$. By this way, more details about eavesdropping method taken by Eve can be found. Moreover, one can easily see that the transformed equations do not depend on V_M. We redraw the physical regions delimited by the new curves for different values of detection efficiency and electronic noise as shown in Figure 7. We obtain the same rule as in Figure 5 that the physical region gradually decreases as the detection efficiency increases and the electronic noise decreases. Secure and unsecure regions under the realistic detection condition are shown in Figure 8. The cyan curve with the secret key rate of zero represents the boundary of two regions. Although the parameters T_y and ε_y are unknown, they are confined to the curve $V_y^{B_1} = 1 + T_y\varepsilon_y$, which can be estimated by randomly measuring the phase quadrature in Bob's side, meaning that T_y and ε_y cannot be set simultaneously in other physical places outside this curve. We can see that Eve essentially changes the value of the parameter $C_y^{B_1}$ by controlling the value of T_y. For a constant value of $V_y^{B_1}$, we can calculate the minimum secret key rate by scanning T_y or ε_y in the physical region. As shown in Figure 8, the curve corresponding to the minimum secret key rate is divided into three parts. The red curve part overlaps with the left boundary of the black solid curve which corresponds to the black solid curve in Figure 6d. As the value of $V_y^{B_1}$ increases, the worst-case T_y and ε_y (green curve part) gradually separate from the black solid curve, meaning that the secret key rate of the protocol is not always monotonically decreasing as ε_y increases or T_y decreases, but still lie in the secure region. The blue curve represents the part where the minimum secret key rate is less than zero. We also find that this minimum secret key rate is equal to the minimum secret key rate that was obtained by scanning $C_y^{B_1}$ (corresponding to the red solid line of Figure 6d) when other parameter values are set to be consistent.

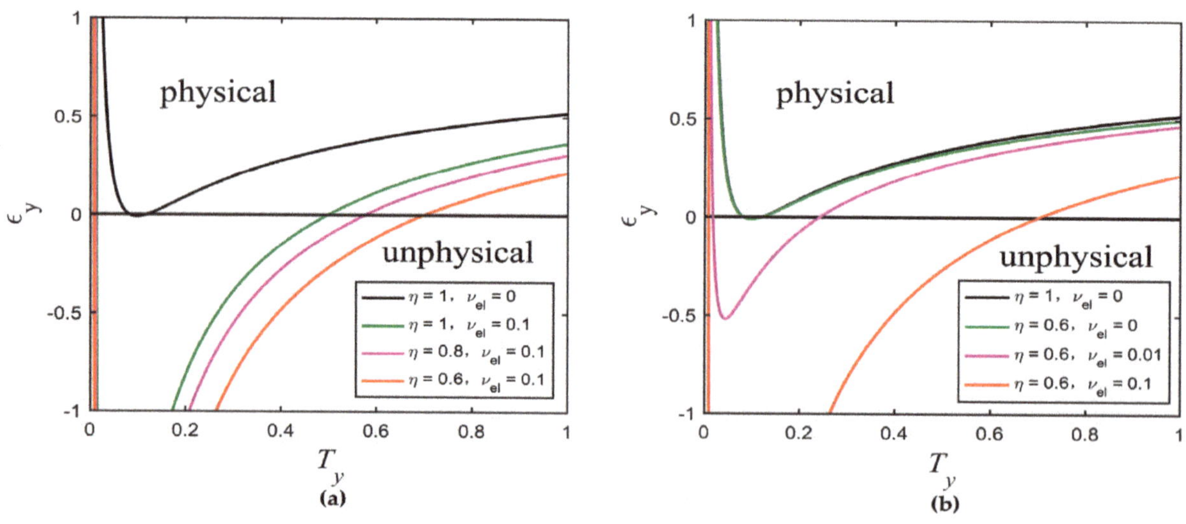

Figure 7. Comparison among physical regions delimited by the new curves of the UD protocol. **(a)** Changes of the physical region according to different values of η (v_{el} remains constant); **(b)** Changes of the physical region according to different values of v_{el} (η remains constant). The other parameters are $\beta = 0.99$, $T_x = 0.1$, $\varepsilon_x = 0.01$, and $V_M = 3$.

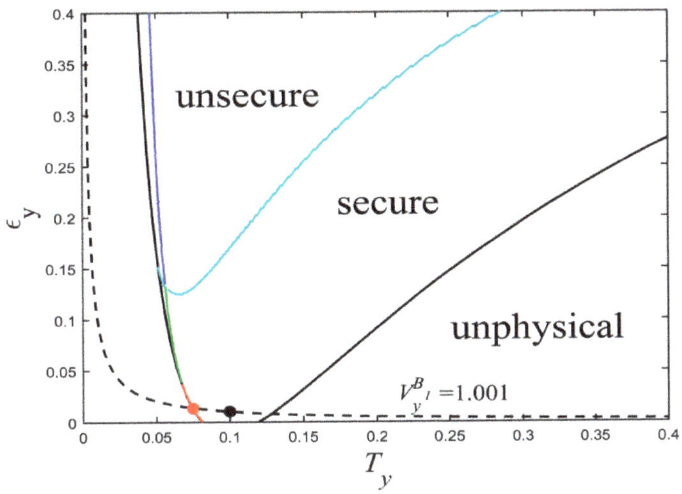

Figure 8. Secure and unsecure regions of the UD protocol under realistic detection condition. The parameters are set to $\beta = 0.99$, $V_M = 3$, $T_x = 0.1$, $\varepsilon_x = 0.01$, $\eta = 0.6$, and $v_{el} = 0.1$.

In typical communication channels, the value of $V_y^{B_1}$ can be estimated by setting $V_y^{B_1} \approx 1 + T_x\varepsilon_x = 1.001$, which is plotted with the black dashed line of Figure 8. At the black point, the conditions $T_x = T_y$ and $\varepsilon_x = \varepsilon_y$ are satisfied. The red point represents the worst-case T_y and ε_y, which is the intersection of the red line and black dashed line. Because Eve can distinguishes T_y, ε_y from T_x, ε_x by measuring coherent states sent by Alice, she can arbitrarily change the values of both T_y and ε_y, while keeps $V_y^{B_1}$ unchanged, eventually, obtains more information. If Alice and Bob use T_x and ε_x to estimate T_y and ε_y (black point), then this will underestimate the ability of the eavesdropper Eve and provide security loopholes. Therefore, here we should consider the minimum secret key rate (red point).

In Figure 9, the curves representing the maximal tolerable excess noise versus the channel losses under ideal and realistic detection conditions are shown. We observe that the UD protocol has a lower tolerance to the excess noise than the SY protocol. However, the UD protocol reduces the complexity of the experiment and still provides a reasonable secure region (all of the parameters are set under the actual conditions).

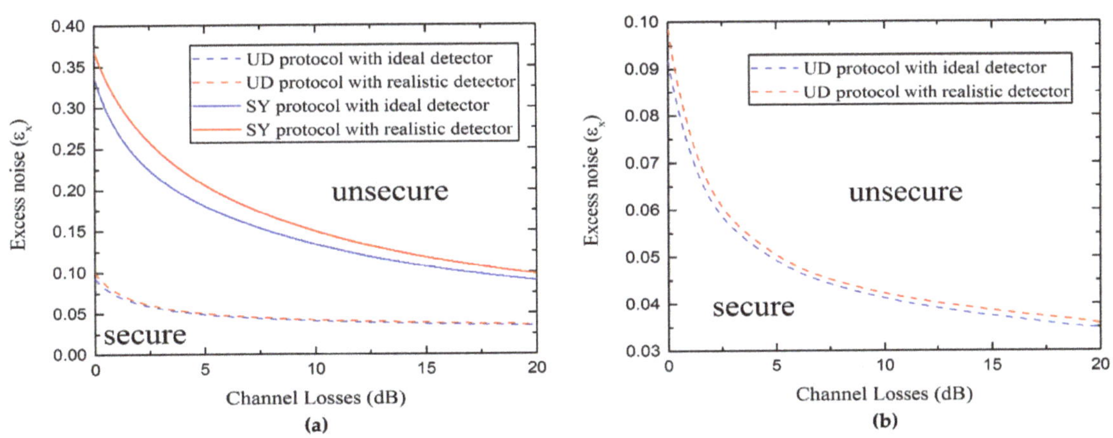

Figure 9. (**a**) Comparison between secure and unsecure regions for the SY coherent-state protocol and UD coherent-state protocol under different detection conditions; (**b**) Secure and unsecure regions of the UD protocol using an ideal homodyne detector ($\eta = 1, v_{el} = 0$) and a realistic one ($\eta = 0.6, v_{el} = 0.1$). Here we consider $V_M = 3$, $\beta = 0.99$, and the estimated value $V_y^{B_1} \approx 1 + T_x\varepsilon_x$.

In addition, from Figure 9b, it is not difficult to find out that the realistic Bob's detection can slightly increase the secure region of the UD protocol. This effect can be explained by considering the fact that the noise added on Bob's side not only affects Alice's and Bob's mutual information, but also decreases Eve's information in reverse reconciliation. Due to the detection at Bob's side, which can be controlled and observed by Bob, the noise added on Bob's side could be considered as a believable noise not controlled by the eavesdropper Eve. Moreover, it is found that there is an optimal noise χ_{hom} (characterized by the detection efficiency η and electronic noise v_{el}) that Bob needs to add to maximize the secret key rate for each channel loss. Then, we can effectively improve the secret key rate and increase the transmission distance by adding proper noise to Bob's side, as we show in Figure 10.

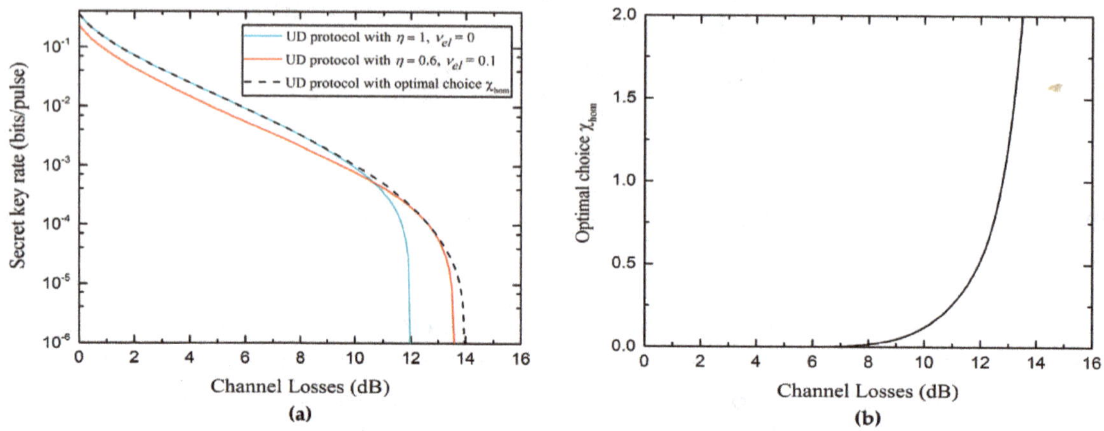

Figure 10. (**a**) Minimum secret key rate as a function of the channel losses; (**b**) Optimal choice of χ_{hom} that maximizes the secret key rate in (**a**). The other parameters are $\beta = 0.99$, $\varepsilon_x = 0.04$, $V_M = 3$, and $V_y^{B_1} \approx 1 + T_x\varepsilon_x$.

4. Conclusions

In this paper, we have proven the equivalence between the EB scheme and the PM scheme of the UD CV-QKD protocol, and investigated the physical and secure regions of the SY coherent-state protocol based on the Heisenberg uncertainty relation. It was shown that the realistic detection condition in UD protocol results in an excess pseudo physical region, which corresponds to the unphysical attack of Eve. In order to ensure the physicality, we should select the physical region delimited by the ideal curve. We also found that a realistic Bob's detection improves the resistance of the protocol to the channel excess noise, therefore, the performance in terms of the secret key rates and transmission distances of the UD coherent-state protocol can be improved by adding an optimal noise to the reconciliation side. Overall, the results confirm the potential of a long-distance secure communication through the usage of the UD CV-QKD protocol.

Acknowledgments: This work was supported by the Key Project of the Ministry of Science and Technology of China (2016YFA0301403), the National Natural Science Foundation of China (NSFC) (Grants No. 11504219, No. 61378010), the Shanxi 1331KSC, and the Program for the Outstanding Innovative Teams of Higher Learning Institutions of Shanxi.

Author Contributions: Pu Wang conceived the study, performed theoretical calculations and numerical simulations and drafted the article. Xuyang Wang designed the conception of the study, discussed the results, checked the draft and critically reviewed the manuscript. Yongmin Li proposed and supervised the project, checked the draft and provided a critical revision of the manuscript. All authors have read and approved the final manuscript.

References

1. Gisin, N.; Ribordy, G.; Tittel, W.; Zbinden, H. Quantum cryptography. *Rev. Mod. Phys.* **2002**, *74*, 145–195. [CrossRef]

2. Scarani, V.; Bechmann-Pasquinucci, H.; Cerf, N.J.; Dušek, M.; Lütkenhaus, N.; Peev, M. The security of practical quantum key distribution. *Rev. Mod. Phys.* **2009**, *81*, 1301–1350. [CrossRef]
3. Cerf, N.J.; Levy, M.; Van Assche, G. Quantum distribution of gaussian keys using squeezed states. *Phys. Rev. A* **2001**, *63*, 052311. [CrossRef]
4. Silberhorn, C.; Ralph, T.C.; Lütkenhaus, N.; Leuchs, G. Continuous variable quantum cryptography: Beating the 3 db loss limit. *Phys. Rev. Lett.* **2002**, *89*, 167901. [CrossRef]
5. Weedbrook, C.; Lance, A.M.; Bowen, W.P.; Symul, T.; Ralph, T.C.; Lam, P.K. Quantum cryptography without switching. *Phys. Rev. Lett.* **2004**, *93*, 170504. [CrossRef]
6. Fossier, S.; Diamanti, E.; Debuisschert, T.; Villing, A.; Tualle-Brouri, R.; Grangier, P. Field test of a continuous-variable quantum key distribution prototype. *New J. Phys.* **2009**, *11*, 045023. [CrossRef]
7. Leverrier, A.; Grangier, P. Continuous-variable quantum-key-distribution protocols with a non-gaussian modulation. *Phys. Rev. A* **2011**, *83*, 042312. [CrossRef]
8. Madsen, L.S.; Usenko, V.C.; Lassen, M.; Filip, R.; Andersen, U.L. Continuous variable quantum key distribution with modulated entangled states. *Nat. Commun.* **2012**, *3*, 1083. [CrossRef]
9. Wang, X.Y.; Bai, Z.L.; Du, P.Y.; Li, Y.M.; Peng, K.C. Ultrastable fiber-based time-domain balanced homodyne detector for quantum communication. *Chin. Phys. Lett.* **2012**, *29*, 124202. [CrossRef]
10. Weedbrook, C.; Pirandola, S.; García-Patrón, R.; Cerf, N.J.; Ralph, T.C.; Shapiro, J.H.; Lloyd, S. Gaussian quantum information. *Rev. Mod. Phys.* **2012**, *84*, 621–669. [CrossRef]
11. Wang, X.Y.; Bai, Z.L.; Wang, S.F.; Li, Y.M.; Peng, K.C. Four-state modulation continuous variable quantum key distribution over a 30-km fiber and analysis of excess noise. *Chin. Phys. Lett.* **2013**, *30*, 010305. [CrossRef]
12. Gehring, T.; Handchen, V.; Duhme, J.; Furrer, F.; Franz, T.; Pacher, C.; Werner, R.F.; Schnabel, R. Implementation of continuous-variable quantum key distribution with composable and one-sided-device-independent security against coherent attacks. *Nat. Commun.* **2015**, *6*, 8795. [CrossRef] [PubMed]
13. Zhang, Y.; Li, Z.; Weedbrook, C.; Marshall, K.; Pirandola, S.; Yu, S.; Guo, H. Noiseless linear amplifiers in entanglement-based continuous-variable quantum key distribution. *Entropy* **2015**, *17*, 4547. [CrossRef]
14. Li, H.S.; Wang, C.; Huang, P.; Huang, D.; Wang, T.; Zeng, G.H. Practical continuous-variable quantum key distribution without finite sampling bandwidth effects. *Opt. Express* **2016**, *24*, 20481–20493. [CrossRef]
15. Bai, D.Y.; Huang, P.; Ma, H.X.; Wang, T.; Zeng, G.H. Performance improvement of plug-and-play dual-phase-modulated quantum key distribution by using a noiseless amplifier. *Entropy* **2017**, *19*, 546. [CrossRef]
16. Bai, Z.L.; Yang, S.S.; Li, Y.M. High-efficiency reconciliation for continuous variable quantum key distribution. *Jpn. J. Appl. Phys.* **2017**, *56*, 044401. [CrossRef]
17. Grosshans, F.; Grangier, P. Continuous variable quantum cryptography using coherent states. *Phys. Rev. Lett.* **2002**, *88*, 057902. [CrossRef]
18. Grosshans, F.; Van Assche, G.; Wenger, J.; Brouri, R.; Cerf, N.J.; Grangier, P. Quantum key distribution using gaussian-modulated coherent states. *Nature* **2003**, *421*, 238–241. [CrossRef]
19. Iblisdir, S.; Van Assche, G.; Cerf, N.J. Security of quantum key distribution with coherent states and homodyne detection. *Phys. Rev. Lett.* **2004**, *93*, 170502. [CrossRef]
20. Grosshans, F. Collective attacks and unconditional security in continuous variable quantum key distribution. *Phys. Rev. Lett.* **2005**, *94*, 020504. [CrossRef] [PubMed]
21. Lance, A.M.; Symul, T.; Sharma, V.; Weedbrook, C.; Ralph, T.C.; Lam, P.K. No-switching quantum key distribution using broadband modulated coherent light. *Phys. Rev. Lett.* **2005**, *95*, 180503. [CrossRef] [PubMed]
22. Lodewyck, J.; Bloch, M.; Garcia-Patron, R.; Fossier, S.; Karpov, E.; Diamanti, E.; Debuisschert, T.; Cerf, N.J.; Tualle-Brouri, R.; McLaughlin, S.W.; et al. Quantum key distribution over 25 km with an all-fiber continuous-variable system. *Phys. Rev. A* **2007**, *76*, 042503. [CrossRef]
23. Qi, B.; Huang, L.L.; Qian, L.; Lo, H.K. Experimental study on the gaussian-modulated coherent-state quantum key distribution over standard telecommunication fibers. *Phys. Rev. A* **2007**, *76*, 052323. [CrossRef]
24. Yang, S.S.; Bai, Z.L.; Wang, X.Y.; Li, Y.M. FPGA-based implementation of size-adaptive privacy amplification in quantum key distribution. *Photonics J.* **2017**, *9*, 7600308. [CrossRef]
25. Jouguet, P.; Kunz-Jacques, S.; Leverrier, A.; Grangier, P.; Diamanti, E. Experimental demonstration of long-distance continuous-variable quantum key distribution. *Nat. Photonics* **2013**, *7*, 378–381. [CrossRef]

26. Li, Y.M.; Wang, X.Y.; Bai, Z.L.; Liu, W.Y.; Yang, S.S.; Peng, K.C. Continuous variable quantum key distribution. *Chin. Phys. B* **2017**, *26*, 040303. [CrossRef]
27. Liu, W.Y.; Wang, X.Y.; Wang, N.; Du, S.N.; Li, Y.M. Imperfect state preparation in continuous-variable quantum key distribution. *Phys. Rev. A* **2017**, *96*, 042312. [CrossRef]
28. Usenko, V.C.; Grosshans, F. Unidimensional continuous-variable quantum key distribution. *Phys. Rev. A* **2015**, *92*, 062337. [CrossRef]
29. Wang, X.Y.; Liu, W.Y.; Wang, P.; Li, Y.M. Experimental study on all-fiber-based unidimensional continuous-variable quantum key distribution. *Phys. Rev. A* **2017**, *95*, 062330. [CrossRef]
30. Wang, P.; Wang, X.Y.; Li, J.Q.; Li, Y.M. Finite-size analysis of unidimensional continuous-variable quantum key distribution under realistic conditions. *Opt. Express* **2017**, *25*, 27995–28009. [CrossRef]
31. Braunstein, S.L.; van Loock, P. Quantum information with continuous variables. *Rev. Mod. Phys.* **2005**, *77*, 513–577. [CrossRef]
32. Serafini, A. Detecting entanglement by symplectic uncertainty relations. *J. Opt. Soc. Am. B* **2007**, *24*, 347–354. [CrossRef]
33. Grosshans, F.; Cerf, N.J.; Wenger, J.; Tualle-Brouri, R.; Grangier, P. Virtual entanglement and reconciliation protocols for quantum cryptography with continuous variables. *Quantum Inform. Comput.* **2003**, *3*, 535–552.
34. Ben-Israel, A.; Greville, T.N.E. *Generalized Inverses: Theory and Applications*, 2nd ed.; Springer: New York, NY, USA, 2003.
35. Serafini, A.; Paris, M.G.A.; Illuminati, F.; Siena, S.D. Quantifying decoherence in continuous variable systems. *J. Opt. B* **2005**, *7*, R19. [CrossRef]
36. Garcia-Patron, R.; Cerf, N.J. Continuous-variable quantum key distribution protocols over noisy channels. *Phys. Rev. Lett.* **2009**, *102*, 130501. [CrossRef]
37. Milicevic, M.; Feng, C.; Zhang, L.M.; Gulak, P.G. Key reconciliation with low-density parity-check codes for long-distance quantum cryptography. *arXiv* **2017**, arXiv:1702.07740.

10

A New Mechanism of Open System Evolution and its Entropy using Unitary Transformations in Noncomposite Qudit Systems

Julio A. López-Saldívar [1,2,*], Octavio Castaños [1], Margarita A. Man'ko [3] and Vladimir I. Man'ko [2,3,4]

1 Instituto de Ciencias Nucleares, Universidad Nacional Autónoma de México, Apdo. Postal 70-543, Ciudad de México 04510, Mexico
2 Moscow Institute of Physics and Technology, State University, Institutskii Per. 9, Dolgoprudnyi, Moscow 141700, Russia
3 Lebedev Physical Institute, Leninskii Prospect 53, Moscow 119991, Russia
4 Department of Physics, Tomsk State University, Lenin Avenue 36, Tomsk 634050, Russia
* Correspondence: julio.lopez.8303@gmail.com

Abstract: The evolution of an open system is usually associated with the interaction of the system with an environment. A new method to study the open-type system evolution of a qubit (two-level atom) state is established. This evolution is determined by a unitary transformation applied to the qutrit (three-level atom) state, which defines the qubit subsystems. This procedure can be used to obtain different qubit quantum channels employing unitary transformations into the qutrit system. In particular, we study the phase damping and spontaneous-emission quantum channels. In addition, we mention a proposal for quasiunitary transforms of qubits, in view of the unitary transform of the total qutrit system. The experimental realization is also addressed. The probability representation of the evolution and its information-entropic characteristics are considered.

Keywords: entropy; open systems; unitary evolution; qubit; qutrit

---

1. Introduction

The open system evolution of a qudit state is known to be the result of interactions with an environment. Usually, the states of the complete system are thought to evolve by a unitary transformation in the Hilbert space $\hat{H} = \hat{H}_q \otimes \hat{H}_{env}$, then the density operator of the composite system leads us, using the partial tracing procedure, to the density operator of the subsystem $\hat{\rho}_q$ (qudit), and its evolution is induced by the unitary evolution of the complete system. In this picture, the qubit state dynamics needs the structure of the Hilbert space $\hat{H}$ corresponding to the presence of two subsystems, qudit and environment [1]. In this work, we suggest a new mechanism to study the open system evolution, which does not demand the complete system to have a subsystem.

We show that for any system without subsystems, there exist a unitary evolution, which due to hidden correlations in the system, evolves according to the Gorini–Kossakowski–Sudarshan–Lindblad equation [2–5]. We demonstrate this picture using the example of a qutrit (complete system without subsystems), where the open-like evolution is available for their associated qubits.

In previous works [6–10], a new method to define different qubit density matrices from a qudit system was established. This procedure uses the occupation probabilities and transition probability amplitudes for different levels of a qudit system and groups them as if there exists two levels only. This is done by mapping the qudit density matrix to the closest higher even-dimensional density matrix. The partial trace operation then is enacted on the resulting matrix in order to obtain well-defined qubit density matrices.

The obtained qubits have been used to define a new geometric representation of the d-dimensional qudit states through d Bloch vectors [10] associated with the generated qubits. Furthermore, it has been possible to describe quantum phenomena as the entanglement on a two-qubit system in terms of standard probabilities [9].

The evolution of a qutrit density matrix can provide the quantum channel, which maps the initial state $\hat{\rho}_a$ onto the density matrix $\hat{\rho}'_a$. The proposed open-type evolution establishes a new mechanism, which will need a special state preparation and a specific unitary operation for the qutrit system, as we will show later on. The experimental possibilities by which one can realize this new mechanism are related to superconducting circuit devices [11,12].

Most quantum computing processes consider a set of pure qubit states, which are transformed by unitary operators, also called gates, that are used to implement different computing algorithms. In this article, instead, we have density matrices (which might be describing a mixed state) of larger qudit systems. The definition of a set of qubit states from a qudit system is similar to the ideas established in [13], where the emulation of a spin system was obtained from qudit states, and in [14], where the quantum logic of qubits was simplified by the use of a higher dimensional Hilbert space; and in general, with all the procedures that make use of larger Hilbert spaces. In this work, we demonstrate that subsystems of qubits defined by larger systems can be used in quantum information. A principal foundation of quantum computation is the study of quantum channels. These channels are linked to unitary transformations of the qubit density matrix. There exist several channels that can describe the interaction between a quantum system and its environment such as the bit-flip, depolarization, spontaneous emission, phase, and amplitude damping channels. For this, the study of quantum channels has been of relevance in the error correction theory of quantum computation [15,16].

Here, we present different examples of quantum channels, which act on the associated qubits to qudit states. These quantum channels have the advantage of being represented as unitary transformations acting in the qudit system, providing the possibility to study the qubits as if they were interacting with an environment.

On the other hand, the study of the interaction of three-level systems with electromagnetic fields has led to the discovery of important phenomena, such as the presence of dark states [17] together with black resonances [18] and electromagnetically-induced transparency [19–21]. This is important to our objectives as in some cases, the herein proposed qubit quantum channels can be obtained by a unitary transformation of dark states, suggesting the possibility of checking our results experimentally.

The work is organized as follows: In Section 2, a review of the qubit density matrices that are associated with a qutrit state is given. Furthermore, the association of a unitary transform of the qutrit to the nonunitary transformations of the qubits is studied. In Section 3, the definitions of the qubit phase damping and spontaneous-emission quantum channels are reviewed. Later, the unitary transformations of a qutrit system are explicitly given, which yields the phase damping and spontaneous-emission channels on the associated qubits. A way to obtain a quasi-unitary transformation on the qubits is also explored. The change of entropy associated with the nonunitary evolution of the qubits is discussed in Section 4. Finally, some concluding remarks are given.

2. Nonunitary Evolution for the Qubit Decomposition of Qutrit States

In a previous work [10], we showed the existence of six different qubit states associated with a general qutrit density matrix:

$$\hat{\rho} = \begin{pmatrix} \rho_{11} & \rho_{12} & \rho_{13} \\ \rho_{21} & \rho_{22} & \rho_{23} \\ \rho_{31} & \rho_{32} & \rho_{33} \end{pmatrix}.$$

To define these states, different maps of $\hat{\rho}$ to a 4×4 density matrix, with one row and one column equal to zero (in such a way that ensures an eigenvalue equal to zero), were used. Then, the partial trace of the resulting 4×4 matrix was performed as if it was describing a two-qubit system. The obtained qubit partial density operators can be explicitly written as:

$$\hat{\rho}_1 = \begin{pmatrix} 1-\rho_{33} & \rho_{13} \\ \rho_{31} & \rho_{33} \end{pmatrix}, \quad \hat{\rho}_2 = \begin{pmatrix} 1-\rho_{22} & \rho_{12} \\ \rho_{21} & \rho_{22} \end{pmatrix}, \quad \hat{\rho}_3 = \begin{pmatrix} \rho_{11} & \rho_{13} \\ \rho_{31} & 1-\rho_{11} \end{pmatrix},$$
$$\hat{\rho}_4 = \begin{pmatrix} \rho_{22} & \rho_{23} \\ \rho_{32} & 1-\rho_{22} \end{pmatrix}, \quad \hat{\rho}_5 = \begin{pmatrix} \rho_{11} & \rho_{12} \\ \rho_{21} & 1-\rho_{11} \end{pmatrix}, \quad \hat{\rho}_6 = \begin{pmatrix} 1-\rho_{33} & \rho_{23} \\ \rho_{32} & \rho_{33} \end{pmatrix}. \tag{1}$$

The qubit states can be characterized in different sets by their corresponding von Neumann entropy $S_k = -\mathrm{Tr}\,\rho_k \ln \rho_k$, with $k = 1, 2, \ldots, 6$. These qubits correspond to the reduction of the three-level system to different two-level systems by the summation of the population probabilities of two levels into one.

When the qutrit state is transformed using a general three-dimensional unitary matrix $\hat{U}$, i.e., $\hat{\rho}' = \hat{U}^\dagger \hat{\rho}\, \hat{U}$, the qubits in Equation (1) are transformed in a nonunitary way. The transformed qubit density matrices can be written by the following expressions:

$$\begin{aligned}
\hat{\rho}'_1 &= \frac{1}{D}\begin{pmatrix} D - M_{3,1}N_{1,3} + M_{2,1}N_{2,3} - M_{1,1}N_{3,3} & M_{3,3}N_{1,3} - M_{2,3}N_{2,3} + M_{1,3}N_{3,3} \\ M_{3,1}N_{1,1} - M_{2,1}N_{2,1} + M_{1,1}N_{3,1} & M_{3,1}N_{1,3} - M_{2,1}N_{2,3} + M_{1,1}N_{3,3} \end{pmatrix}, \\
\hat{\rho}'_2 &= \frac{1}{D}\begin{pmatrix} D + M_{3,2}N_{1,2} - M_{2,2}N_{2,2} + M_{1,2}N_{3,2} & M_{3,3}N_{1,2} - M_{2,3}N_{2,2} + M_{1,3}N_{3,2} \\ -M_{3,2}N_{1,1} + M_{2,2}N_{2,1} - M_{1,2}N_{3,1} & -M_{3,2}N_{1,2} + M_{2,2}N_{2,2} - M_{1,2}N_{3,2} \end{pmatrix}, \\
\hat{\rho}'_3 &= \frac{1}{D}\begin{pmatrix} M_{3,3}N_{1,1} - M_{2,3}N_{2,1} + M_{1,3}N_{3,1} & M_{3,3}N_{1,3} - M_{2,3}N_{2,3} + M_{1,3}N_{3,3} \\ M_{3,1}N_{1,1} - M_{2,1}N_{2,1} + M_{1,1}N_{3,1} & D - M_{3,3}N_{1,1} + M_{2,3}N_{2,1} - M_{1,3}N_{3,1} \end{pmatrix}, \\
\hat{\rho}'_4 &= \frac{1}{D}\begin{pmatrix} -M_{3,2}N_{1,2} + M_{2,2}N_{2,2} - M_{1,2}N_{3,2} & -M_{3,2}N_{1,3} + M_{2,2}N_{2,3} - M_{1,2}N_{3,3} \\ M_{3,1}N_{1,2} - M_{2,1}N_{2,2} + M_{1,1}N_{3,2} & D + M_{3,2}N_{1,2} - M_{2,2}N_{2,2} + M_{1,2}N_{3,2} \end{pmatrix}, \\
\hat{\rho}'_5 &= \frac{1}{D}\begin{pmatrix} M_{3,3}N_{1,1} - M_{2,3}N_{2,1} + M_{1,3}N_{3,1} & M_{3,3}N_{1,2} - M_{2,3}N_{2,2} + M_{1,3}N_{3,2} \\ -M_{3,2}N_{1,1} + M_{2,2}N_{2,1} - M_{1,2}N_{3,1} & D - M_{3,3}N_{1,1} + M_{2,3}N_{2,1} - M_{1,3}N_{3,1} \end{pmatrix}, \\
\hat{\rho}'_6 &= \frac{1}{D}\begin{pmatrix} D - M_{3,1}N_{1,3} + M_{2,1}N_{2,3} - M_{1,1}N_{3,3} & -M_{3,2}N_{1,3} + M_{2,2}N_{2,3} - M_{1,2}N_{3,3} \\ M_{3,1}N_{1,2} - M_{2,1}N_{2,2} + M_{1,1}N_{3,2} & M_{3,1}N_{1,3} - M_{2,1}N_{2,3} + M_{1,1}N_{3,3} \end{pmatrix},
\end{aligned} \tag{2}$$

where $N_{jk} = (\hat{\rho}\hat{U})_{jk}$, D is the determinant of $\hat{U}$, and M_{jk} are the components of the minors of matrix $\hat{U}$, i.e., its elements are the determinants after eliminating the $(4-j)^{\text{th}}$ row and $(4-k)^{\text{th}}$ column of $\hat{U}$. The transformed states are characterized into different sets by their corresponding transformed entropies $S'_k = -\mathrm{Tr}\,\rho'_k \ln \rho'_k$. We emphasize that the resulting qubit density matrices are associated, in general, with a nonunitary evolution of the original qubits. This fact establishes a new mechanism to obtain the open-like system evolution in a noncomposite qutrit system. Additionally, this procedure can be extended to any qudit system, in view of the general definition of the qubit density matrices obtained from a qudit system [10].

In [9], we discussed that a two-qubit density matrix with one of its rows and columns equal to zero describes separable states, if one of the off-diagonal terms is equal to zero, for example, the state:

$$\hat{\rho} = \begin{pmatrix} \rho_{11} & \rho_{12} & \rho_{13} & 0 \\ \rho_{21} & \rho_{22} & \rho_{23} & 0 \\ \rho_{31} & \rho_{32} & \rho_{33} & 0 \\ 0 & 0 & 0 & 0 \end{pmatrix}$$

is separable iff $\rho_{23} = 0$. To show this, one can consider the previous density matrix to be in the standard two-qubit representation $|00\rangle$, $|01\rangle$, $|10\rangle$, and $|11\rangle$. It can be seen that the partial transpose operation [22] implies the change $\rho_{12} \leftrightarrow \rho_{21}$, and for this reason, the eigenvalues of $\hat{\rho}$ with $\rho_{23} = 0$ are equal to the eigenvalues of its partial transpose. As the partial transpose is a nonnegative operator, then the system is separable. The separability implies the invariance of the partial density matrices under local unitary transformations. As this two-qubit density matrix has a pair of row-column with a diagonal term equal to zero, the correspondence with a qutrit density matrix can be made. On the other hand, the correspondence between two-qubit local unitary transformations and qutrit unitary transformations can be made in the same way, e.g., by eliminating one row and one column of the two-qubit local transformation. This procedure allows us to define different unitary transformations that almost leave the qubits in Expression (1) invariant.

3. Phase Damping and Spontaneous-Emission Channels

It is known that the interaction of a qubit system with an environment leads to several physical phenomena such as dissipation and decoherence in the qubit subsystem; an example of these interactions is the phase damping channel. In this channel, the evolution of the qubit plus environment $(|\cdots\rangle_q|\cdots\rangle_e)$ is given by a unitary transformation $\hat{T}$, which acts differently if the qubit is in the ground or excited state, according to the following rules: $\hat{T}(|0\rangle_q|0\rangle_e) = \sqrt{1-p}|0\rangle_q|0\rangle_e + \sqrt{p}|0\rangle_q|1\rangle_e$ and $\hat{T}(|1\rangle_q|0\rangle_e) = \sqrt{1-p}|0\rangle_q|0\rangle_e + \sqrt{p}|0\rangle_q|2\rangle_e$ with p being a probability, i.e., the environment subsystem goes to a superposition of the states $(|0\rangle_e, |1\rangle_e)$, or to $(|0\rangle_e, |2\rangle_e)$, if the environment is in $|0\rangle_e$, or $|1\rangle_e$, respectively [15,23]. This two-qubit unitary transformations result in a nonunitary change when the partial trace over the environment subsystem is taken:

$$\begin{pmatrix} 1-\rho_{22} & \rho_{12} \\ \rho_{12}^* & \rho_{22} \end{pmatrix} \to \begin{pmatrix} 1-\rho_{22} & \rho_{12}(1-p) \\ \rho_{12}^*(1-p) & \rho_{22} \end{pmatrix}.$$

When the map is applied a very large number of times ($\to \infty$), it is straightforward that the initial state tends to the completely decoherent state:

$$\begin{pmatrix} 1-\rho_{22} & \rho_{12} \\ \rho_{12}^* & \rho_{22} \end{pmatrix} \to \begin{pmatrix} 1-\rho_{22} & 0 \\ 0 & \rho_{22} \end{pmatrix},$$

with an exponential convergence.

The other example is the spontaneous-emission (also called the amplitude-damping) quantum channel. In this channel, the dynamics of the qubit system plus the environment is determined by a unitary transform $\hat{T}$, which only acts if the qubit system is in the excited state $|1\rangle_q$, according to the following rules: $\hat{T}(|0\rangle_q|0\rangle_e) = |0\rangle_q|0\rangle_e$ and $\hat{T}(|1\rangle_q|0\rangle_e) = \sqrt{1-p}|1\rangle_q|0\rangle_e + \sqrt{p}|0\rangle_q|1\rangle_e$, where p is the probability [15,23]. This channel then defines a nonunitary evolution over the qubit subsystem, which transforms the qubit density matrix as follows:

$$\begin{pmatrix} 1-\rho_{22} & \rho_{12} \\ \rho_{12}^* & \rho_{22} \end{pmatrix} \to \begin{pmatrix} 1-(1-p)\rho_{22} & \rho_{12}\sqrt{1-p} \\ \rho_{12}^*\sqrt{1-p} & (1-p)\rho_{22} \end{pmatrix}.$$

If this channel is applied a very large number of times ($\to \infty$), the density matrix converges to a ground state, i.e.,

$$\begin{pmatrix} 1-\rho_{22} & \rho_{12} \\ \rho_{12}^* & \rho_{22} \end{pmatrix} \to \begin{pmatrix} 1 & 0 \\ 0 & 0 \end{pmatrix}.$$

In addition to these examples, there exists another type of quantum channel defined in the theory of interaction between a quantum system and an environment, which can be considered [15,23].

It is possible to demonstrate that phase damping and spontaneous-emission quantum channels for qubits $\hat{\rho}_1, \ldots, \hat{\rho}_6$ in Equation (1) can be obtained by the use of particular unitary transformations of a qutrit state $\hat{\rho}$. To justify this, we assumed a two-qubit quantum system where one of the levels cannot be populated, i.e., the 4×4 density matrix has an eigenvalue equal to zero, e.g.,

$$\hat{\rho} = \begin{pmatrix} \rho_{11} & \rho_{12} & \rho_{13} & 0 \\ \rho_{21} & \rho_{22} & 0 & 0 \\ \rho_{31} & 0 & \rho_{33} & 0 \\ 0 & 0 & 0 & 0 \end{pmatrix}; \tag{3}$$

it is clear that this density matrix is separable since $\rho_{23} = \rho_{32}^* = 0$. The partial density matrices can be operated locally by unitary transformations of the form $\hat{u}_1 \otimes \hat{u}_2$. When only one of the qubits is operated, i.e., when the unitary matrix corresponds to a controlled operation [15]: $\hat{u}_1 = \hat{I}$ or $\hat{u}_2 = \hat{I}$. If $\hat{u}_2 = \hat{I}$, then the unitary transformation only operates over the second qubit,

$$\hat{u} = \begin{pmatrix} u_{11} & u_{12} & 0 & 0 \\ u_{21} & u_{22} & 0 & 0 \\ 0 & 0 & u_{11} & u_{12} \\ 0 & 0 & u_{21} & u_{22} \end{pmatrix}. \tag{4}$$

By means of this type of unitary matrix, one can define an operation in the qutrit system that approximately only affects $\hat{\rho}_2$. This is done by ignoring the fourth row and the fourth column of (3); the resulting qutrit state is then operated by the unitary matrix resulting from the elimination of the fourth row and the fourth column of Equation (4). For the operator to be still unitary, the $(3,3)$ entry must be replaced by one. Following these and other analogous arguments, we study the application of the unitary transforms:

$$\hat{U}_1 = \begin{pmatrix} u_{11} & u_{12} & 0 \\ u_{21} & u_{22} & 0 \\ 0 & 0 & 1 \end{pmatrix}, \quad \hat{U}_2 = \begin{pmatrix} u_{11} & 0 & u_{12} \\ 0 & 1 & 0 \\ u_{21} & 0 & u_{22} \end{pmatrix}, \quad \hat{U}_3 = \begin{pmatrix} 1 & 0 & 0 \\ 0 & u_{11} & u_{12} \\ 0 & u_{21} & u_{22} \end{pmatrix} \tag{5}$$

on the qutrit density matrices:

$$\hat{\sigma}_1 = \begin{pmatrix} \rho_{11} & \rho_{12} & \rho_{13} \\ \rho_{21} & \rho_{22} & 0 \\ \rho_{31} & 0 & \rho_{33} \end{pmatrix}, \quad \hat{\sigma}_2 = \begin{pmatrix} \rho_{11} & 0 & \rho_{13} \\ 0 & \rho_{22} & \rho_{23} \\ \rho_{31} & \rho_{32} & \rho_{33} \end{pmatrix}, \quad \hat{\sigma}_3 = \begin{pmatrix} \rho_{11} & \rho_{12} & 0 \\ \rho_{21} & \rho_{22} & \rho_{23} \\ 0 & \rho_{23} & \rho_{33} \end{pmatrix}. \tag{6}$$

The unitary transformations in Equation (5) can be enacted on any of the density matrices in Equation (6), which define a nonunitary transformation of the qubits defined in Equation (1). These qubit transformations are found by the substitution of Equations (5) and (6) into Equation (2), e.g., the unitary transformation $\hat{U}_1^\dagger \hat{\sigma}_1 \hat{U}_1$ results in the following transformations of the qubits:

$$\begin{aligned}
\hat{\rho}_1' &= \begin{pmatrix} 1-\rho_{33} & \rho_{13}\,u_{11}^* \\ \rho_{31}\,u_{11} & \rho_{33} \end{pmatrix}, \\
\hat{\rho}_2' &= \begin{pmatrix} 1-u_{12}^*(\hat{\sigma}_1\hat{U}_1)_{12} - u_{22}^*(\hat{\sigma}_1\hat{U}_1)_{22} & u_{11}^*(\hat{\sigma}_1\hat{U}_1)_{12} + u_{21}^*(\hat{\sigma}_1\hat{U}_1)_{22} \\ u_{12}^*(\hat{\sigma}_1\hat{U}_1)_{11} + u_{22}^*(\hat{\sigma}_1\hat{U}_1)_{21} & u_{12}^*(\hat{\sigma}_1\hat{U}_1)_{12} + u_{22}^*(\hat{\sigma}_1\hat{U}_1)_{22} \end{pmatrix}, \\
\hat{\rho}_3' &= \begin{pmatrix} u_{11}^*(\hat{\sigma}_1\hat{U}_1)_{11} + u_{21}^*(\hat{\sigma}_1\hat{U}_1)_{21} & \rho_{13}\,u_{11}^* \\ \rho_{31}\,u_{11} & 1-u_{11}^*(\hat{\sigma}_1\hat{U}_1)_{11} - u_{21}^*(\hat{\sigma}_1\hat{U}_1)_{21} \end{pmatrix}, \\
\hat{\rho}_4' &= \begin{pmatrix} u_{12}^*(\hat{\sigma}_1\hat{U}_1)_{12} + u_{22}^*(\hat{\sigma}_1\hat{U}_1)_{22} & \rho_{13}u_{12}^* \\ \rho_{31}u_{12} & 1-u_{12}^*(\hat{\sigma}_1\hat{U}_1)_{12} - u_{22}^*(\hat{\sigma}_1\hat{U}_1)_{22} \end{pmatrix}, \\
\hat{\rho}_5' &= \begin{pmatrix} u_{11}^*(\hat{\sigma}_1\hat{U}_1)_{11} + u_{21}^*(\hat{\sigma}_1\hat{U}_1)_{21} & u_{11}^*(\hat{\sigma}_1\hat{U}_1)_{12} + u_{21}^*(\hat{\sigma}_1\hat{U}_1)_{22} \\ u_{12}^*(\hat{\sigma}_1\hat{U}_1)_{11}(r11u11+r12u21) + u_{22}^*(\hat{\sigma}_1\hat{U}_1)_{21} & 1-u_{11}^*(\hat{\sigma}_1\hat{U}_1)_{11} - u_{21}^*(\hat{\sigma}_1\hat{U}_1)_{21} \end{pmatrix}, \\
\hat{\rho}_6' &= \begin{pmatrix} 1-\rho_{33} & \rho_{13}\,u_{12}^* \\ \rho_{31}\,u_{12} & \rho_{33} \end{pmatrix}.
\end{aligned}$$

From these results, one can notice that the transformed qubits $\hat{\rho}'_1$ and $\hat{\rho}'_6$ correspond to the phase damping channel of $\hat{\rho}_1$ with different damping parameters. Furthermore, the qubit states $\hat{\rho}'_2$, $\hat{\rho}'_5$ can be seen as quasi-unitary transformations of the initial states $\hat{\rho}_2$, $\hat{\rho}_5$, respectively. In a similar way, one can obtain all the possible unitary transformations of the density matrices in Equation (6). These transformations lead to the identification of two types of quantum channels: the phase damping and a quasi-unitary operation described below.

The unitary transformation over the density matrices $\hat{\sigma}_1$, $\hat{\sigma}_2$, and $\hat{\sigma}_3$ results in a change over their associated qubits $\hat{\rho}_1, \dots, \hat{\rho}_6$, to $\hat{\rho}'_1, \dots, \hat{\rho}'_6$, which denote the qubits after the transformation. We have found the following interesting expressions:

$$\begin{aligned}
\hat{U}_1^\dagger \hat{\sigma}_1 \hat{U}_1 \Rightarrow \hat{\rho}'_1 &= \begin{pmatrix} 1-\rho_{33} & u_{11}^*\rho_{13} \\ u_{11}\rho_{31} & \rho_{33} \end{pmatrix}, \quad \hat{\rho}'_6 = \begin{pmatrix} 1-\rho_{33} & u_{12}^*\rho_{13} \\ u_{12}\rho_{31} & \rho_{33} \end{pmatrix}; \\
\hat{U}_2^\dagger \hat{\sigma}_1 \hat{U}_2 \Rightarrow \hat{\rho}'_2 &= \begin{pmatrix} 1-\rho_{22} & u_{11}^*\rho_{12} \\ u_{11}\rho_{21} & \rho_{22} \end{pmatrix}, \quad \hat{\rho}'_4 = \begin{pmatrix} 1-\rho_{33} & u_{12}^*\rho_{12} \\ u_{12}\rho_{21} & \rho_{33} \end{pmatrix}; \\
\hat{U}_2^\dagger \hat{\sigma}_2 \hat{U}_2 \Rightarrow \hat{\rho}'_2 &= \begin{pmatrix} 1-\rho_{22} & u_{21}^*\rho_{32} \\ u_{21}\rho_{23} & \rho_{22} \end{pmatrix}, \quad \hat{\rho}'_4 = \begin{pmatrix} \rho_{22} & u_{22}\rho_{23} \\ u_{22}^*\rho_{32} & 1-\rho_{22} \end{pmatrix}; \\
\hat{U}_3^\dagger \hat{\sigma}_2 \hat{U}_3 \Rightarrow \hat{\rho}'_3 &= \begin{pmatrix} \rho_{11} & u_{22}\rho_{13} \\ u_{22}^*\rho_{31} & 1-\rho_{11} \end{pmatrix}, \quad \hat{\rho}'_5 = \begin{pmatrix} \rho_{11} & u_{21}\rho_{13} \\ u_{21}^*\rho_{31} & 1-\rho_{11} \end{pmatrix}; \\
\hat{U}_1^\dagger \hat{\sigma}_3 \hat{U}_1 \Rightarrow \hat{\rho}'_1 &= \begin{pmatrix} 1-\rho_{33} & u_{21}^*\rho_{23} \\ u_{21}\rho_{32} & \rho_{33} \end{pmatrix}, \quad \hat{\rho}'_6 = \begin{pmatrix} 1-\rho_{33} & u_{22}^*\rho_{23} \\ u_{22}\rho_{32} & \rho_{33} \end{pmatrix}; \\
\hat{U}_3^\dagger \hat{\sigma}_3 \hat{U}_3 \Rightarrow \hat{\rho}'_3 &= \begin{pmatrix} \rho_{11} & u_{12}\rho_{12} \\ u_{12}^*\rho_{21} & 1-\rho_{11} \end{pmatrix}, \quad \hat{\rho}'_5 = \begin{pmatrix} \rho_{11} & u_{11}\rho_{12} \\ u_{11}^*\rho_{21} & 1-\rho_{11} \end{pmatrix}.
\end{aligned} \tag{7}$$

In most of the cases, the resulting qubits $\hat{\rho}'_j$ correspond to the phase damping quantum channel of $\hat{\rho}_j$, as can be seen in Expression (8). In this channel, the probability amplitudes given by the original off-diagonal terms of the qubits are multiplied by a number. The damping parameters are associated with different entries of the unitary transformation u_{jk}, which in general are complex numbers. When the unitary transformation correspond to a real matrix, then the expression for the standard phase damping map is obtained. As you can see in Equation (8), in some cases, the unitary transformations leads to the quantum channel of another qubit, e.g., after the application of $\hat{U}_1$ to $\hat{\sigma}_1$, the qubit $\hat{\rho}'_6$ is the phase damping channel of $\hat{\rho}_1$. Furthermore, in some other cases, the obtained density matrices correspond to transformations similar to the phase damping channel of matrices outside the ones in Equation (1), e.g., $\hat{\rho}'_4$ after the application of $\hat{U}_2$ to $\hat{\sigma}_1$. Although these matrices seem unrelated, they have the same form as the phase damping channel. In the case of $\hat{U}$ being a rotation matrix with a time-dependent angle $\theta = \omega t$, the original qubit states can be recovered at the time $t = 2\pi l/\omega, l = 0, 1, 2, \dots$.

The unitary transformations ($\hat{U}_1$, $\hat{U}_2$, $\hat{U}_3$) previously described can also lead to quasi-unitary transformations of the qubits. In particular, for the unitary transformation $\hat{U}_1^\dagger \hat{\sigma}_1 \hat{U}_1$, one gets the quasi-unitary transformations:

$$\begin{aligned}
\hat{\rho}'_2 &= \hat{U}^\dagger \hat{\rho}_2 \hat{U} + \rho_{33} \begin{pmatrix} |u_{12}|^2 & -u_{11}^* u_{12} \\ -u_{11} u_{12}^* & -|u_{12}|^2 \end{pmatrix}, \\
\hat{\rho}'_5 &= \hat{U}^\dagger \hat{\rho}_5 \hat{U} + \rho_{33} \begin{pmatrix} -|u_{21}|^2 & -u_{21}^* u_{22} \\ -u_{21} u_{22}^* & |u_{21}|^2 \end{pmatrix},
\end{aligned} \tag{8}$$

with $\hat{\mathcal{U}} = \begin{pmatrix} u_{11} & u_{12} \\ u_{21} & u_{22} \end{pmatrix}$ being a two-dimensional unitary transformation. For the other qubits, one can also define quasi-unitary transformations as follows:

(a) From the qutrit unitary transformation $\hat{U}_1^\dagger \hat{\sigma}_3 \hat{U}_1$,

$$\begin{aligned} \hat{\rho}_2' &= \hat{\mathcal{U}}^\dagger \hat{\rho}_2 \hat{\mathcal{U}} + \rho_{33} \begin{pmatrix} -|u_{12}|^2 & u_{11}^* u_{12} \\ u_{11} u_{12}^* & |u_{12}|^2 \end{pmatrix}, \\ \hat{\rho}_5' &= \hat{\mathcal{U}}^\dagger \hat{\rho}_5 \hat{\mathcal{U}} + \rho_{33} \begin{pmatrix} |u_{21}|^2 & u_{21}^* u_{22} \\ u_{21} u_{22}^* & -|u_{21}|^2 \end{pmatrix}, \end{aligned} \tag{9}$$

(b) For the transformation $\hat{U}_2^\dagger \hat{\sigma}_1 \hat{U}_2$,

$$\begin{aligned} \hat{\rho}_1' &= \hat{\mathcal{U}}^\dagger \hat{\rho}_1 \hat{\mathcal{U}} + \rho_{22} \begin{pmatrix} |u_{12}|^2 & -u_{11}^* u_{12} \\ -u_{11} u_{12}^* & -|u_{12}|^2 \end{pmatrix}, \\ \hat{\rho}_3' &= \hat{\mathcal{U}}^\dagger \hat{\rho}_3 \hat{\mathcal{U}} + \rho_{22} \begin{pmatrix} -|u_{21}|^2 & -u_{21}^* u_{22} \\ -u_{21} u_{22}^* & |u_{21}|^2 \end{pmatrix}. \end{aligned} \tag{10}$$

(c) For the transformation $\hat{U}_2^\dagger \hat{\sigma}_2 \hat{U}_2$,

$$\begin{aligned} \hat{\rho}_1' &= \hat{\mathcal{U}}^\dagger \hat{\rho}_1 \hat{\mathcal{U}} + \rho_{22} \begin{pmatrix} |u_{12}|^2 & -u_{11}^* u_{12} \\ -u_{11} u_{12}^* & -|u_{12}|^2 \end{pmatrix}, \\ \hat{\rho}_3' &= \hat{\mathcal{U}}^\dagger \hat{\rho}_3 \hat{\mathcal{U}} + \rho_{22} \begin{pmatrix} -|u_{21}|^2 & -u_{21}^* u_{22} \\ -u_{21} u_{22}^* & |u_{21}|^2 \end{pmatrix}, \end{aligned} \tag{11}$$

(d) From $\hat{U}_3^\dagger \hat{\sigma}_2 \hat{U}_3$,

$$\begin{aligned} \hat{\rho}_4' &= \hat{\mathcal{U}}^\dagger \hat{\rho}_4 \hat{\mathcal{U}} + \rho_{11} \begin{pmatrix} -|u_{21}|^2 & -u_{12}^* u_{22} \\ -u_{21} u_{22}^* & |u_{21}|^2 \end{pmatrix}, \\ \hat{\rho}_6' &= \hat{\mathcal{U}}^\dagger \hat{\rho}_6 \hat{\mathcal{U}} + \rho_{11} \begin{pmatrix} -|u_{12}|^2 & u_{11}^* u_{12} \\ u_{11} u_{12}^* & |u_{12}|^2 \end{pmatrix}. \end{aligned} \tag{12}$$

(e) Finally, for $\hat{U}_3^\dagger \hat{\sigma}_3 \hat{U}_3$,

$$\begin{aligned} \hat{\rho}_4' &= \hat{\mathcal{U}}^\dagger \hat{\rho}_4 \hat{\mathcal{U}} + \rho_{11} \begin{pmatrix} -|u_{21}|^2 & -u_{12}^* u_{22} \\ -u_{21} u_{22}^* & |u_{21}|^2 \end{pmatrix}, \\ \hat{\rho}_6' &= \hat{\mathcal{U}}^\dagger \hat{\rho}_6 \hat{\mathcal{U}} + \rho_{11} \begin{pmatrix} -|u_{12}|^2 & u_{11}^* u_{12} \\ u_{11} u_{12}^* & |u_{12}|^2 \end{pmatrix}, \end{aligned} \tag{13}$$

For all the cases, $\hat{\mathcal{U}}$ is a two-dimensional unitary transformation.

As in the phase-damping case, one can think of a rotation matrix with a time-dependent angle $\theta = \omega t$ as the unitary operation, i.e.,

$$\hat{\mathcal{U}} = \begin{pmatrix} \cos(\omega t) & -\sin(\omega t) \\ \sin(\omega t) & \cos(\omega t) \end{pmatrix},$$

which, in the case where $t \approx 0$, results in the following transformations:

$$\hat{\rho}'_j = \hat{\mathcal{U}}^\dagger \hat{\rho}_j \hat{\mathcal{U}} - \rho_{kk}\,\omega\, t\, \hat{\sigma}_x + \mathcal{O}(t^2)\,, \tag{14}$$

where $\hat{\sigma}_x$ is the Pauli matrix and ρ_{kk} is a diagonal component of $\hat{\rho}$, which depends on j. Its value is $k = 2$ for $j = 1, 3$, $k = 3$ for $j = 2, 5$, and $k = 1$ for $j = 4, 6$. It is necessary to point out that, for $\hat{\rho}'_5$ associated with $\hat{U}_1^\dagger \hat{\sigma}_3 \hat{U}_1$, we need to replace ρ_{33} with $-\rho_{33}$ in Equation (14).

In the case where the density matrices correspond to states, where one of the accessible levels is not occupied, i.e.,

$$\hat{\sigma}_4 = \begin{pmatrix} \rho_{11} & \rho_{12} & 0 \\ \rho_{21} & \rho_{22} & 0 \\ 0 & 0 & 0 \end{pmatrix}, \quad \hat{\sigma}_5 = \begin{pmatrix} \rho_{11} & 0 & \rho_{13} \\ 0 & 0 & 0 \\ \rho_{31} & 0 & \rho_{33} \end{pmatrix}, \quad \hat{\sigma}_6 = \begin{pmatrix} 0 & 0 & 0 \\ 0 & \rho_{22} & \rho_{23} \\ 0 & \rho_{32} & \rho_{33} \end{pmatrix}, \tag{15}$$

we obtain the expressions:

$$\begin{aligned}
\hat{U}_2^\dagger \hat{\sigma}_4 \hat{U}_2 \Rightarrow \hat{\rho}'_5 &= \begin{pmatrix} \rho_{11}|u_{11}|^2 & \rho_{12}u_{11}^* \\ \rho_{21}u_{11} & 1-\rho_{11}|u_{11}|^2 \end{pmatrix}, \quad \hat{\rho}'_6 = \begin{pmatrix} 1-\rho_{11}|u_{12}|^2 & \rho_{21}u_{12} \\ \rho_{12}u_{12}^* & \rho_{11}|u_{12}|^2 \end{pmatrix}, \\
\hat{U}_3^\dagger \hat{\sigma}_4 \hat{U}_3 \Rightarrow \hat{\rho}'_1 &= \begin{pmatrix} 1-\rho_{22}|u_{12}|^2 & \rho_{12}u_{12} \\ \rho_{21}u_{12}^* & \rho_{22}|u_{12}|^2 \end{pmatrix}, \quad \hat{\rho}'_2 = \begin{pmatrix} 1-\rho_{22}|u_{11}|^2 & \rho_{12}u_{11} \\ \rho_{21}u_{11}^* & \rho_{22}|u_{11}|^2 \end{pmatrix}, \\
\hat{U}_1^\dagger \hat{\sigma}_5 \hat{U}_1 \Rightarrow \hat{\rho}'_3 &= \begin{pmatrix} \rho_{11}|u_{11}|^2 & \rho_{13}u_{11}^* \\ \rho_{31}u_{11} & 1-\rho_{11}|u_{11}|^2 \end{pmatrix}, \quad \hat{\rho}'_4 = \begin{pmatrix} \rho_{11}|u_{12}|^2 & \rho_{13}u_{12}^* \\ \rho_{31}u_{12} & 1-\rho_{11}|u_{12}|^2 \end{pmatrix}, \\
\hat{U}_3^\dagger \hat{\sigma}_5 \hat{U}_3 \Rightarrow \hat{\rho}'_1 &= \begin{pmatrix} 1-\rho_{33}|u_{22}|^2 & \rho_{13}u_{22} \\ \rho_{31}u_{22}^* & \rho_{33}|u_{22}|^2 \end{pmatrix}, \quad \hat{\rho}'_2 = \begin{pmatrix} 1-\rho_{33}|u_{21}|^2 & \rho_{13}u_{21} \\ \rho_{31}u_{21}^* & \rho_{33}|u_{21}|^2 \end{pmatrix}, \\
\hat{U}_1^\dagger \hat{\sigma}_6 \hat{U}_1 \Rightarrow \hat{\rho}'_3 &= \begin{pmatrix} \rho_{22}|u_{21}|^2 & \rho_{23}u_{21}^* \\ \rho_{32}u_{21} & 1-\rho_{22}|u_{21}|^2 \end{pmatrix}, \quad \hat{\rho}'_4 = \begin{pmatrix} \rho_{22}|u_{22}|^2 & \rho_{23}u_{22}^* \\ \rho_{32}u_{22} & 1-\rho_{22}|u_{22}|^2 \end{pmatrix}, \\
\hat{U}_2^\dagger \hat{\sigma}_6 \hat{U}_2 \Rightarrow \hat{\rho}'_5 &= \begin{pmatrix} \rho_{33}|u_{21}|^2 & \rho_{32}u_{21}^* \\ \rho_{23}u_{21} & 1-\rho_{33}|u_{21}|^2 \end{pmatrix}, \quad \hat{\rho}'_6 = \begin{pmatrix} 1-\rho_{33}|u_{22}|^2 & \rho_{23}u_{22} \\ \rho_{32}u_{22}^* & \rho_{33}|u_{22}|^2 \end{pmatrix}.
\end{aligned} \tag{16}$$

These transformations in many of the cases can represent the spontaneous-emission quantum channel. As in the other examples studied above, when the unitary matrices are rotated by angle $\theta = \omega t$, the original qubit systems can be recovered at times $t = 2\pi l/\omega; l = 0, 1, 2, \ldots$. It is important to mention that the states represented by Equation (15) correspond to three-level systems, where one of the levels is a dark state, and then only two of the levels can be populated, which have been experimentally obtained [24]. These kinds of systems have been of relevance as they can be created by two-photon processes in a three-level system [25] or by the adiabatic variation of the Rabi frequencies associated with the transitions between the three states [26]. For example, to obtain the state $\hat{\sigma}_4$, one can think of an atomic Λ-type three-level system ($|1\rangle$, $|2\rangle$, $|3\rangle$), which interacts with an environment [26]; see Figure 1. The Hamiltonian associated with this system can be written in the form:

$$\hat{H} = \begin{pmatrix} \omega_1 & 0 & \omega_{13} \\ 0 & \omega_2 & \omega_{23} \\ \omega_{13} & \omega_{23} & 0 \end{pmatrix},$$

where $\omega_{1,2}$ are the energies of the states $|1\rangle, |2\rangle$, respectively. By considering the energy of the ground state $|3\rangle$ equal to zero, ω_{13} and ω_{23} are the transition energies. Taking the zero energy in the ground state $|3\rangle$, we can make the replacements $\omega_{13} \to \omega_1 e^{-i\omega_1 t}$ and $\omega_{23} \to \omega_2 e^{-i\omega_2 t}$. The time evolution of the density matrix can be obtained by the expression:

$$\frac{d}{dt}\hat{\rho} = i[\hat{\rho}, \hat{H}] + \hat{\rho}' , \tag{17}$$

where the matrix $\hat{\rho}'$ is given by the interaction of the original density matrix with the environment:

$$\hat{\rho}' = \begin{pmatrix} \gamma_{31}\rho_{33} & -\gamma'\rho_{12} & -\gamma_1\rho_{13} \\ -\gamma'\rho_{21} & \gamma_{32}\rho_{33} & -\gamma_2\rho_{23} \\ -\gamma_1\rho_{31} & -\gamma_2\rho_{32} & -\gamma\rho_{33} \end{pmatrix},$$

where the parameters γ_{31}, γ_{32}, and γ are the spontaneous-emission rates, which must satisfy $\gamma = \gamma_{31} + \gamma_{32}$, and the relaxation terms for the coherence components are named γ_1 and γ_2, which also satisfy $\gamma' = \gamma_1 + \gamma_2$. The resulting differential Equation (17) can be reduced by considering that the variation of the parameters ρ_{13}, ρ_{23}, and ρ_{33} over time is smaller compared to the spontaneous emission and decoherence terms γ_{31} and γ_{32}; this is called the adiabatic hypothesis. Under this hypothesis, it is possible to obtain a state with $\rho_{13} = \rho_{23} = \rho_{33} = 0$, as the solution of the evolution of the density matrix $\hat{\sigma}_4$ discussed above.

Another way to obtain these types of systems is the case where the environmental interaction is neglected, i.e., $\hat{\rho}' = 0$ in Equation (17). The corresponding Schrödinger equation is $i\,\frac{d|\psi\rangle}{dt} = \hat{H}|\psi\rangle$, with $|\psi\rangle = a_1(t)e^{-i\omega_1 t}|1\rangle + a_2(t)e^{-i\omega_2 t}|2\rangle + a_3(t)|3\rangle$, which in view of the initial conditions $a_1(0) = \frac{\omega_2}{\sqrt{\omega_1^2+\omega_2^2}}$, $a_2(0) = -\frac{\omega_1}{\sqrt{\omega_1^2+\omega_2^2}}$, $a_3(0) = 0$ leads to the solution:

$$a_1(t) = \frac{\omega_2}{\sqrt{\omega_1^2+\omega_2^2}}, \quad a_2(t) = -\frac{\omega_1}{\sqrt{\omega_1^2+\omega_2^2}}; \quad a_3(t) = 0,$$

so the level $|3\rangle$ is never populated.

The density matrices $\hat{\sigma}_5$ and $\hat{\sigma}_6$ can be obtained by means of analogous procedures applied to the V and Ξ configurations of the three-level system depicted in Figure 1.

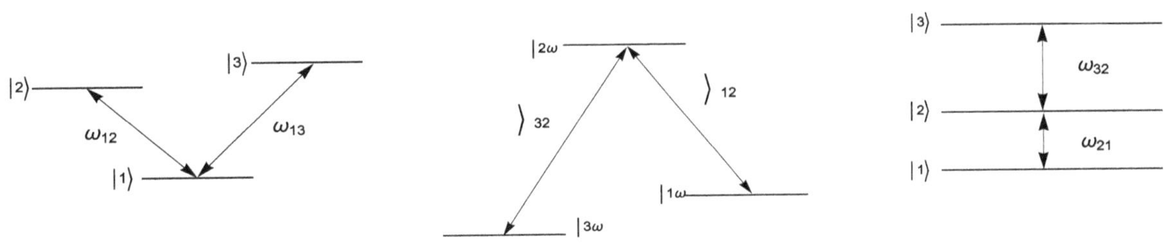

Figure 1. State configurations for the V- (**left**), the Λ- (**center**), and the Ξ-level (**right**) systems.

It is also important to mention that the unitary transformations defined by the matrices $\hat{U}_1$, $\hat{U}_2$, and $\hat{U}_3$ in Equation (5) can be generated experimentally by different proposed mechanisms, such as sliding mode control [27], adiabatic passage [28–30], and the robust control scheme [31,32]. We want to emphasize that the resulting quasi-unitary evolutions and the different quantum channels obtained in our work can have applications in quantum computing and quantum information theories. We think so because the quasi-unitary operations discussed here could be used as approximations to the standard quantum gates, and furthermore, the obtained quantum channels could also be used in the quantum correction algorithms found in the literature.

4. Probability Representation of the Qubit-State Evolution

In the quantum tomographic approach of qubit states [33,34], the states are identified with tomographic probability distributions. In the case of the minimal number of probability parameters, the density matrix of the qubit (spin-1/2) state reads [6]:

$$\hat{\rho} = \begin{pmatrix} p_3 & p_1 - 1/2 - i(p_2 - 1/2) \\ p_1 - 1/2 + i(p_2 - 1/2) & 1 - p_3 \end{pmatrix}, \quad \sum_{j=1}^{3} \left(p_j - \frac{1}{4}\right)^2 \leq \frac{1}{4}, \tag{18}$$

where $0 \leq p_k, \leq 1$ with $k = 1,2,3$ are the probabilities to obtain the value $+1/2$ in the x, y, z axis, respectively. Thus, any qubit state can be identified through the probabilities p_1, p_2, and p_3, i.e., given the density operator, one can get the set $\hat{\rho} \leftrightarrow p_1, p_2, p_3$ and vice versa. In the case of qubits (1) associated with the qutrit state, the evolution of the probabilities after the unitary operation of the qutrit is determined by Equation (2). For example, we have a probabilistic representation corresponding to $\hat{\rho}_5'$ in the first formula of Equation (17), i.e.,

$$p_3 \to p_3|u_{11}|^2, \quad p_1 - 1/2 - i(p_2 - 1/2) \to (p_1 - 1/2 - i(p_2 - 1/2))u_{11}^* \,. \tag{19}$$

The change of probabilities can be characterized by the evolution of the Tsallis and Shannon entropies. For example, in (19), the unitary matrix parameter u_{11} determines the evolution of the Shannon entropy related to a coin probability distribution $(p_3, 1 - p_3)$ (assume that we have two nonideal classical coins I and II in such a game as coin flipping, coin tossing, or heads (up, $\oplus$) or tails (down, $\ominus$), which is the practice of throwing a coin in the air and checking which side is showing when it lands, in order to choose between two alternatives P_k or $(1 - P_k)$; $k = 1,2$). This evolution is of the form:

$$S(\hat{U}) = -p_3|u_{11}|^2 \ln\left(p_3|u_{11}|^2\right) - (1 - p_3|u_{11}|^2) \ln\left(1 - p_3|u_{11}|^2\right) .$$

This entropy, as a function of the unitary evolution applied to the qutrit state, characterizes some aspects of the open dynamics of qubits. We point out that, as for p_3, there exist other classical entropic characteristics associated with the evolution of p_1 and p_2 given by Equation (19).

5. Concluding Remarks

A new mechanism to study the open system evolution of a noncomposite qudit system was established. As an example of the general procedure, we considered a qutrit system. Associated with the qutrit system, one can define different qubit density matrices, which evolve in an open-like way when a unitary transformation is enacted on the qutrit.

The application of the resulting transformations for the qubits within the qutrit was also discussed. The quasi-unitary transformations obtained here might be used as an approximation to quantum gates, whereas the quantum channels could be employed in quantum correction protocols.

Different types of quantum channels can be observed using the qubit decomposition of a qutrit system. In particular, the phase damping and the spontaneous-emission channels were obtained using a unitary transformation acting on specific qutrit density matrices. The phase damping channel was obtained when a unitary transformation of the density matrix with one off-diagonal term equal to zero was performed. A spontaneous-emission channel can be observed by unitary transformations acting over a dark state, i.e., a three-level state where one of the levels cannot be populated.

In addition to these channels, quasi-unitary transformations of the qubit states can be defined. This was also done by the application of a unitary matrix to the generic qutrit state.

The entropy evolution of the tomographic-probability distributions determined by the system of qubits was discussed.

We can extend our analysis to other qudit systems without subsystems since, o an arbitrary spin-j density matrix and the spin unitary evolution, one can associate the smaller spin $j' < j$ evolution.

The possible experimental implementation of the procedure was also addressed, given that there exist several proposed ways to generate the unitary transformations such as by sliding mode control [27], adiabatic passage [28–30], or the robust control scheme [31,32].

Author Contributions: The original idea was given by J.A.L.-S. All authors contributed equally to the conception, design, and methodology of this study. All authors contributed equally to the analysis of the results and the conclusions. All authors contributed equally to the final writing of the manuscript.

References

1. Weiss, U. *Quantum Dissipative Systems*; World Scientific: Singapore, 1993.
2. Kossakowski, A. On quantum statistical mechanics of non-Hamiltonian systems. *Rep. Math. Phys.* **1972**, *3*, 247–274. [CrossRef]
3. Ingarden, R.S.; Kossakowski, A. On the connection of nonequilibrium information thermodynamics with non-Hamiltonian quantum mechanics of open systems. *Ann. Phys.* **1975**, *89*, 451–485. [CrossRef]
4. Gorini, V.; Kossakowski, A.; Sudarshan, E.C.G. Completely positive dynamical semigroups of *N*-level systems. *J. Math. Phys.* **1976**, *17*, 821–825. [CrossRef]
5. Lindblad, G. On the generators of quantum dynamical semigroups. *Commun. Math. Phys.* **1976**, *48*, 119–130. [CrossRef]
6. Chernega, V.N.; Man'ko, O.V.; Man'ko, V.I. Triangle Geometry of the Qubit State in the Probability Representation Expressed in Terms of the Triada of Malevich's Squares. *J. Russ. Laser Res.* **2017**, *38*, 141–149. [CrossRef]
7. Chernega, V.N.; Man'ko, O.V.; Man'ko, V.I. Probability Representation of Quantum Observables and Quantum States. *J. Russ. Laser Res.* **2017**, *38*, 324–333. [CrossRef]
8. Chernega, V.N.; Man'ko, O.V.; Man'ko, V.I. Triangle Geometry for Qutrit States in the Probability Representation. *J. Russ. Laser Res.* **2017**, *38*, 416–425. [CrossRef]
9. López-Saldívar, J.A.; Castaños, O.; Nahmad-Achar, E.; López-Peña , R.; Man'ko, V.I.; Man'ko, M.A. Geometry and Entanglement of Two-Qubit States in the Quantum Probabilistic Representation. *Entropy* **2018**, *20*, 630. [CrossRef]
10. López-Saldívar, J.A.; Castaños, O.; Man'ko, M.A.; Man'ko, V.I. Qubit representation of qudit states: Correlations and state reconstruction. *Quantum Inf. Process.* **2019**, *18*, 210. [CrossRef]
11. Devoret, M.H.; Wallraff, A.; Martinis, J.M. Superconducting Qubits: A Short Review. *arXiv* **2004**, arXiv:cond-mat/0411174v1.
12. Devoret, M.H.; Schoelkopf, R.J. Superconducting Circuits for Quantum Information: An Outlook. *Science* **2013**, *339*, 1169–1174. [CrossRef]
13. Neeley, M.; Ansmann, M.; Bialczak, R.C.; Hofheinz, M.; Lucero, E.; O'Connell, A.D.; Sank, D.; Wang, H.; Wenner, J.; Cleland, A.N.; et al. Emulation of a quantum spin with a superconducting phase qudit. *Science* **2009**, *325*, 722–725. [CrossRef]
14. Lanyon, B.P.; Barbieri, M.; Almeida, M.P.; Jennewein, T.; Ralph, T.C.; Resch, K.J.; Pryde, G.J.; O'Brien, J.L.; Gilchrist, A.; White, A.G. Simplifying quantum logic using higher-dimensional Hilbert spaces. *Nat. Phys.* **2009**, *5*, 134–140. [CrossRef]
15. Chuang, I.L.; Nielsen, M.A. *Quantum Computation and Quantum Information*; Cambridge University Press: Cambridge, UK, 2000.
16. Terhal, B.M. Quantum Error Correction for Quantum Memories. *Rev. Mod. Phys.* **2015**, *87*, 307. [CrossRef]
17. Arimondo, E.; Orriols, G. Nonabsorbing atomic coherences by coherent two-photon transitions in a three-level optical pumping. *Lett. Nuovo C.* **1976**, *17*, 333–338. [CrossRef]
18. Dalibard, J.; Reynaud, S.; Cohen-Tannoudji, C. La cascade radiative de l'atome habillé. In *Interaction of Radiation with Matter*; A Volume in Honour of Adriano Gozzini; Scuola Normale Superiore: Pisa, Italy, 1987.
19. Harris, S.E.; Field, J.E.; Imamoglu, A. Nonlinear Optical Processes Using Electromagnetically Induced Transparency. *Phys. Rev. Lett.* **1990**, *64*, 1107. [CrossRef]
20. Harris, S.E. Electromagnetically induced transparency with matched pulses. *Phys. Rev. Lett.* **1993**, *70*, 552. [CrossRef]
21. Fleischhauer, M.; Imamoglu, A.; Marangos, J.P. Electromagnetically induced transparency: Optics in coherent media. *Rev. Mod. Phys.* **2005**, *77*, 633.

[CrossRef]
22. Horodecki, M.; Horodecki, P.; Horodecki, R. Separability of mixed states: Necessary and sufficient conditions. *Phys. Lett. A* **1996**, *223*, 1–8. [CrossRef]
23. Caruso, F.; Giovannetti, V.; Lupo, C.; Mancini, S. Quantum channels and memory effects. *Rev. Mod. Phys.* **2014**, *86*, 1203. [CrossRef]
24. Alzetta, G.; Gozzini, A.; Moi, L.; Orriols, G. An Experimental Method for the Observation of R.F. Transitions and Laser Beat Resonances in Oriented Na Vapour. *Il Nuovo C. B* **1976**, *36*, 5–20. [CrossRef]
25. Brewer, R.G.; Hahn, E.L. Coherent two-photon processes: Transient and steady-state cases. *Phys. Rev. A* **1975**, *11*, 1641. [CrossRef]
26. Zanon-Willette, T.; de Clercq, E.; Arimondo, E. Ultrahigh-resolution spectroscopy with atomic or molecular dark resonances: Exact steady-state line shapes and asymptotic profiles in the adiabatic pulsed regime. *Phys. Rev. A* **2011**, *84*, 062502. [CrossRef]
27. Dong, D.; Petersen, I.R. Sliding mode control of quantum systems. *New J. Phys.* **2009**, *11*, 105033. [CrossRef]
28. Chen, J.-M.; Liang, L.-M.; Li, C.-Z.; Deng, Z.-J. Arbitrary state controlled-unitary gate between two remote atomic qubits via adiabatic passage. *Opt. Commun.* **2009**, *282*, 4020–4024. [CrossRef]
29. Guérin, S.; Hakobyan, V.; Jauslin, H.R. Optimal adiabatic passage by shaped pulses: Efficiency and robustness. *Phys. Rev. A* **2011**, *84*, 013423. [CrossRef]
30. Torosov, B.T.; Guérin, S.; Vitanov, N.V. High-Fidelity Adiabatic Passage by Composite Sequences of Chirped Pulses. *Phys. Rev. Lett.* **2011**, *106*, 233001. [CrossRef]
31. Zhang, J.; Greenman, L.; Deng, X.; Whaley, K.B. Robust Control Pulses Design for Electron Shuttling in Solid-State Devices. *IEEE Trans. Control Syst. Technol.* **2014**, *22*, 2354–2359. [CrossRef]
32. Wu, C.; Qi, B.; Chen, C.; Dong, D. Robust Learning Control Design for Quantum Unitary Transformations. *IEEE Trans. Cybern.* **2017**, *47*, 4405–4417. [CrossRef]
33. Dodonov, V.V.; Man'ko, V.I. Positive distribution description for spin states. *Phys. Lett. A* **1997**, *229*, 335–339. [CrossRef]
34. Man'ko, V.I.; Man'ko, O.V. Spin state tomography. *J. Exp. Theor. Phys.* **1997**, *85*, 430–434. [CrossRef]

11

Quantum Quantifiers for an Atom System Interacting with a Quantum Field Based on Pseudoharmonic Oscillator States

Bahaaudin Mohammadnoor Raffah [1] and Kamal Berrada [2,*]

[1] Department of Physics, Faculty of Sciences, King Abdulaziz University, Jeddah 21589, Saudi Arabia; braffah@kau.edu.sa

[2] Department of Physics, College of Science, Al Imam Mohammad Ibn Saud Islamic University (IMSIU), Riyadh 11623, Saudi Arabia

* Correspondence: berradakamal@ymail.com

Abstract: We develop a useful model considering an atom-field system interaction in the framework of pseudoharmonic oscillators. We examine qualitatively the different physical quantities for a two-level atom (TLA) system interacting with a quantized coherent field in the context of photon-added coherent states of pseudoharmonic oscillators. Using these coherent states, we solve the model that exhibits the interaction between the TLA and field associated with these kinds of potentials. We analyze the temporal evolution of the entanglement, statistical properties, geometric phase and squeezing entropies. Finally, we show the relationship between the physical quantities and their dynamics in terms of the physical parameters.

Keywords: pseudoharmonic oscillators; entanglement; von Neumann entropy; geometric phase; nonclassicality; squeezing entropies; quantum dynamics; photon-added coherent states

---

1. Introduction

Recently, the outgrowth and development of quantum information processing (QIP) have been supplied to enhance a large knowledge-base and increase the literature background of the quantum entanglement phenomenon, which is responsible for the implantation of the most tasks of QIP [1–4]. The importance of quantum entanglement in various applications of QIP has led to the examination and realization of high-dimensional systems and provided the significance of this kind of correlation in many-body quantum systems [5]. In recent years, various optical devices have been suggested to realize and generate the quantum entanglement, such as NMR systems [6], beam splitters [7], nanoresonators [8] and cavity QED [9]. Moreover, the generation of this kind of correlation actually emerges as an objective in the quantum experimental implementation when examining the non-classicality effects in quantum mechanics. Several attempts have been made to measure the quantum entanglement among particles and fields. Entanglement between atoms and photons has been treated and examined at optical frequencies with atoms [10] and electron spins [11], to interface stationary and flying qubits [12], to perform quantum communication [13] and to implement nodes for quantum repeaters [14] and networks [15].

The geometric phase (GP) is an example of the features of quantum mechanics that could remain overlooked by almost two generations of physicists. A considerable understanding of the formal description of quantum mechanics has been achieved after Berry's discovery [16–20] of a geometric feature related to the dynamics of a quantum system in the adiabatic and cyclic unitary evolution of non-degenerate states. There are plenty of generalizations including nonadiabatic [17], non-cyclic and even nonunitary evolution of the quantum state. Berry has demonstrated that the wave function of a quantum system retains a memory of its evolution in its complex phase argument, which apart from the usual dynamical contribution, only depends on the geometry of the path traversed by the system. Known as the GP factor, this contribution originates from the very heart of the structure of quantum mechanics. The GP is attractive for the implementations of fault-tolerant quantum computation [21–25]. The idea is to exploit this inherent robustness provided by the topological properties of some quantum systems as a means of constructing built-in fault-tolerant quantum logic gates.

Squeezed states in quantized electromagnetic fields have attracted much attention and exhibited in several interesting works in the literature [26]. This squeezing physical concept has been extended to atomic systems [27] considering the definition used for the radiation field. In this context, the atomic squeezing has obtained a great deal of interest and provided many potential applications [28–30]. The atom-photon interactions are utilized to describe the conditions under which the squeezing effect will exist [31]. The appearance of atomic squeezing in a system of three-level atoms placed in a two-mode cavity is analyzed through the effective dipole-dipole interaction between atoms [32]. The model of spin-squeezed atoms, which is based on the Raman scattering with a strong laser pulse, was used to determine the transfer of the change of the correlation between the atom and the light [33]. Atomic squeezing under collective emission was considered to introduce a method for controlling the temporal behavior of the squeezing factor and characterizing the collective emission by the influence of the squeezing effect [34]. The squeezing effect in optimal and nonlinear spin states has been examined in [35,36], respectively. The relationship between the atomic spin squeezing and bosonic quadrature was introduced [37]. Moreover, the experimental realization for an ensemble of V-type atoms was reported [38,39]. In all these cases, the atomic squeezing has been treated in the framework of the Heisenberg uncertainty relations.

The Jaynes–Cummings (JC) model has received much interest, and various axes in different branches of the optical physics both theoretically and experimentally have been developed. The JC model has seen its real practice by exploiting the experimental step in the electrodynamics cavities. In order to understand the physical phenomena through that model, it is important to include the external noises on the studied quantum system [40–43]. Interestingly, it is shown that the noises that lead to the loss of energy have a significant impact on experimental progress in realistic physical situations. On the other hand, the noises that lead to destruction of the coherence in the system state also play a crucial role in those fields.

Coherent states play a crucial role in various physical branches [44,45], which are introduced as an eigenvector of the lowering operator for quantum harmonic oscillators [46]. These states exhibit physical properties like the classical electromagnetic field. In this context, the classical trajectory is used to determine the center of the coherent states' wave packet for the harmonic oscillator potentials. There are other coherent states for nonlinear quantum electromagnetic fields, called nonclassical states, which are antibunching and sub-Poissonian statistics, squeezing and high order squeezing [47,48]. When the nonclassical quantum effects are taken into account, the classical limit and nonclassical limit of the radiation fields are determined by the ordinary coherent state.

The pseudoharmonic oscillator (PHO) potentials have attracted much attention, and more insights are being obtained on different physical subjects [49]. The PHO can be considered in a certain sense as an intermediate potential between the harmonic oscillator (HO) potential (an ideal potential) and anharmonic potentials (the more realistic potentials) [50]. A comparative analysis of potentials considering three-dimensional harmonic oscillator potential (HO-3D) and PHO was introduced in [51]. It is claimed that like the coherent states (CSs) for the HO, the CSs for the PHO can be helpful in the theory of quantum information [50]. In this context, it is shown that even if the HO-3D can be considered as a limit oscillator of the PHO, it is possible to find a harmonic limit that leads the obtained formulae for the PHO in the CSs formalism to the corresponding well-known formulae for the Glauber coherent state of the HO-1D (referring to a coherent state of a quantum simple harmonic oscillator) [52,53]. In fact, it is shown that apart from their theoretical merit (by contributing to a better understanding of the behavior and properties of the PHO), the formalism of the CSs of the PHO may have also a practical importance (by using it in the quantum information theory and practice) [50].

In the present manuscript, we consider the coherent states associated with the pseudoharmonic oscillator potentials and propose a new model of the atom-field in the framework of these kinds of potentials. We investigate the dynamical behavior of the atomic inversion, photons' distribution, geometric phase, degree of entanglement and atomic squeezing for the quantum system, which will be described in the next section. The paper is organized as follows: In Section 2, we describe our Hamiltonian model and provide an exact form for the ket state of the system using the Schrödinger picture. Section 3 describes the quantum quantifiers considered in this manuscript. In Section 4, we show the numerical results and discuss the variation of the population inversion, entanglement, geometric phase and atomic squeezing. Finally, some conclusions are given in Section 5.

2. Physics Model and Dynamics

The PHO is considered as an anharmonic potential [51,54], which plays a similar role as the HO potential, and it admits exact mathematical studies. The PHO potential can be utilized in some cases as an intermediate oscillator between the HO and more anharmonic oscillators, e.g., Morse oscillator [55,56], Pöschl–Teller oscillator [57], which are more realistic. Similarly to the HO-1D (and a few other quantum systems, e.g., the Morse potential, as well as Poschl–Teller), the PHO potential accepts the building of coherent states [49,58]. Generally, the coherent states are of special importance due to their remarkable mathematical properties and interesting physical applications, especially in quantum optics [59] and also in quantum information theory [60]. The excitation on CSs can be considered as one of the possible generalizations of CSs. These states may be useful in the optical communications field, which employes the nonclassical signal beams, usually mixed with thermal noise [61]. On the other hand, the statistical properties of the CSs are useful in quantum optics and quantum electronics. A new class of states has been introduced, which are generated by the successive action of the raising operator on the Klauder–Perelomov coherent states of the PHO, and we have shown the important nonclassical properties such states possess [61].

The PHO effective potential has the form [58]:

$$V(r) = \frac{M\omega^2}{8} r_j^2 \left(\frac{r}{r_j} - \frac{r_j}{r} \right)^2 + \frac{M\omega^2}{4} \left(r_j^2 - r_o^2 \right), \tag{1}$$

where ω presents the angular frequency, M defines the reduced mass and r_j is the equilibrium distance depending on the rotational quantum j; the parameter α is given by:

$$r_j = \left\{ \frac{2\hbar}{M\omega} \sqrt{\alpha^2 - \frac{1}{4}} \right\}, \quad \alpha = \sqrt{\left(j + \frac{1}{2} \right)^2 + \left(\frac{M\omega}{2\hbar} r_0^2 \right)^2}. \tag{2}$$

It is shown that the bounded states for the PHO are associated with the dynamical $SU(1,1)$ group [49]. The $su(1,1)$ Lie algebra is of great interest in quantum optics because it can characterize many kinds of quantum optical systems [62,63]. It has recently been utilized for investigating the nonclassical properties of light in quantum optical systems [45]. In particular, the bosonic realization of $su(1,1)$ describes the degenerate and non-degenerate parametric amplifiers [64]. The squeezed states and nonlinear CSs of photons have been considered in terms of the $su(1,1)$ Lie algebra and the CSs associated with this algebra [64]. The photon-added coherent states of the pseudoharmonic oscillator (PA-PHOCSs) are expanded as [65]:

$$|z, m, k\rangle = \frac{1}{\sqrt{C_k\left(|z|^2\right)}} \sum_{n=0}^{\infty} \frac{z^n}{\sqrt{R(n,k,m)}} |n + m, k\rangle, \tag{3}$$

where:

$$R(n,k,m) = \frac{\Gamma(2k)\,\{\Gamma(n+1)\}^2}{\Gamma(n+m+2k)\Gamma(n+m+1)}, \tag{4}$$

and:

$$C_k\left(|z|^2\right) = \sum_{n=0}^{\infty} \frac{z^{2n}}{R(n,k,m)}. \tag{5}$$

where m denotes the excited or the number of added photons and k is the Bragmann index. The energy spectrum of the PHO is identical to the HO-1D energy spectrum, up to a translation in the energy scale. It was demonstrated [66,67] that two specific elements of the positive discrete series of the $SU(1,1)$ group unitary irreducible representations (for $k = 1/4$ and $k = 3/4$) reduce the corresponding representation spaces to the Hilbert space of the HO-1D.

The JC model is considered as one of the simplest models to describe the interaction between matter and radiation [68]. This model provides a considerably richer way to investigate the dynamical behavior of the physical phenomena that occur in the atom-field systems. In the rotating wave approximation limit, the model allows an explicit solution, which may be proven empirically. Here, we consider a TLA system interacting with PA-PHOCSs where the coupling term is dependent on the time,

$$H_I(t) = \lambda(t)\left(\hat{a}\sqrt{\hat{a}^{\dagger}\hat{a}}\,|0\rangle\langle 1| + \sqrt{\hat{a}^{\dagger}\hat{a}}\hat{a}^{\dagger}\,|1\rangle\langle 0|\right), \tag{6}$$

where $|1\rangle$ (respectively $|0\rangle$) defines the lower (respectively upper) level of the two-level atom (TLA), $\hat{a}$ (two-level atom $\hat{a}^{\dagger}$) is correspondent to the annihilation (respectively creation) operator of the quantum field and $\lambda(t) = g\sin^2(t)$ is the coupling parameter, where in the case of constant coupling between the TLA and the field, it can be obtained at $\lambda(t) = g$. The time-dependent coupling $\lambda(t)$ is assumed to be a sine function. In this context, the transient regime of the coupling varies rapidly with time. The generalization from the constant coupling λ to arbitrary time-dependent coupling $\lambda(t)$ gives the possibility to model various new physical situations not discussed before. A realization of particular interest is when $\lambda(t)$ may be the time-dependent alignment or orientation of the atomic/molecular dipole moment using a laser pulse [69] and the motion of the atom through the cavity. Theoretical examination of a cavity-quantum electrodynamics (QED) system monitored by utilizing bichromatic adiabatic passage under the influence of a dissipative environment [70], where the authors have analyzed the generation of a controlled Fock number state inside the cavity by a traveling atom, encounters the time-dependent effect and the delays of the Rabi frequencies of the laser fields and cavity.

We suppose that the TLA starts from the upper state $|0\rangle$, and the quantum field is prepared in the PA-PHOCS, $|z,m,k\rangle$; hence, the quantum state of the combined system is written as:

$$|\varpi(0)\rangle = |\varpi_A(0)\rangle \otimes |z,m,k\rangle\,. \tag{7}$$

The ket state vector for any later time is written as:

$$|\varpi(t)\rangle = \exp\left\{-i\int_0^t H_I(T)dT\right\}|\varpi(0)\rangle = \sum_{n=0}^{\infty}\{X_n(t)|n,u\rangle + Y_n(t)|n+1,l\rangle\}\,. \tag{8}$$

The time-dependent functions X_n and Y_n are given by:

$$\begin{aligned} X_n(t) &= Q_n\cos(f(t)(n+m+1)) \\ Y_n(t) &= Q_n\sin(f(t)(n+m+1)), \end{aligned}$$

where $|z,m,k\rangle$ in Equation (7) can be written as:

$$|z,m,k\rangle = \sum_{n=0}^{\infty} Q_n|n\rangle,$$

and:

$$f(t) = \begin{cases} gt & for \quad \lambda(t) = g \\ g(\frac{t}{2} - \frac{\sin(2t)}{4}) & for \quad \lambda(t) = g\sin^2(t). \end{cases}$$

Once the wave function has been analytically obtained, it can be employed to analyze and discuss many physical features of the whole system and subsystems.

3. Quantum Quantifiers

In this section, we define and give a brief discussion of the different physical quantities. The atomic inversion is introduced as the probability difference of getting the TLA system in the upper and lower levels:

$$S_z(t) = \sum_{n=0}^{\infty} \left\{|X_n(t)|^2 - |Y_n(t)|^2\right\}. \tag{9}$$

When the field is defined in a Glauber state at $t = 0$, the atomic inversion exhibits a collapse-revival feature during the time-evolution [71]. The origin of this phenomenon is dependent on the photon distribution of the field, and it is experimentally realizable through ionization detectors as the atomic beam leaving the cavity [72].

To examine the dynamical behavior of the entanglement for the TLA-field state, we introduce the von Neumann entropy as a measure, which is defined as [73]:

$$S_A(t) = -Tr\{\rho_A \ln \rho_A\} = -\sum_{j=1}^{2} \mu_j \ln \mu_j, \tag{10}$$

where ρ_A (respectively ρ_F) presents the TLA (respectively field) density operator, obtained by making the trace over the quantum field (respectively TLA) element basis, i.e., $\rho_A = \text{Tr}_F\left(|\varpi(t)\rangle\langle\varpi(t)|\right)$ and μ_j denotes the eigenvalues of the TLA (respectively field) density operator. This entropy function changes from zero value for a factorizable state to one for a maximally-entangled state.

In order to analyze the photons' distribution, we utilize Mandel's parameter, which is considered as an accurate measure for the statistical properties of the quantum field. It is defined in terms of the average photon number of the field state as [74,75]:

$$M_P = \frac{\langle N^2\rangle - \langle N\rangle^2 - \langle N\rangle}{\langle N\rangle}, \tag{11}$$

where:

$$\langle N^i\rangle = \sum_{n=0}^{\infty} \left\{ n^i |X_n(t)|^2 + (n+1)^i |Y_n(t)|^2\right\}, \quad i = 1,2. \tag{12}$$

The M_P parameter determines the statistical properties of the field state, where $(-1 \le M_P < 0)$ corresponds to the sub-Poissonian photon distribution, $M_P > 0$ is for super-Poissonian distribution and $Q_P = 0$ is for the Poissonian distribution (semi-classical states).

The evolution of the quantum system is described as noncyclic, when the initial and final states are considered different. The initial and the final vector ket states are not connected through a complex scalar factor. If we assume that the initial ket state $|\varpi(0)\rangle$ evolves to $|\varpi(t)\rangle$ and the scalar product $M(t) = \langle\varpi(0)|\varpi(t)\rangle$ is expressed by a real number ℓ, where $M(t) = Re^{i\phi}$, consequently, the noncyclic phase is given by the angle ϕ. The cyclic geometric phase is considered as a particular case of the noncyclic phase, and it can be obtained by taking $R = 1$. The Pancharatnam phase includes the geometric phase (GP) and dynamical phase and is defined as [76]:

$$\Phi_G(t) = \arg(\langle \varpi(0) | \varpi(t) \rangle). \tag{13}$$

The Heisenberg uncertainty relation (HUR) is introduced to examine the squeezing entropy, which is described by Pauli matrices σ_x, σ_y and σ_z for the TLA in the framework of the quantum field as:

$$\Delta\sigma_x \Delta\sigma_y \geq \frac{1}{2} |\langle \sigma_z \rangle|, \tag{14}$$

where $\Delta\sigma_\alpha = \sqrt{\langle \sigma_\alpha^2 \rangle - \langle \sigma_\alpha \rangle^2}$. If σ_α verifies the condition:

$$V(\sigma_\alpha) = \Delta\sigma_\alpha - \sqrt{\frac{|\langle \sigma_z \rangle|}{2}} < 0, \qquad \alpha = x, y. \tag{15}$$

then, the atomic dipole fluctuation in σ_α will be squeezed.

For sets of complementary observable in an even-dimensional Hilbert space, an optimal entropic uncertainty relation has been studied through the quantum entropy theory [77],

$$\sum_{k=1}^{N+1} H(\sigma_k) \geq \frac{N}{2} \ln(\frac{N}{2}) + (1 + \frac{N}{2}) \ln(1 + \frac{N}{2}), \tag{16}$$

with $H(\sigma_k)$ giving the information entropy corresponding to the variable S_k. For the general criterion of the squeezing, we employ entropic uncertainty relation (EUR) defined in Equation (16) in terms of the information entropy to examine the squeezing for the considered JC model. For the TLA state, ρ_A, the information entropies corresponding to the operators σ_x, σ_y and σ_z are given by:

$$\begin{aligned} H(\sigma_x) \quad = \quad & -\left\{\frac{1}{2} + \Re\left[\rho_{lu}(t)\right]\right\} \ln\left\{\frac{1}{2} + \Re\left[\rho_{lu}(t)\right]\right\} \\ & -\left\{\frac{1}{2} - \Re\left[\rho_{lu}(t)\right]\right\} \ln\left\{\frac{1}{2} - \Re\left[\rho_{lu}(t)\right]\right\}, \end{aligned} \tag{17}$$

$$\begin{aligned} H(\sigma_y) \quad = \quad & -\left\{\frac{1}{2} + \Im\left[\rho_{lu}(t)\right]\right\} \ln\left\{\frac{1}{2} + \Im\left[\rho_{lu}(t)\right]\right\} \\ & -\left\{\frac{1}{2} - \Im\left[\rho_{lu}(t)\right]\right\} \ln\left\{\frac{1}{2} - \Im\left[\rho_{lu}(t)\right]\right\}, \end{aligned} \tag{18}$$

$$H(\sigma_z) = -\rho_{Uu}(t) \ln \rho_{Uu}(t) - \rho_{ll}(t) \ln \rho_{ll}(t), \tag{19}$$

where $\rho_{Uu}(t) = \sum_{n=0}^{\infty} |X_n(t)|^2$, $\rho_{ll}(t) = \sum_{n=0}^{\infty} |Y_n(t)|^2$ and $\rho_{lu}(t) = \sum_{n=0}^{\infty} X_n(t) Y_n^*(t)$. For a TLA $N = 2$, then $0 \leq H(\sigma_\alpha) \leq \ln 2$, while from Equation (16), we obtain that the information entropies corresponding to the operators σ_x, σ_y and σ_z verify:

$$H(\sigma_x) + H(\sigma_y) \geq 2 \ln 2 - H(\sigma_z). \tag{20}$$

The aforementioned inequality may be also given as:

$$\delta H(\sigma_x) \delta H(\sigma_y) \geq \frac{4}{\delta |H(\sigma_z)|}, \tag{21}$$

where:

$$\delta H(\sigma_\alpha) = \exp[H(\sigma_\alpha)]. \tag{22}$$

The EUR, described by Equation (21), evidences the impossibility of knowledge of simultaneous information about the observables σ_x and σ_y, where the uncertainty of the polarization component σ_x (respectively σ_y) is measured by $\delta H(\sigma_x)$ (respectively $\delta H(\sigma_y)$).

Let us now introduce the squeezing of the TLA using EUR defined in Equation (21), which is called squeezing entropy [77]. The component fluctuations σ_α ($\alpha = x$ or y) of the TLA are said to be squeezed if the entropy $H(\sigma_\alpha)$ of σ_α verifies the inequality,

$$E(\sigma_\alpha) = \delta H(\sigma_\alpha) - \frac{2}{\sqrt{|\delta H(\sigma_z)|}} < 0, \quad \alpha = x, y. \tag{23}$$

4. Numerical Results and Discussion

In Figure 1, for a TLA initially defined in an upper-level state and the quantum field in PA-PHOCSs, we display the variation of the population inversion S_z dimensionless for the scaled time gt with respect to various physical parameters. We compare the effects of parameters k and m in both cases with and without the time effect. We can see that the population inversion makes period oscillations during the time-evolution. The atomic inversion after suddenly decreasing to its minimum value at the beginning of the interaction increases to a maximum value for each periodicity in the case of $m = 0$. When $m \neq 0$, the behavior of the atomic inversion makes rapid oscillations exhibiting local minima and local maxima in each periodicity. Moreover, as is seen, the parameter k has an impact on the temporal evolution of the atomic inversion only in the absence of photon excitation $k = 0$ and leads to a decrease in the amount of S_z by a suitable choice k. On the other hand, the existence of time-dependent coupling influence leads to a reduction of the oscillations of the S_z during the time-evolution.

The numerical results of the von Neumann entropy have been shown versus the time gt in Figure 2 for various values of the photon-added number in the absence and existence of the time-dependent coupling influence when the TLA initially stated in the upper level and the quantum field is in PA-PHOCSs. The dashed line (red) is for $m = 0$, and the solid line (blue) is for $m = 10$. Figure 1c,d presents the temporal evolution of the entanglement for $\lambda(t) = g$ and $\lambda(t) = g\sin^2(t)$, respectively, in the case of $k = 3/4$. Generally, the von Neumann is a periodic function with sudden death and sudden birth entanglement phenomenon during the time-evolution. In the ideal case in which no atomic motion is considered, von Neumann entropy suddenly increases from zero to its maximum value, then it decays to zero for each periodic time interval, whereas when the time-dependent effect is considered, S_A attains rapid oscillations due to the fluctuations during the interaction, but also leads to an enhancement or reduction of the degree of entanglement. This shows that the quantum field system can help to stabilize the temporal evolution of the entanglement. This behavior is due only to the influence of the kind of coupling term via the generalized parameter $\lambda(t)$ and the photon-added number m. Moreover, it is found that as the parameter m increases, the structure of the oscillations becomes very complex for different values of k, whereas the coupling effect leads to the disappearance of these structures of the von Neumann entropy, i.e., the periodicity time increases and the oscillations are more transparent and accompanied by an increase in the lifetime of sudden death entanglement phenomenon. From Figures 1 and 2, an interesting relationship can be seen between the dynamical behavior of the population inversion and the quantum entanglement.

Figure 3 refers to the effect of the parameters m and k on the time evolution of Mandel's M_P parameter defined by Equation (11) when the field is initially defined in PA-PHOCSs in the absence and existence of the time-dependent coupling influence. Generally, Mandel's parameter makes periodic oscillations, exhibiting a sub-Poissonian and Poissonian distribution at $m = 0$ for different values of k. Whereas, for $m \neq 0$, the field statistics tend to fluctuate around the sub-Poissonian distribution. Interestingly, we obtain that the choice of the initial parameter k only influences the photon statistics of the quantum field in the absence of the excited photons. When the time-dependent effect is considered, Mandel's parameter keeps its behavior with an increase in the periodic time interval during the evolution.

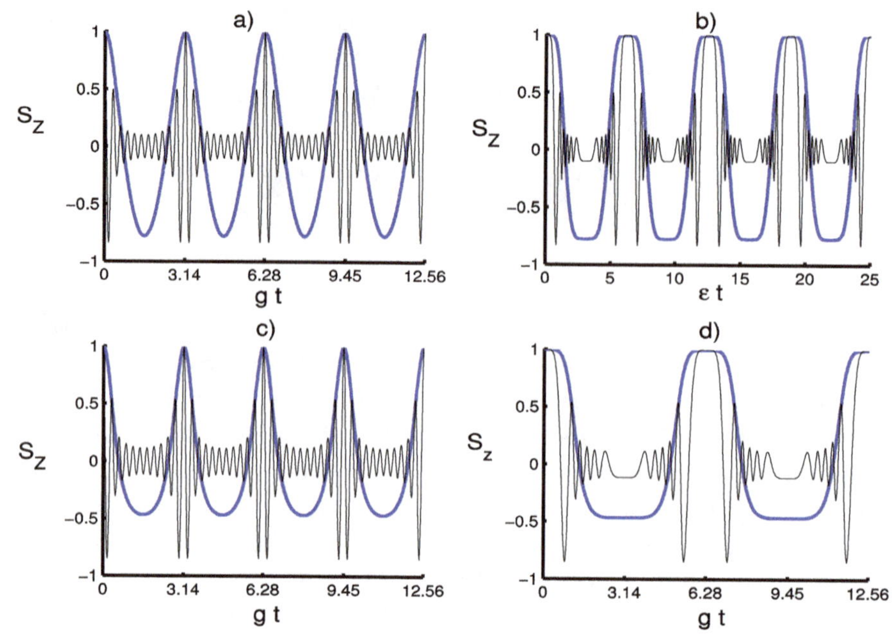

Figure 1. Population inversion for the two-level atom (TLA) initially defined in its excited state, and the field is in photon-added coherent states of the pseudoharmonic oscillator (PA-PHOCSs) for $z = 0.5$. (**a**) The solid line (blue) is for $(k, m) = (\frac{1}{4}, 0)$, and the dashed line (red) is for $(k, m) = (\frac{1}{4}, 10)$. (**b**) The solid line (blue) is for $(k, m) = (\frac{1}{4}, 0)$, and the solid line (red) is for $(k, m) = (\frac{1}{4}, 10)$. (**c**) the solid line is for $(k, m) = (\frac{3}{4}, 0)$ and the dashed line $(k, m) = (\frac{3}{4}, 10)$. (**d**) The solid line (blue) is for $(k, m) = (\frac{3}{4}, 0)$ and the solid line (red) is for $(k, m) = (\frac{3}{4}, 10)$. In (**a**,**c**), we consider the case of constant coupling $\lambda(t) = g$ and in (**b**,**d**), the time dependent coupling $\lambda(t) = g \sin^2(t)$.

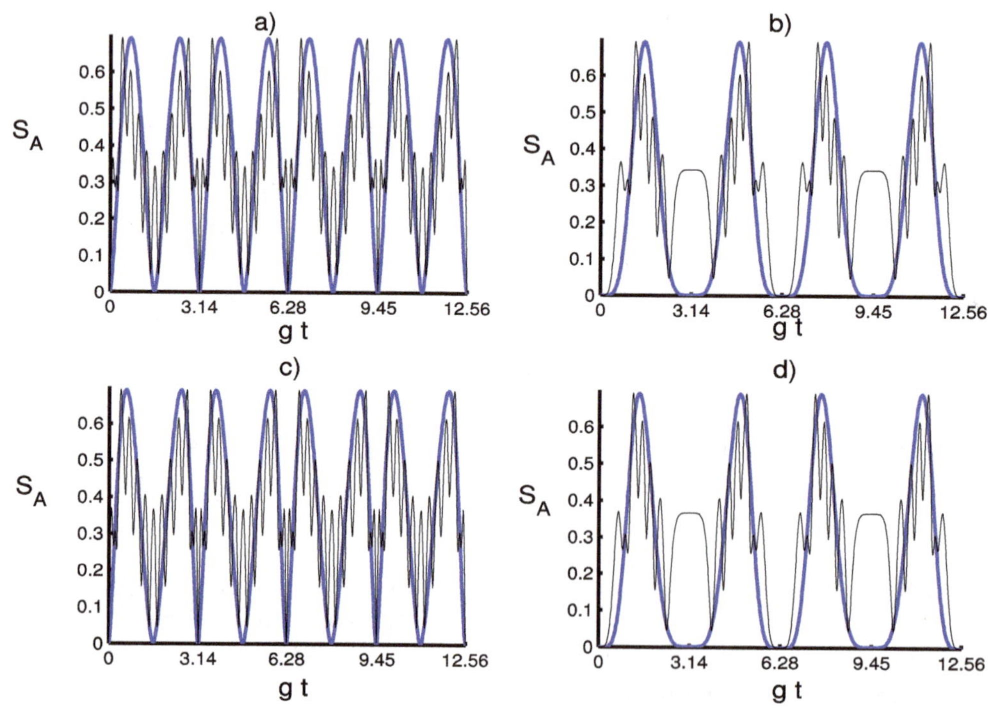

Figure 2. von Neumann entropy S_A with the same conditions of Figure 1. (**a**) The solid line (blue) is for $(k, m) = (\frac{1}{4}, 0)$, and the dashed line (red) is for $(k, m) = (\frac{1}{4}, 10)$. (**b**) The solid line (blue) is for $(k, m) = (\frac{1}{4}, 0)$, and the solid line (red) is for $(k, m) = (\frac{1}{4}, 10)$. (**c**) the solid line is for $(k, m) = (\frac{3}{4}, 0)$ and the dashed line $(k, m) = (\frac{3}{4}, 10)$. (**d**) The solid line (blue) is for $(k, m) = (\frac{3}{4}, 0)$ and the solid line (red) is for $(k, m) = (\frac{3}{4}, 10)$.

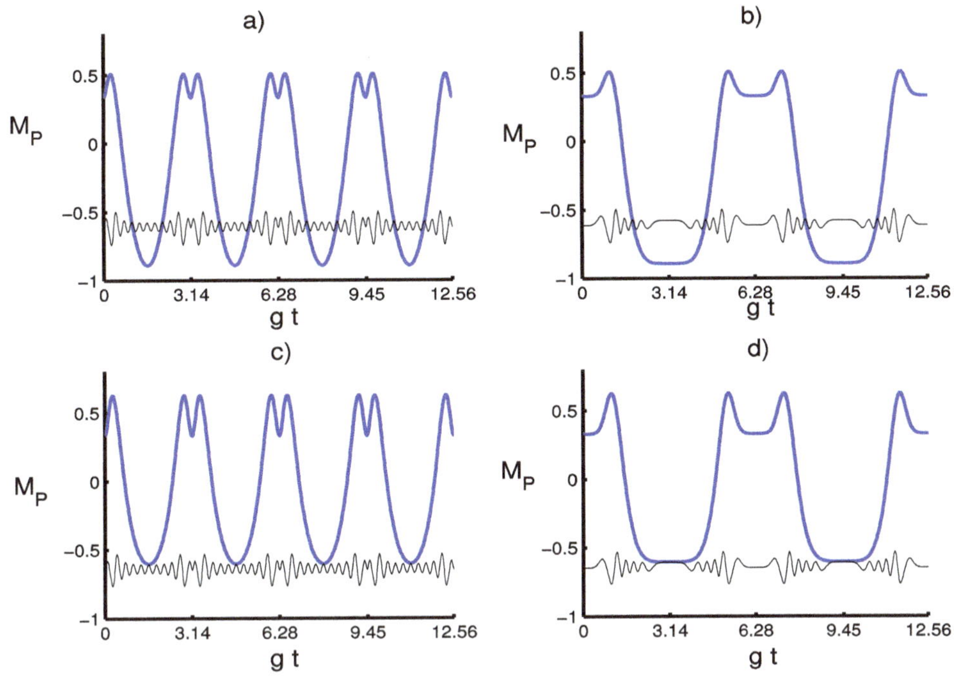

Figure 3. Parameter M_P with the same situation as in Figure 1. (**a**) The solid line (blue) is for $(k, m) = (\frac{1}{4}, 0)$, and the dashed line (red) is for $(k, m) = (\frac{1}{4}, 10)$. (**b**) The solid line (blue) is for $(k, m) = (\frac{1}{4}, 0)$, and the solid line (red) is for $(k, m) = (\frac{1}{4}, 10)$. (**c**) the solid line is for $(k, m) = (\frac{3}{4}, 0)$ and the dashed line $(k, m) = (\frac{3}{4}, 10)$. (**d**) The solid line (blue) is for $(k, m) = (\frac{3}{4}, 0)$ and the solid line (red) is for $(k, m) = (\frac{3}{4}, 10)$.

Let us investigate the main results on the variation of the GP for the whole system state $|\psi\rangle$ with respect to the physical parameters in the presence and absence of the time-depending coupling influence. To understand the impact of the parameters' effects on Φ_G, we display the dynamical behavior of the Φ_G in Figure 4 with respect to different values of k and m. Generally, it can be seen that the GP provides a periodic behavior, exhibiting collapse and revival phenomena. The duration of these phenomena strictly depends on the excited number m and coupling term $\lambda(t)$, where the atomic motion leads to an increase in the periodicity time of the GP. On the other hand, for large values of m, the GP is unaffected by the parameter k, and the result seems to be similar for both cases $k = 1/4$ and $k = 3/4$. From the obtained results, we find that the control and the stabilization of the system dynamics highly benefit from the combination of the quantum field and coupling term parameters.

We now examine the dynamical behavior of the atomic squeezing with regard to the physical parameters. In Figures 5 and 6, we plot the time-evolution of $E(\sigma_x)$ and $E(\sigma_y)$ versus the dimensionless time gt, respectively, with respect to different values of the parameters m and k for both cases $\lambda(t) = g$ and $\lambda(t) = g\sin^2(t)$. We find that the atomic squeezing provides periodic oscillations, where $E(\sigma_x)$ and $E(\sigma_y)$ remain unchanged under the parameter k as the added photon number m obtains large values. This shows that the enhancement and loss of squeezing are due to the physical properties of the quantum field. Interestingly, the atomic motion leads to an increase in the time periodicity of the squeezing entropies. On the other hand, it seems that the squeezing occurs only in the variable y and no squeezing in x, where the increase in m would be accompanied by an increase in $E(\sigma_y)$ and enhance the squeezing effect during the time-evolution. In a nutshell, the obtained results provide that the effect of the initial parameters k and m on the physical quantities seems to be the same in the existence and absence of the atomic motion influence, showing a monotonic relationship between these quantifiers with respect to the initial parameters.

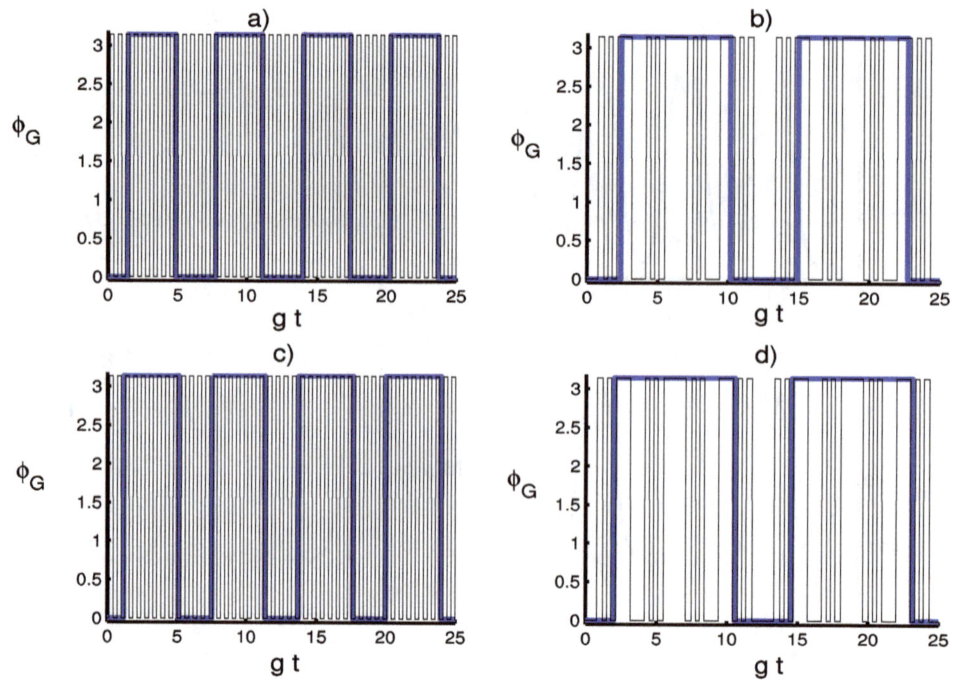

Figure 4. Geometric phase ϕ_P with the same situation as in Figure 1. (**a**) The solid line (blue) is for $(k, m) = (\frac{1}{4}, 0)$, and the dashed line (red) is for $(k, m) = (\frac{1}{4}, 10)$. (**b**) The solid line (blue) is for $(k, m) = (\frac{1}{4}, 0)$, and the solid line (red) is for $(k, m) = (\frac{1}{4}, 10)$. (**c**) the solid line is for $(k, m) = (\frac{3}{4}, 0)$ and the dashed line $(k, m) = (\frac{3}{4}, 10)$. (**d**) The solid line (blue) is for $(k, m) = (\frac{3}{4}, 0)$ and the solid line (red) is for $(k, m) = (\frac{3}{4}, 10)$.

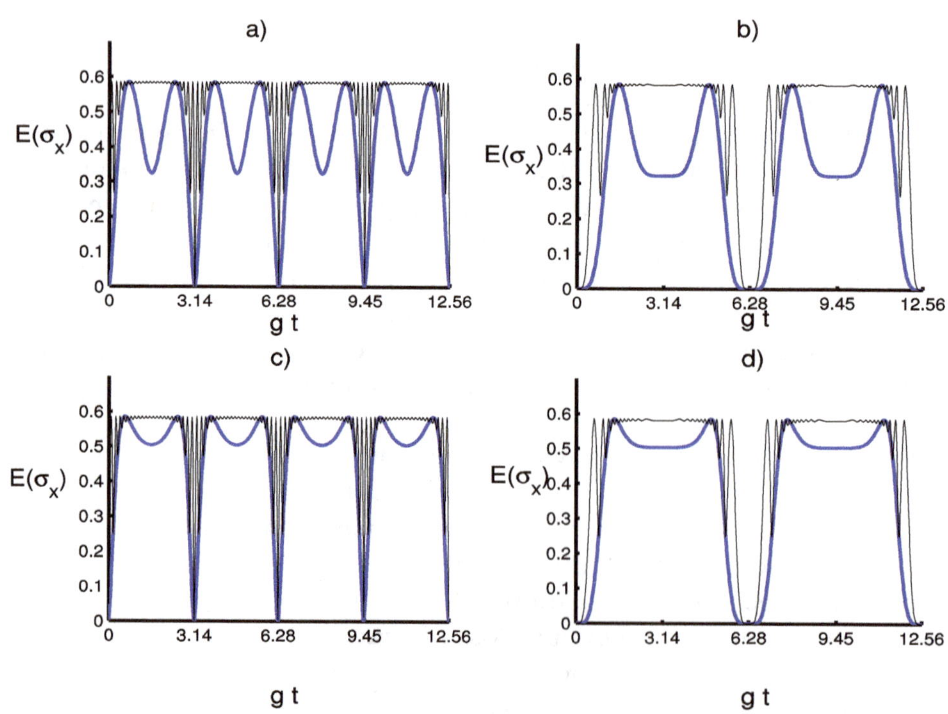

Figure 5. Entropy squeezing component $E(\sigma_x)$ with the same conditions of Figure 1. (**a**) The solid line (blue) is for $(k, m) = (\frac{1}{4}, 0)$, and the dashed line (red) is for $(k, m) = (\frac{1}{4}, 10)$. (**b**) The solid line (blue) is for $(k, m) = (\frac{1}{4}, 0)$, and the solid line (red) is for $(k, m) = (\frac{1}{4}, 10)$. (**c**) the solid line is for $(k, m) = (\frac{3}{4}, 0)$ and the dashed line $(k, m) = (\frac{3}{4}, 10)$. (**d**) The solid line (blue) is for $(k, m) = (\frac{3}{4}, 0)$ and the solid line (red) is for $(k, m) = (\frac{3}{4}, 10)$.

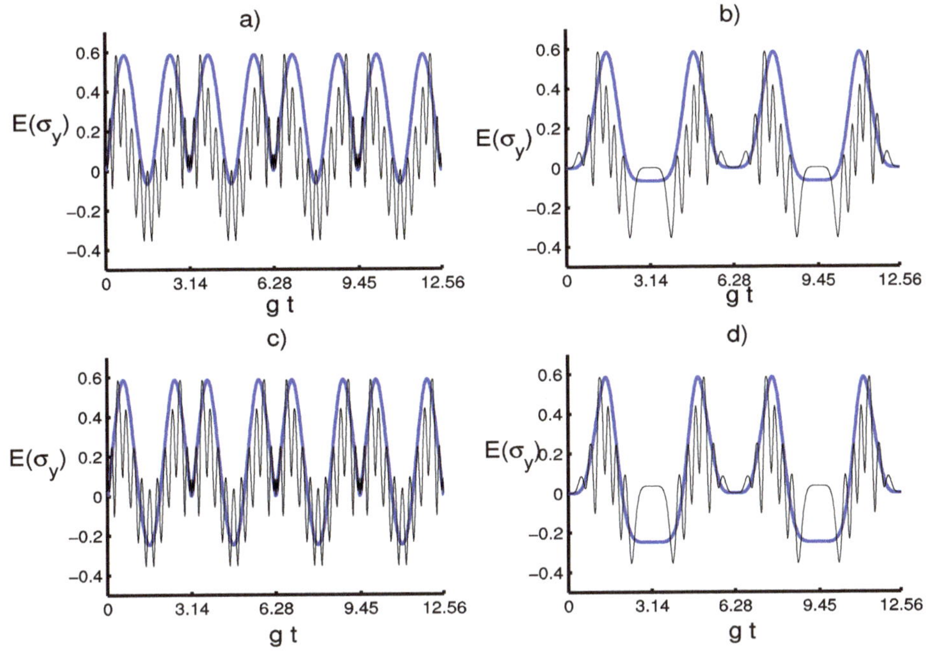

Figure 6. Entropy squeezing component $E(\sigma_y)$ with the same conditions of Figure 1. **(a)** The solid line (blue) is for $(k, m) = (\frac{1}{4}, 0)$, and the dashed line (red) is for $(k, m) = (\frac{1}{4}, 10)$. **(b)** The solid line (blue) is for $(k, m) = (\frac{1}{4}, 0)$, and the solid line (red) is for $(k, m) = (\frac{1}{4}, 10)$. **(c)** the solid line is for $(k, m) = (\frac{3}{4}, 0)$ and the dashed line $(k, m) = (\frac{3}{4}, 10)$. **(d)** The solid line (blue) is for $(k, m) = (\frac{3}{4}, 0)$ and the solid line (red) is for $(k, m) = (\frac{3}{4}, 10)$.

5. Conclusions

We have developed a JC model considering the interaction between a TLA and a quantum field in the framework of pseudoharmonic oscillator potentials. We have shown the necessary optimal conditions that are appropriate for empirical implementation to execute various tasks of quantum computational and information technologies. We have examined qualitatively various quantum quantifiers in terms of the initial parameters during the time-evolution with and without time-dependent coupling, considering the quantum entanglement, geometric phase, nonclassicality and atomic squeezing. Furthermore, we have displayed the relationship between the different physical quantities in terms of the initial parameters during the evolution. We have shown that the change of the parameters strongly influences the dynamical behavior of the quantifiers. The obtained results confirm that the considered quantum system is helpful to withstand the effect of noises on the physical quantities by a suitable choice of the initial parameters. The result suggests future study, considering that the initial mixed state under the effect of the finite-temperature environments on the quantifiers could be pondered.

Author Contributions: B.M.R. and K.B. shared the concept, revision, enriched the research point, conducted the theoretical calculations and writing the manuscript. All authors shared equally the revision of the final version.

References

1. Nielsen, M.A.; Chuang, I.L. *Quantum Computation and Information*; Cambridge University Press: Cambridge, UK, 2000.
2. Alber, G.; Beth, T.; Horodecki, M.; Horodecki, P.; Horodecki, R.; Rötteler, M.; Weinfurter, H.; Zeilinger, R.A. *Quantum Information*; Springer: Berlin, Germany, 2001; Chapter 5.

3. Horodecki, R.; Horodecki, P.; Horodecki, M.; Horodecki, K. Quantum entanglement. *Rev. Mod. Phys.* **2009**, *81*, 865–942. [CrossRef]
4. Joo, J.; Munro, W.J.; Spiller, T.P. Quantum metrology with entangled coherent states. *Phys. Rev. Lett.* **2011**, *107*, 083601. [CrossRef] [PubMed]
5. Amico, L.; Fazio, R.; Osterloh, A.; Vedral, V. Entanglement in many-body systems. *Rev. Mod. Phys.* **2008**, *80*, 517–576. [CrossRef]
6. Gershenfeld, N.A.; Chuang, I.L. Bulk spin-resonance quantum computation. *Science* **1997**, *275*, 350–356. [CrossRef] [PubMed]
7. Tóth, G.; Simon, C.; Cirac, J.I. Entanglement detection based on interference and particle counting. *Phys. Rev. A* **2003**, *68*, 062310. [CrossRef]
8. Sete, E.A.; Eleuch, H.; Ooi, C.R. Light-to-matter entanglement transfer in optomechanics. *JOSA B* **2014**, *31*, 2821–2828. [CrossRef]
9. Zheng, S.B.; Guo, G.C. Efficient scheme for two-atom entanglement and quantum information processing in cavity QED. *Phys. Rev. Lett.* **2000**, *85*, 2392–2395. [CrossRef] [PubMed]
10. Blinov, B.B.; Moehring, D.L.; Duan, L.M.; Monroe, C. Observation of entanglement between a single trapped atom and a single photon. *Nature* **2004**, *428*, 153–157. [CrossRef] [PubMed]
11. Togan, E.; Chu, Y.; Trifonov, A.S.; Jiang, L.; Maze, J.; Childress, L.; Dutt, M.G.; Sørensen, A.S.; Hemmer, P.R.; Zibrov, A.S.; et al. Quantum entanglement between an optical photon and a solid-state spin qubit. *Nature* **2010**, *466*, 730–734. [CrossRef] [PubMed]
12. Wilk, T.; Webster, S.C.; Kuhn, A.; Rempe, G. Single-atom single-photon quantum interface. *Science* **2007**, *317*, 488–490. [CrossRef] [PubMed]
13. Olmschenk, S.; Matsukevich, D.N.; Maunz, P.; Hayes, D.; Duan, L.M.; Monroe, C. Quantum teleportation between distant matter qubits. *Science* **2009**, *323*, 486–489. [CrossRef] [PubMed]
14. Yuan, Z.S.; Chen, Y.A.; Zhao, B.; Chen, S.; Schmiedmayer, J.; Pan, J.W. Experimental demonstration of a BDCZ quantum repeater node. *Nature* **2008**, *454*, 1098–1101. [CrossRef] [PubMed]
15. Ritter, S.; Nölleke, C.; Hahn, C.; Reiserer, A.; Neuzner, A.; Uphoff, M.; Mücke, M.; Figueroa, E.; Bochmann, J.; Rempe, G. An elementary quantum network of single atoms in optical cavities. *Nature* **2012**, *484*, 195–200. [CrossRef] [PubMed]
16. Berry, M.V. Quantal phase factors accompanying adiabatic changes. *Proc. R. Soc. Lond. A* **1984**, *392*, 45–57. [CrossRef]
17. Aharonov, Y.; Anandan, J. Phase change during a cyclic quantum evolution. *Phys. Rev. Lett.* **1987**, *58*, 1593–1596. [CrossRef] [PubMed]
18. Simon, B. Holonomy, the quantum adiabatic theorem, and Berry's phase. *Phys. Rev. Lett.* **1983**, *51*, 2167–2170. [CrossRef]
19. Anandan, J.; Stodolsky, L. Some geometrical considerations of Berry's phase. *Phys. Rev. D* **1987**, *35*, 2597–2600. [CrossRef]
20. Samuel, J.; Bhandari, R. General setting for Berry's phase. *Phys. Rev. Lett.* **1988**, *60*, 2339–2342. [CrossRef] [PubMed]
21. Liu, T.; Cao, X.Z.; Su, Q.P.; Xiong, S.J.; Yang, C.P. Multi-target-qubit unconventional geometric phase gate in a multi-cavity system. *Sci. Rep.* **2016**, *6*, 21562. [CrossRef] [PubMed]
22. Feng, X.L.; Wang, Z.; Wu, C.; Kwek, L.C.; Lai, C.H.; Oh, C.H. Scheme for unconventional geometric quantum computation in cavity QED. *Phys. Rev. A* **2007**, *75*, 052312. [CrossRef]
23. Xiang-Bin, W.; Keiji, M. Nonadiabatic conditional geometric phase shift with NMR. *Phys. Rev. Lett.* **2001**, *87*, 097901. [CrossRef] [PubMed]
24. Zhu, S.L.; Wang, Z.D. Implementation of universal quantum gates based on nonadiabatic geometric phases. *Phys. Rev. Lett.* **2002**, *96*, 289901. [CrossRef]
25. Falci, G.; Fazio, R.; Palma, G.M.; Siewert, J.; Vedral, V. Detection of geometric phases in superconducting nanocircuits. *Nature* **2000**, *407*, 355–358. [CrossRef] [PubMed]
26. Drummond, P.D.; Ficek, Z. *Quantum Squeezing*; Springer: Berlin, Germany, 2004.
27. Wodkiewicz, K. Reduced quantum fluctuations in the Josephson junction. *Phys. Rev. B* **1985**, *32*, 4750. [CrossRef]
28. Agarwal, G.S.; Puri, R.R. Cooperative behavior of atoms irradiated by broadband squeezed light. *Phys. Rev. A* **1990**, *41*, 3782–3791. [CrossRef] [PubMed]

29. Ashraf, M.M.; Razmi, M.S.K. Atomic-dipole squeezing and emission spectra of the nondegenerate two-photon Jaynes-Cummings model. *Phys. Rev. A* **1992**, *45*, 8121–8128. [CrossRef] [PubMed]
30. Kitagawa, M.; Ueda, M. Squeezed spin states. *Phys. Rev. A* **1993**, *47*, 5138. [CrossRef] [PubMed]
31. Civitarese, O.; Reboiro, M. Atomic squeezing in three level atoms. *Phys. Lett. A* **2006**, *357*, 224–228. [CrossRef]
32. Civitarese, O.; Reboiro, M.; Rebón, L.; Tielas, D. Atomic squeezing in three-level atoms with effective dipole–dipole atomic interaction. *Phys. Lett. A* **2010**, *374*, 2117–2121. [CrossRef]
33. Poulsen, U.V.; Mølmer, K. Squeezed light from spin-squeezed atoms. *Phys. Rev. Lett.* **2001**, *87*, 123601. [CrossRef] [PubMed]
34. Yukalov, V.I.; Yukalova, E.P. Atomic squeezing under collective emission. *Phys. Rev. A* **2004**, *70*, 053828. [CrossRef]
35. Wang, X. Spin squeezing in nonlinear spin-coherent states. *J. Opt. B Quantum Semiclass. Opt.* **2001**, *3*, 93–96. [CrossRef]
36. Rojo, A.G. Optimally squeezed spin states. *Phys. Rev. A* **2003**, 68, 013807. [CrossRef]
37. Wang, X.; Sanders, B.C. Relations between bosonic quadrature squeezing and atomic spin squeezing. *Phys. Rev. A* **2003**, *68*, 033821. [CrossRef]
38. Dicke, R.H. Coherence in spontaneous radiation processes. *Phys. Rev.* **1954**, *93*, 99–110. [CrossRef]
39. El-Orany, F.A.; Wahiddin, M.R.B.; Obada, A.S. Single-atom entropy squeezing for two two-level atoms interacting with a single-mode radiation field. *Opt. Commun.* **2008**, *281*, 2854–2863. [CrossRef]
40. Barnett, S.M.; Knight, P.L. Dissipation in a fundamental model of quantum optical resonance. *Phys. Rev. A* **1986**, *33*, 2444–2448. [CrossRef]
41. Puri, R.R.; Agarwal, G.S. Finite-Q cavity electrodynamics: Dynamical and statistical aspects. *Phys. Rev. A* **1987**, *35*, 3433–3449. [CrossRef]
42. Eiselt, J.; Risken, H. Quasiprobability distributions for the Jaynes-Cummings model with cavity damping. *Phys. Rev. A* **1991**, *43*, 346–360. [CrossRef] [PubMed]
43. Abdel-Khalek, S.; Obada, A.S. New features of Wehrl entropy and Wehrl PD of a single Cooper-pair box placed inside a dissipative cavity. *Ann. Phys.* **2010**, *325*, 2542–2549. [CrossRef]
44. Klauder, J.R.; Skagerstam, B.S. *Coherent States-Applications in Physics and Mathematical Physics*; World Scientific: Singapore, 1985.
45. Zhang, W.M.; Gilmore, R. Coherent states: Theory and some applications. *Rev. Mod. Phys.* **1990**, *62*, 867–927. [CrossRef]
46. Glauber, R.J. The quantum theory of optical coherence. *Phys. Rev.* **1963**, *130*, 2529–2539. [CrossRef]
47. Walls, D.F. Squeezed states of light. *Nature* **1983**, *306*, 141–146. [CrossRef]
48. Loudon, R.; Knight, P.L. Squeezed light. *J. Mod. Opt.* **1987**, *34*, 709–759. [CrossRef]
49. Popov, D. Barut-Girardello coherent states of the pseudoharmonic oscillator. *J. Phys. A Math. Gen.* **2001**, *34*, 5283.
50. Popov, D.; Sajfert, V.; Zaharie, I. Pseudoharmonic oscillator and their associated Gazeau–Klauder coherent states. *Phys. A Stat. Mech. Appl.* **2008**, *387*, 4459–4474. [CrossRef]
51. Sage, M.; Goodisman, J. Improving on the conventional presentation of molecular vibrations: Advantages of the pseudoharmonic potential and the direct construction of potential energy curves. *Am. J. Phys.* **1985**, *53*, 350–355. [CrossRef]
52. Gazeau, J.P.; Klauder, J.R. Coherent states for systems with discrete and continuous spectrum. *J. Phys. A Math. Gen.* **1999**, *32*, 123–132. [CrossRef]
53. Klauder, J.R.; Penson, K.A.; Sixdeniers, J.M. Constructing coherent states through solutions of Stieltjes and Hausdorff moment problems. *Phys. Rev. A* **2001**, *64*, 013817. [CrossRef]
54. Gol'dman, I.I.; Krivchenko, V.D.; Kogan, V.I.; Galitskiy, V.M. *Problems in Quantum Mechanics*; Infosearch: London, UK, 1960.
55. Roy, B.; Roy, P. Gazeau–Klauder coherent state for the Morse potential and some of its properties. *Phys. Lett. A* **2002**, *296*, 187–191. [CrossRef]
56. Fakhri, H.; Chenaghlou, A. Barut–Girardello coherent states for the Morse potential. *Phys. Lett. A* **2003**, *310*, 1–8. [CrossRef]
57. Popov, D. Gazeau–Klauder quasi-coherent states for the Morse oscillator. *Phys. Lett. A* **2003**, *316*, 369–381. [CrossRef]
58. Popov, D.; Davidovic, D.M.; Arsenovic, D.; Saifert, V. P-function of the pseudo harmonic oscillator in terms of Klauder-Perelomov coherent states. *Acta Phys. Slovaca* **2006**, *56*, 445–453.

59. Walls, D.F.; Milburn, G.J. *Quantum Optics*; Springer: Berlin, Germany, 1994.
60. Wang, X.; Sanders, B.C.; Pan, S.H. Entangled coherent states for systems with SU (2) and SU (1, 1) symmetries. *J. Phys. A Math. Gen.* **2000**, *33*, 7451–7467. [CrossRef]
61. Popov, D.; Pop, N.; Luminosu, I.; Chiriţoiu, V. Density matrix approach of the excitation on coherent states of the pseudoharmonic oscillator. *EPL Europhys. Lett.* **2009**, *87*, 44003. [CrossRef]
62. Mojaveri, B.; Dehghani, A. Generalized su (1, 1) coherent states for pseudo harmonic oscillator and their nonclassical properties. *Eur. Phys. J. D* **2013**, *67*, 179. [CrossRef]
63. Perelomov, A.M. Coherent states for arbitrary Lie group. *Commun. Math. Phys.* **1972**, *26*, 222–236.
64. Wodkiewicz, K.; Eberly, J.H. Coherent states, squeezed fluctuations, and the SU (2) am SU (1, 1) groups in quantum-optics applications. *JOSA B* **1985**, *2*, 458–466. [CrossRef]
65. Popov, D.; Pop, N.; Sajfert, V. Excitation on the Coherent States of Pseudoharmonic Oscillator. *AIP Conf. Proc.* **2009**, *1131*, 61–66.
66. Perelomov, A.M. *Generalized Coherent States and Their Applications*; Springer: Berlin, Germany, 1986.
67. Gerry, C.C.; Silverman, S. Path integral for coherent states of the dynamical group SU (1, 1). *J. Math. Phys.* **1982**, *23*, 1995–2003. [CrossRef]
68. Janes, E.T.; Cummings, F.W. Comparison of quantum and semiclassical radiation theories with application to the beam maser. *Proc. IEEE* **1963**, *51*, 89–109. [CrossRef]
69. Phoenix, S.J.; Knight, P.L. Comment on "Collapse and revival of the state vector in the Jaynes-Cummings model: An example of state preparation by a quantum apparatus". *Phys. Rev. Lett.* **1991**, *66*, 2833. [CrossRef] [PubMed]
70. Friedrich, B.; Herschbach, D. Alignment and trapping of molecules in intense laser fields. *Phys. Rev. Lett.* **1995**, *74*, 4623–4626. [CrossRef] [PubMed]
71. Rempe, G.; Walther, H.; Klein, N. Observation of quantum collapse and revival in a one-atom maser. *Phys. Rev. Lett.* **1987**, *58*, 353–356. [CrossRef] [PubMed]
72. Scully, M.O.; Zubairy, M.S. *Quantum Optics*; Cambridge University Press: Cambridge, UK, 1997.
73. Von Neumann, J. *Mathematical Foundations of Quantum Mechanics*; Princeton University Press: Princeton, NJ, USA, 1955.
74. Berrada, K.; El Baz, M.; Hassouni, Y. On the construction of generalized su (1, 1) coherent states. *Rep. Math. Phys.* **2011**, *68*, 23–35. [CrossRef]
75. Abdel-Khalek, S.; Berrada, K.; Ooi, C.R. Beam splitter entangler for nonlinear bosonic fields. *Laser Phys.* **2012**, *22*, 1449–1454. [CrossRef]
76. Pancharatnam, S. The adiabatic phase and pancharatnam's phase for polarized light. *Proc. Indian Acad. Sci.* **1956**, *44*, 247–262.
77. Fang, M.F.; Zhou, P.; Swain, S. Entropy squeezing for a two-level atom. *J. Mod. Opt.* **2000**, *47*, 1043–1053. [CrossRef]

12

Hypergraph Contextuality

Mladen Pavičić

Center of Excellence for Advanced Materials and Sensors, Research Unit Photonics and Quantum Optics, Institute Ruder Bošković, Zagreb 10000, Croatia; mpavicic@irb.hr

Abstract: Quantum contextuality is a source of quantum computational power and a theoretical delimiter between classical and quantum structures. It has been substantiated by numerous experiments and prompted generation of state independent contextual sets, that is, sets of quantum observables capable of revealing quantum contextuality for any quantum state of a given dimension. There are two major classes of state-independent contextual sets—the Kochen-Specker ones and the operator-based ones. In this paper, we present a third, hypergraph-based class of contextual sets. Hypergraph inequalities serve as a measure of contextuality. We limit ourselves to qutrits and obtain thousands of 3-dim contextual sets. The simplest of them involves only 5 quantum observables, thus enabling a straightforward implementation. They also enable establishing new entropic contextualities.

Keywords: quantum contextuality; hypergraph contextuality; MMP hypergraphs; operator contextuality; qutrits; Yu-Oh contextuality; Bengtsson-Blanchfield-Cabello contextuality; Xu-Chen-Su contextuality; entropic contextuality

---

1. Introduction

Recently, quantum contextuality found applications in quantum communication [1,2], quantum computation [3,4], quantum nonlocality [5] and lattice theory [6,7]. This has prompted experimental implementation with photons [8–19], classical light [20–23], neutrons [24–26], trapped ions [27], solid state molecular nuclear spins [28] and superconducting quantum systems [29].

Quantum contextuality, which the aforementioned citations refer to, precludes assignments of predetermined values to dense sets of projection operators, and in our approach we shall keep to this feature of the considered contextual sets. Contextual theoretical models and experimental tests involve additional subtle issues, such as the possibility of classical noncontextual hidden variable models that can reproduce quantum mechanical predictions, up to arbitrary precision [30] or a generalization and redefinition of noncontextuality [31,32]. These elaborations are outside the scope of the present paper, though, since it is primarily focused on contextuality, which finds applications within quantum computation versus noncontextuality, which is inherent in the current classical binary computation. That means that we consider classical models with predetermined binary values, which can be assigned to measurement outcomes of classical observables, which underlie the latter computation, versus quantum models that do not allow for such values and underlie quantum computation. As for the direct relevance of our results to quantum computation, we point out that the hypergraph presented in Figure 2 of Reference [3]—from which the contextual "magic" of quantum computation has been derived—is the kind of hypergraph contextual sets we present in this paper. However, the hypergraph is from a 4-dim Hilbert space, so, we will not elaborate on it in this paper.

We give a pedestrian overview of our approach, methods and results, as well as their background in the last few paragraphs of this introduction, describing the organization of the paper.

A class of state-independent contextual (SIC) [33] sets that have been elaborated on the most in the literature are the Kochen-Specker (KS) sets [34–48]. They boil down to a list of n-dim vectors and their n-tuples of orthogonalities, such that one cannot assign definite binary values to them.

Recently, different SIC sets have been designed and/or considered by Yu and Oh [49], Bengtsson, Blanchfield and Cabello [33], Xu, Chen and Su [50], Ramanathan and Horodecki [51], and Cabello, Kleinmann and Budroni [52]. They all make use of operators defined by vectors that define their sets. You and Oh construct rather involved expression of state/vector defined 3×3 operators that eventually reduces to a multiple of a unit operator while the other authors make use of projectors whose expressions also reduce to a multiple of a unit operator. Therefore, we call their sets the *operator-based contextuality sets* and assume that they form an *operator contextuality class*. All the sets make use of a particular list of 3-dim vectors and their orthogonal doublets and triplets, such that a given expression of definite binary variables has an upper bound which is lower than that of a corresponding quantum expression. The last two References [51,52] also provide us with the necessary and sufficient conditions for an SIC set in any dimension.

The difference between the KS contextuality and the operator contextuality is that KS statistics include measured values of all vectors from each n-tuple, while the statistics of measurements are built on values obtained via operators defined by possibly less than n vectors from each n-tuple.

In this paper, we blend the two aforementioned contextualities so as to arrive at hypergraph one. We consider hypergraphs with 3-dim vectors in which some of those vectors that belong to only one triplet are dropped, as in the observable approach, and generate smaller hypergraphs from them, such that one cannot assign definite binary values to them, as in the KS approach. We call our present approach the McKay-Megill-Pavičić hypergraph (MMPH) approach. MMPH non-binary sets directly provide us with noncontextual inequalities. On the other hand, via our algorithms and programs we obtain thousands of smaller MMPH sets which can serve for various applications as, for example, to generate new entropic tests of contextuality or new operator-based contextual sets.

The smallest MMPH non-binary set we obtain is a pentagon with five vectors (vertices) cyclically connected with 5 pairs of orthogonality (edges). It corresponds to the pentagram from Reference [53], implemented in [15,20,23]. The difference is that the pentagram inequality is state dependent, while the MMPH pentagon inequality is state independent. More specifically, in Reference [53], one obtains a nonclassical inequality by means of the projections of five pentagram vectors at a chosen sixth vector directed along a fivefold symmetry axis of the pentagram. By our method, one gets a nonclassical inequality between the maximum sum of possible assignments of 1, representing classical measurement clicks and the sum of probabilities of obtaining quantum measurement clicks.

Entropic test of contextuality for pentagram/pentagon has been formulated in Reference [54] following Reference [55]. It can be straightforwardly reformulated for the other MMPH non-binary sets we obtained.

The paper is organized as follows.

In Section 2.1 we present the hypergraph formalism and define n-dim MMPH set and n-dim MMPH binary and non-binary sets as well as *filled* MMPH set. We explain how vertices and edges in an n-MMPH set correspond to vectors and their orthogonalities, that is, m-tuples ($2 \leq m \leq n$) of mutually orthogonal vectors, respectively.

In Section 2.2 we give the KS theorem and a definition of a KS set and prove that a KS set is a special non-binary set. In Definition 3 we define a *critical* KS set, that is, the one which would stop being a KS set if we removed any of its edges. Then we introduce known KS sets to compare them with operator defined sets. In particular, we start with Conway-Kochen, Bub, Peres and original Kochen-Specker's sets. We show that the number of vectors they are characterised with in the original papers and most of the subsequent ones, as well as in books—that is, 31, 33, 33, and 117, respectively—are not critical. That, actually, enables the whole approach presented in this paper. We show that the aforementioned authors dropped the vectors that are contained in only one triplet. If we took all the stripped vectors into account, that is, if we formed filled sets, we would get 51, 49, 57 and 192 vectors, respectively.

These sets are critical and the majority of researchers assumed that their stripped versions are critical too and so they did not try to use them as a source of smaller non-classical 3-dim sets.

Next, we connect and compare KS sets with operator-based sets, in particular YU-Oh's 13 vector set whose filled version has 25 vectors and 16 triplets—we denote it as 25–16. In Figure 1, we show Yu-Oh's 25-16 as a subgraph of Peres' 57–40. In Figure 2, we show how 25–16 can be stripped of vectors contained in only one triplet, so as to arrive at the original Yu-Oh's 13-16 set. Equations (1)–(6) and their comments explain how Yu and Oh defined their operators with the help of the 13 vectors and how they used them to arrive, via Equation (4), at the inequality defined by Equation (6). We then used the operator expression given by Equation (4) to test 50 sets smaller and bigger than the 13–16 but did not obtain an analogous result. Some of the sets are shown in Figure 3.

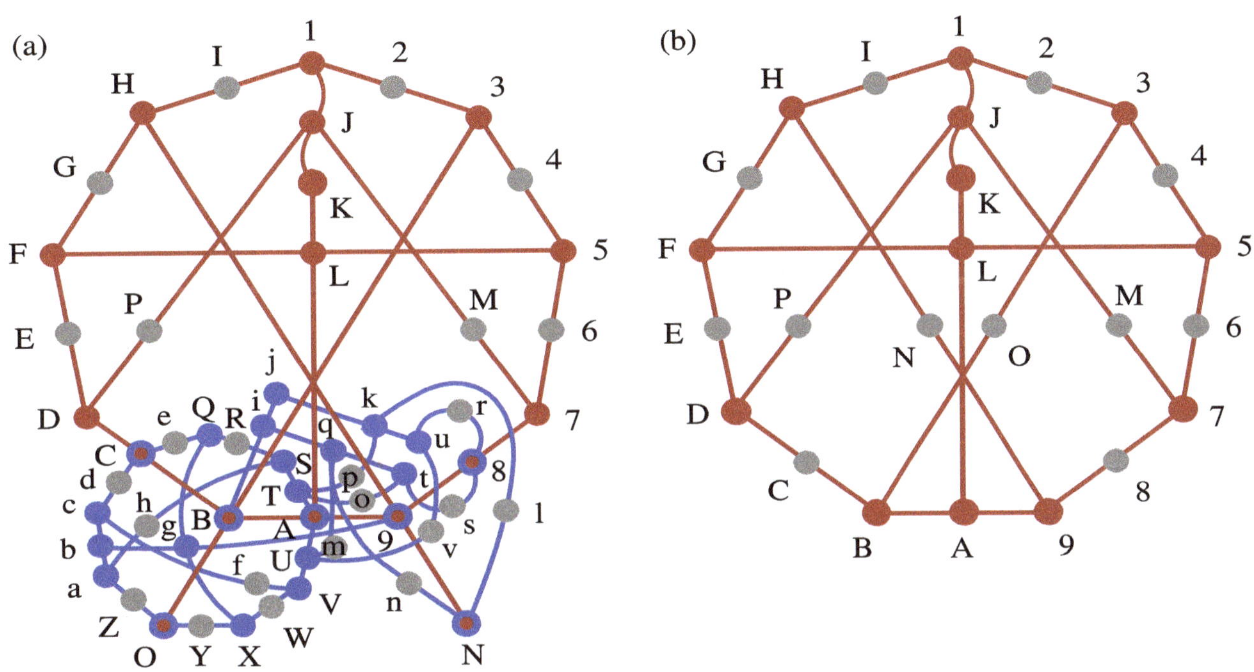

Figure 1. (**a**) Peres' KS set 57-40 in the MMPH representation and containing the full scale Yu-Oh set (drawn in red); (**b**) The full scale Yu-Oh non-KS set 25–16; Vertices (vectors) that share only one edge (triplet) are given as gray dots. See text.

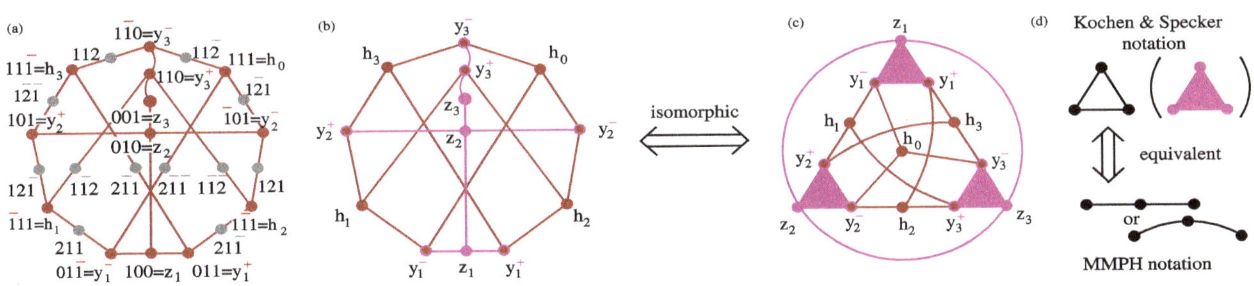

Figure 2. (**a**) An MMPH subgraph of Peres' KS MMPH; (**b**) Yu-Oh's reduction of (**a**); (**c**) Yu-Oh's Figure 2 from [49]; (**d**) Yu and Oh adopted a mixture of Kochen & Specker notation [56]; Cf. (Figure 19 in the [46],) (the triangles in (**c**)) and MMPH notation (the circle in (**c**)).

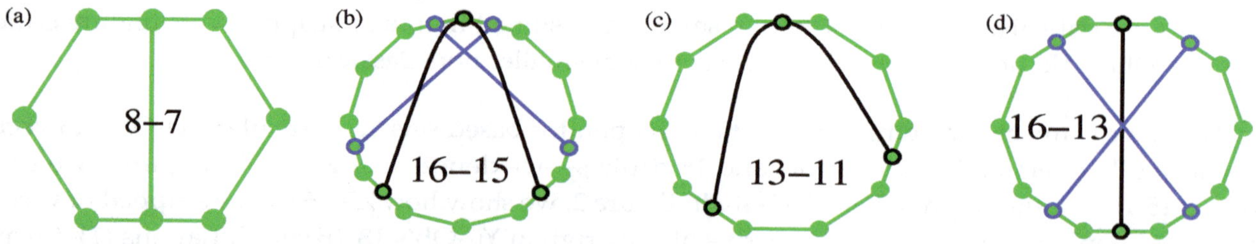

Figure 3. (**a**) Hexagon MMPH from the KS set 192(117)–118 (Figure 6(ii) in the [35]) where it appears in 15 instances; (**b**) a symmetric subgraph of Peres' MMPH with a non-diagonal $\hat{L}$; (**c**) an asymmetric subgraph of Peres' MMPH with a diagonal $\hat{L}$ and $\langle\hat{L}\rangle < Max[C]$; (**d**) a constructed symmetric MMPH with a diagonal $\hat{L}$ and $\langle\hat{L}\rangle < Max[C]$ but whose full scale version does not have a coordinatization.

In Section 2.3 we give a historical background of stripping the aforementioned vectors that are contained in only one triplet and explain what was behind that "incomplete triplets" issue. Then we give MMPH strings of Conway-Kochen's 31–37, Bub's 33–36, Peres' 33–40 and Kochen-Specker's 117–118 non-critical but still non-binary non-classical MMPH sets and take them as our master sets from which we generate smaller non-binary critical MMPH sets in the next section. However, we stress that any set we obtain by stripping some other number of vertices contained in only one edge from any one of the original four KS sets can serve us as a master set. We give a Peres' 40–40 set as an example.

In Section 2.4 we start with a definition of a critical MMPH non-binary set which differs from that of a critical KS set. If we strip more and more edges from a critical KS set we shall never come to a KS set again. This is not so with MMPH non-binary sets. MMPH non-binary critical sets might properly contain smaller MMPH non-binary critical sets whose number of edges is smaller than the original critical set for at least 2 edges.

Via our algorithms and programs, we obtain thousands of critical sets from our master sets, whose distributions are shown in Figure 4. We say that a collection of MMPH non-binary subgraphs of an MMPH master form its class.

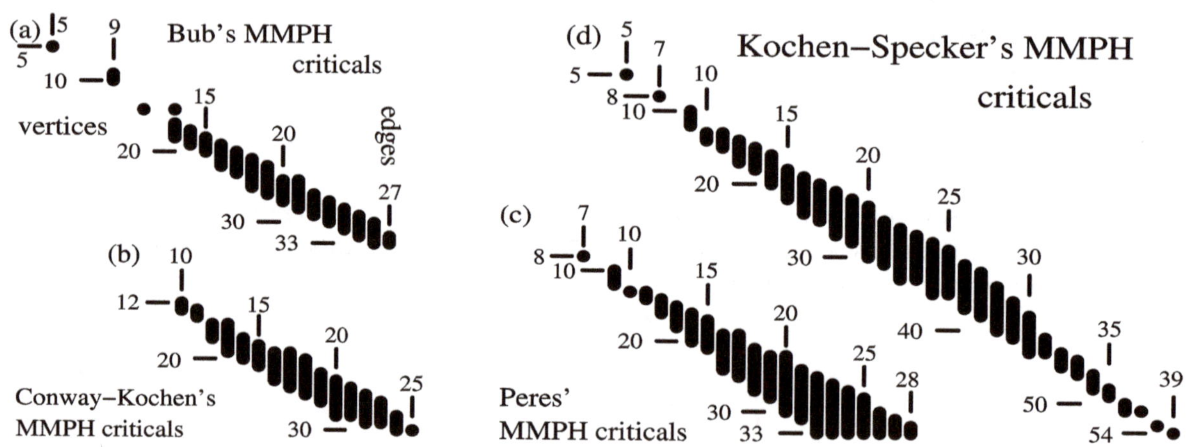

Figure 4. (**a**) Distribution of MMPH non-binary critical sets generated from Bub's MMPH non-binary master set; (**b**) Conway-Kochen's criticals; (**c**) Peres' criticals; (**d**) Kochen-Specker's criticals.

Next we define measurements which can distinguish contextual from non-contextual MMPH sets, that is, non-binary from binary ones. Similar to operator-based contextual measurements, dropped vertices are not considered, that is, clicks obtained at their corresponding out-ports are not taken into account when obtaining the statistics of collected data. So, measurements of MMPH non-binary sets are carried out as for KS sets with triplets, that is, with the $1/3$ probability of detection at each out-port and via *calibrated* detections of a particle or a photon at out-ports of a gate representing a doublet with the $1/2$ probability of getting a click at each of the two considered ports, while ignoring the third one. When a vertex shares a mixture of triplet and doublet edges the probability of detection is $1/p$, where $1/3 \leq p \leq 1/2$. We call detections at all ports notwithstanding whether we include them in our final statistics or not, *uncalibrated* detections—they simply have $1/3$ probability of detection at every port.

To obtain contextual distinguishers of an MMPH set we consider the sum of probabilities of getting clicks for all considered vertices and call it a *quantum hypergraph index*. We distinguish a calibrated quantum hypergraph index, which we denote as HI_q and an uncalibrated one, which we denote as HI_{q-unc}. On the other hand, each MMPH set allows a maximal number of 1s assigned to vertices so as to satisfy the two conditions from Definition 2. We call the number *classical hypergraph index* and denote it as HI_c. Our *weak* contextual distinguisher is the inequality: $HI_q > HI_c$ and the *strong* one is the inequality $HI_{q-unc} > HI_c$. Yu-Oh, Bub, Conway-Kochen and Peres' MMPH non-binary sets as well as others given in the section, like, for example, 13–10, satisfy both inequalities.

We present several small critical MMPH sets in Figures 5 and 6 and discuss their features. We also calculate Yu-Oh's inequalities for several sets different from Yu-Oh's 13–16 set. None of the 50 tested sets satisfy the inequality.

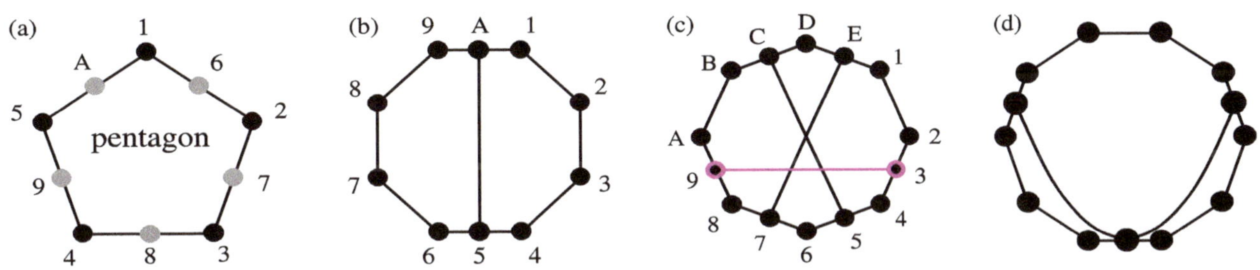

Figure 5. Criticals generated from Bub's master: (**a**) $\overline{\text{subgraph}}$ pentagon 5–5; (**b**) $\overline{\text{subgraph}}$ 10–9; (**c**) standard subgraph 14–11; Critical generated from Peres' master: (**d**) 13–11.

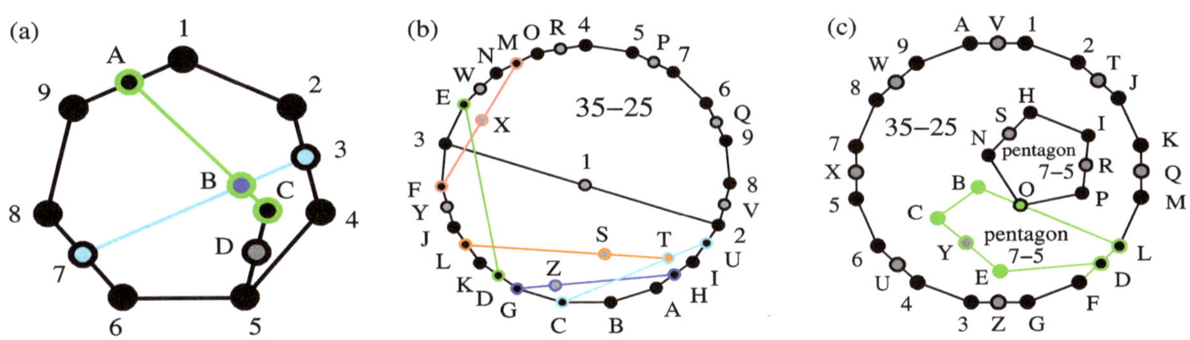

Figure 6. (**a**) Conway-Kochen's MMPH non-binary critical set 13–10; (**b**) Kochen-Specker's 35–25a critical with uncalibrated contextuality; the outer loop is a 19–gon; (**c**) Kochen-Specker's 35–25b critical without uncalibrated contextuality; the outer loop is a 16–gon; See text.

In Section 3 we discuss and reexamine the steps and details of our approach.

2. Results

We consider a set of quantum states represented by vectors in a 3-dim Hilbert space $\mathcal{H}^3$ grouped in triplets of mutually orthogonal vectors. We describe such a set by means of a hypergraph which we call a MMPH. In it, vectors themselves are represented by vertices and mutually orthogonal triplets of them by edges. However, an MMPH itself has a definition which is independent of a possible representation of vertices by means of vectors. For instance, there are MMPHs without a coordinatization, that is,

MMPHs for whose vertices vectors one could assign to, do not exist. Also, edges can contain less than 3 vertices, that is, 2 and form doublets. When a coordinatization exist, that does not mean that a doublet belongs to a 2-dim edge but only that we do not take an existing third vertex/vector into account.

2.1. Formalism

Let us define the hypergraph formalism.

A hypergraph is a pair *v*-*e* where *v* is a set of elements called vertices and *e* is a set of non-empty subsets of *e* called edges. Edge is a set of vertices that are in some sense *related* to each other, in our case *orthogonal* to each other.

The first definition of MMPH was given in Reference [35] where we called them, not hypergraphs, but diagrams. In Reference [46], we gave a definition of an *n*-dim *MMP hypergraph* which required that each edge has at least 3 vertices and that edges that intersect each other in *n*-2 vertices contain at least *n* vertices. The definition of *MMPH* is slightly different.

Definition 1. *An MMPH is an n-dim hypergraph in which*

1. *Every vertex belongs to at least one edge;*
2. *Every edge contains at least 2 vertices;*
3. *Edges that intersect each other in m—2 vertices contain at least m vertices, where* $2 \leq m \leq n$.

Then, in Reference [47] we presented a hypergraph reformulation of the Kochen-Specker theorem [56] from which we derive the following definition of an MMPH non-binary set.

Definition 2. *n-dim MMPH non-binary set,* $n \geq 3$, *is a hypergraph whose each edge contains at least two and at most n vertices to which it is impossible to assign 1s and 0s in such a way that*

1. *No two vertices within any of its edges are both assigned the value 1;*
2. *In any of its edges, not all of the vertices are assigned the value 0.*

An MMPH set to which it is possible to assign 1s and 0s so as to satisfy the above two conditions we call an MMPH binary set.

An MMPH non-binary set with edges of mixed sizes to which vertices are added so as to make all edges of equal size each containing n vertices is called filled MMPH set.

A coordinatization of an MMPH non-binary set means that the vertices of its *filled* MMPH denote *n*-dim vectors in $\mathcal{H}^n$, $n \geq 3$ and that its edges represent orthogonal *n*-tuples, containing vertices corresponding to those mutually orthogonal vectors. Then the vertices of an MMPH set with edges of mixed sizes inherit its coordinatization from the coordinatization of its filled set. In our present approach a coordinatization is automatically assigned to each hypergraph by the very procedure of its generation from master MMPHs as we shall see below.

In the real 3-dim Hilbert space edges form loops of order five (pentagon) or higher as we proved in Reference [35]. For complex vectors our calculations always confirmed this result but we were unable to find an exact proof. Loops of order two are precluded by Definition 1(3).

MMPH are encoded by means of printable ASCII characters organized in a single string, and within it in edges, which are separated by commas; each string ends with a period. Vertices are denoted by one of the following characters: `1 2 ... 9 A B ... Z a b ... z !` " # $ % & ' ( ) * - / : ; < = > ? @ [\ ] ^ _ ' { | } ~ [35]. When all of them are exhausted one reuses them prefixed by '+', then again by '++' and so forth. An MMPH with *k* vertices and *l* edges we denote as a *k*-*l* set. In its graphical representation, vertices are depicted as dots and edges as straight or curved lines connecting orthogonal vertices. In its ASCII string representation (used for computer processing) each MMPH is encoded in a single line followed by assignments of coordinatization to *k* vertices. We handle

MMP hypergraphs by means of algorithms in the programs SHORTD, MMPSTRIP, MMPSUBGRAPH, VECFIND, STATES01, and others [6,35,38,39,57,58].

2.2. KS vs. Operator Contextuality

Let us start with the *Kochen-Specker* theorem and KS sets. Then we shall connect them with the vectors and operators of one type of operator-based contextuality introduced by Yu and Oh.

Theorem 1. (Kochen-Specker [56,59,60]). *In $\mathcal{H}^n$, $n \geq 3$, there are sets of n-tuples of mutually orthogonal vectors to which it is impossible to assign 1 s and 0 s in such a way that*

1. *No two orthogonal vectors are both assigned the value 1;*
2. *In any group of n mutually orthogonal vectors, not all of the vectors are assigned the value 0.*

The sets of such vectors are called KS sets and the vectors themselves are called KS vectors.

There is a one-to-one correspondence between KS n-tuples of vectors and MMPH edges when they are all of their maximal size, as established in Reference [35,46–48] and between KS vectors and MMPH vertices with coordinatization within an MMPH with maximal edges.

Theorem 2. *An n-dim MMPH non-binary set with a coordinatization whose each edge contains n vertices, is a KS set.*

Proof. It follows straightforwardly from the KS theorem, its definition of a KS set and the aforementioned correspondences between its vectors and MMPH vertices. □

In 1988, Asher Peres presented a simple proof of the KS theorem in a 3-dim Hilbert space using real vectors [61]. He implicitly made use of 57 vectors/rays and 40 triplets of mutually orthogonal vectors but seemed to have dropped 24 vectors that appear in only one triplet and called his proof a "33 vector [ray] proof." However, he admitted the role of the remaining vectors, "It can be shown that if a single ray is deleted from the set of 33, the contradiction disappears. It is so even if the deleted ray is not explicitly listed in table 1." ([61], L176, bottom paragraph). From Reference ([61], Table 1) we can reconstruct the 33 vectors within their 40 triplets together with the "non-explicit" 24 vectors and represent them in our MMPH notation, obtaining an MMPH non-binary set with 57 vertices (vectors) and 40 edges (triplets), that is, a 57–40 KS set. We did so in two different ways with two resulting (but isomorphic) hypergraphs in Reference ([6], Figure 4) and Reference ([46], Figure 19). Here we give a third MMPH representation (isomorphic to the previous two) which contains the so-called full scale Yu-Oh set `123,345,567,789,9AB,BCD,DEF,FGH,HI1,1JK,KLA,5LF,JPD,JM7,3OB,HN9.` we elaborate on below. The representation is carried out via our programs SUBGRAPH and LOOP [47].

Peres' 57–40 MMPH KS set reads:

`123,345,567,789,9AB,BCD,DEF,FGH,HI1,1JK,KLA,JM7,3BO,H9N,JPD,FL5,QRS,STA,AUV,VWX, XYO,OZa,abc,cdC,CeQ,Sha,QgX,Vfc,bg9,qmU,Nnq,Bij,jku,klN,ur8,8st,iqt,Tpk,Tot,uvU.`

Its graphical representation is given in Figure 1a.

Notice that gray dots 8,D,N,O in Figure 1b are not gray in Figure 1a and therefore the representation of the original full scale 57–40 Peres KS set (with all gray dots included) by means of the three original Yu-Oh non-KS sets (with gray vertices dropped), as depicted in Figure 1 of [62], apparently does not work. Also, as verified with our program SUBGRAPH, Yu-Oh's set is not a subgraph of Peres' 33–40 set (with all gray dots dropped). On the other hand, Yu-Oh's set cannot be a subgraph of Peres' 57–40 because it lacks gray dots. The full scale Yu-Oh's set 25–16 shown Figure 1b is, of course, a subgraph of the full-scale Peres' 57-40 set as shown in Figure 1a and confirmed by SUBGRAPH.

The arguments that all vertices are indispensable for an experimental implementation of a KS set can be found in Reference ([63], In particular Table on p. 804), Reference ([35], pp. 1583 top, 1588 bottom, and top 1589), and Reference ([64], p. 332, end of the 1st par.). In essence, every n-tuple from the KS Theorem 1 should contain no fewer than n vectors.

Below, the coordinatization of Peres' 57–40 set is obtained via VECFIND [47] from the vector components $0, \pm 1, \sqrt{2}$ (the component $-\sqrt{2}$, used by Peres in Reference [61] is not needed):

1 = {1,$\sqrt{2}$,−1}, 3 = {0,1,$\sqrt{2}$}, 5 = {−1,$\sqrt{2}$,−1}, 7 = {$\sqrt{2}$,1,0}, 8 = {−1,$\sqrt{2}$,0}, 9 = {0,0,1}, A = {0,1,0}, B = {1,0,0}, C = {0,$\sqrt{2}$,1}, D = {0,−1,$\sqrt{2}$}, F = {1,$\sqrt{2}$,1}, H = {$\sqrt{2}$,−1,0}, J = {−1,$\sqrt{2}$,1}, K = {1,0,1}, L = {1,0,−1}, N = {1,$\sqrt{2}$,0}, O = {0,$\sqrt{2}$,−1}, Q = {−1,−1,$\sqrt{2}$}, S = {$\sqrt{2}$,0,1}, T = {−1,0,$\sqrt{2}$}, U = {1,0,$\sqrt{2}$}, V = {$\sqrt{2}$,0,−1}, X = {1,1,$\sqrt{2}$}, a = {−1,1,$\sqrt{2}$}, b = {1,1,0}, c = {1,−1,$\sqrt{2}$}, g = {1,−1,0}, i = {0,1,−1}, j = {0,1,1}, k = {$\sqrt{2}$,−1,1}, q = {$\sqrt{2}$,−1,−1}, t = {$\sqrt{2}$,1,1}, u = {$\sqrt{2}$,1,−1}

The aforementioned Peres' statement, "if a single ray is deleted from the set of 33, the contradiction disappears" amounts to a coarse definition of a vertex-critical KS set: "A KS [set] is termed critical iff it cannot be made smaller by deleting the [vertices]" [65]. However, in KS sets, there are edges whose removal does not remove any vertex (but nevertheless cause a disappearance of the KS property) and, on the other hand, no vertex can be removed from a KS set without removing at least one edge as well, in the sense that all edges/n-tuples should contain n mutually orthogonal vertices/vectors.

Therefore, we adopt a definition of an *edge-critical* KS set [6,46,58] (MMPH sets will require a redefinition of critical sets, as we shall see later on):

Definition 3. *KS sets that do not properly contain any KS subset, meaning that if any of its edges were removed, they would stop being KS sets, are called critical KS sets.*

Hence, the set `13,35,57,79,9AB,BD,DF,FH,H1,1JK,KLA,5LF,JD,J7,3B,H9.` Yu and Oh obtained in Reference [49] cannot be a KS set since it is a subgraph of a critical KS set (Peres' set) and therefore cannot provide a proof of the KS theorem contrary to the claim in the title of Reference [49], as we also show in some detail in Reference ([46], Section XII). But, in Reference [49], Yu and Oh do define a new kind of contextuality which we shall analyse and which we summarize as follows.

Consider the MMPH of the Yu-Oh representation of the MMPH Peres' subgraph, from Figure 1b, shown in Figure 2. They removed all the vertices that share only one edge and which are depicted as gray dots in Figure 2a. Then they define operators by way of the remaining vertices/vectors/rays/states which serve to define filters either for preparation or for detection of arbitrary input or output states, respectively. The procedure goes as follows.

Some of the vectors from Figure 2a are represented as

$$|y_1^-\rangle = \frac{1}{\sqrt{2}}\begin{pmatrix}0\\1\\-1\end{pmatrix}, \quad |h_2\rangle = \frac{1}{\sqrt{3}}\begin{pmatrix}1\\-1\\1\end{pmatrix}, \quad |z_3\rangle = \begin{pmatrix}0\\0\\1\end{pmatrix}, \quad |N\rangle = \frac{1}{\sqrt{6}}\begin{pmatrix}2\\-1\\1\end{pmatrix}. \tag{1}$$

Vectors serve Yu and Oh to define the following operators

$$\hat{A}_i = I - 2|i\rangle\langle i| \tag{2}$$

where $i = 1, \ldots, 13$ correspond to $y_1^-, y_2^-, \ldots, z_3$ and we add $i = 14, \ldots, 25$ corresponding to gray dots in Figure 2a. For instance, for $i = 1, 8, 13, 20$, corresponding to vectors from Equation (1), we have:

$$\hat{A}_1 = \begin{pmatrix}1&0&0\\0&0&1\\0&1&0\end{pmatrix}, \quad \hat{A}_8 = \frac{1}{3}\begin{pmatrix}1&2&-2\\2&1&2\\-2&2&1\end{pmatrix}, \quad \hat{A}_{13} = \begin{pmatrix}1&0&0\\0&1&0\\0&0&-1\end{pmatrix}, \quad \hat{A}_{20} = \frac{1}{3}\begin{pmatrix}-1&2&2\\2&2&1\\-2&1&2\end{pmatrix}. \tag{3}$$

The operators can be combined in the following way:

$$\hat{L}_{13} = \sum_{i}^{13} \hat{A}_i - \frac{1}{4}\sum_{i}^{13}\sum_{j}^{13} \Gamma_{ij}\hat{A}_i\hat{A}_j = \frac{25}{3}I = 8.\dot{3}I, \tag{4}$$

where $\Gamma_{ij} = 1$ whenever corresponding vectors i, j are orthogonal to each other and $\Gamma_{ij} = 0$ when they are not; also $\Gamma_{ii} = 0$. The value $25/3$ is curious since it is also the sum of probabilities of detecting photons in the full scale setup 25–16 shown in Figure 1b. That may be purely accidental. Also, $\hat{L}_{25}$ is not diagonal. Yu and Oh consider neither vectors $|i\rangle$ nor operators $\hat{A}_i$ for $i = 14, \ldots, 25$

The fact that each $\hat{A}_i$ has the spectrum $\{-1, 1, 1\}$ prompted Yu-Oh to calculate the upper bound of a corresponding expression for 13 classical variables with predetermined values -1 and 1:

$$C_{13} = \sum_{i}^{13} a_i - \frac{1}{4}\sum_{i}^{13}\sum_{j}^{13} \Gamma_{ij}a_i a_j \le 8 \tag{5}$$

The inequality

$$\langle \hat{L} \rangle > Max[C] \tag{6}$$

has been verified experimentally [16,21] and also improved theoretically by changing the coefficients in Equations (4) and (5) [66,67]. However, no other set, apart from Yu-Oh's 13–16 itself, with such properties has been found since.

We tested 50 sets and found that $\hat{L}$ of MMPHs without left right symmetry mostly do not have diagonal matrices, although some do, and that $\hat{L}$s of the majority of symmetric MMPHs are also not diagonal; when they are, they are often not multiples of I; for the ones whose $\hat{L}$s are multiples of I we found that they satisfy either $\langle \hat{L} \rangle < Max[C]$ or at most $\langle \hat{L} \rangle = Max[C]$, that is, we have not found instances of Equation (6) being satisfied. We give some examples below.

We should stress here that our definition of a *subgraph* differs from a standard one. The standard definition assumes that a subgraph is a hypergraph contained in a bigger hypergraph as is. In contradistinction, we shall assume that a subgraph might also be a hypergraph obtained from a bigger hypergraph by taking out some edges and connecting the remaining edges together, or simply by taking out some vertices. The latter subgraph we denote as $\overline{\text{subgraph}}$. For instance `123,345,567.` is a standard subgraph of `123,345,567,781.`, while `123,345,561.` and `13,345,567,781.` are its $\overline{\text{subgraphs}}$. Yu-Oh's 13–16 set is a $\overline{\text{subgraph}}$ of Peres' full scale 57–40 set. It is not a subgraph of either Peres' 57–40 or Peres' 33–40.

For a symmetric Kochen & Specker's divided hexagon ([35], Figure 6(ii)) MMPH 8–7, a subgraph of the KS set 117–118 [56], shown in Figure 3a, we obtain $\langle \hat{L}_8 \rangle = Max[C_8] = 9/2$. The contextuality of the set has previously been considered in Reference [68].

From Peres' original KS set, using our programs STATES01, LOOP and VECFIND we can generate arbitrary many subsets. Most of them are asymmetric and their $\hat{L}$s are non-diagonal. Also, many highly symmetric ones, such as, for example, 16–15, shown in Figure 3b with $\hat{L}_{16}$ given in Equation (7), are not diagonal.

$$\hat{L}_{16} = \frac{1}{6}\begin{pmatrix} 57 & 4 & 4 \\ 4 & 54 & 3 \\ 4 & 3 & 60 \end{pmatrix} \tag{7}$$

An example of a non-symmetric 13–11 with a diagonal $\hat{L}$ is given in Figure 3c. It has $\langle \hat{L}_{13} \rangle = 7.5$ and $Max[C_{13}] = 7.75$, that is, $\langle \hat{L} \rangle < Max[C]$.

We might try to construct a symmetric MMPH, for example, the 16–13 one given in Figure 3d. For it we obtain $\langle \hat{L}_{13} \rangle = 9.5$ and $Max[C_{13}] = 9.75$, that is, again $\langle \hat{L} \rangle < Max[C]$. However, the main problem with such constructed MMPHs is that the probability of coming across their filled (full scale) versions with coordinatizations and therefore belonging to the 3-dim Hilbert space is minute, that is, negligible even via automated construction and search on a supercomputer. The full scale version (23–13) of the aforementioned 16–13 apparently does not have a coordinatization either.

We give more examples of $\langle \hat{L} \rangle$ versus $Max[C]$ calculations for other MMPHs in Section 2.4.

2.3. *MMPH Masters*

There are several facts we would like to stress as starting points of our elaboration on the MMPH non-binary sets.

(i) Peres wrote, "It can be shown that if a single ray is deleted from the set of 33, the contradiction disappears. It is so even if the deleted ray is not explicitly listed in Table 1." ([61], L176, bottom paragraph)

Ad *(i)* The first sentence is wrong because MMPH 33–40 set `123,345,47,79,92A,AC,C4,AF,5F,HJ,HL,H7M,NCO,OPQ,QRL,RT,TJ,JPV,VX,XR,Va,La,ce,cT1,cg,FXM,Mhi,ijg,jl,le,ehn,np,pj,nN,gN,t9,t10,t5,ap1,1MO.` is not critical as verified by STATES01. It is also not a KS set but only an MMPH non-binary set. The second sentence is conditionally correct because the full scale MMPH 57–40, `123,345,467,789,92A,ABC,CD4,AEF,5GF,HIJ,HKL,H7M,NCO,OPQ,QRL,RST,TUJ,JPV,VWX,XYR,VZa,Lba,cde,cT1,cfg,FXM,Mhi,ijg,jkl,lme,ehn,nop,pqj,nrN,gsN,tu9,t10,tv5,ap1,1MO.` is a critical KS set but only if assume that with the deleted ray we also delete the edge/triplet it belonged to. (This instance of Peres' 57–40 KS set is isomorphic to the one given above; the sequence of characters is different due to a reshuffling by automated tools we used to obtain 33–40 as a subgraph of 57–40.

(ii) Yu and Oh write, "The KS value assignments to the 13-ray set [13-16] are possible; i.e., no logical contradiction can be extracted by considering conditions 1 and 2 [of Theorem 1]." ([49], p. 3, left column, top)

Ad *(ii)* The claim is provisionally correct, but not because "no logical contradiction can be extracted by considering conditions 1 and 2"—it can be extracted—in 13–16 it is impossible to assign 1s and 0s in such a way that conditions 1 and 2 are satisfied, and not because "value assignments to the 13-ray set are possible"—they are not possible; one cannot assign 1s and 0s to its rays in such a way that conditions 1 and 2 are satisfied—but because the 13–16 set is not a set of triplets and therefore does not satisfy the first part of the KS theorem.

The "incomplete triplets" issue reappears in many papers and books. For instance in Karl Svozil's book [69] in Section 7.4 there is an excellent symmetric figure of Peres' 33–40 set [Figure 7.12], we actually made use of to write down MMPH 57–40 set, but we had to add 24 vertices that were not there; 33 vectors and their corresponding logical proposition were explicitly given, but the remaining 24 vectors were not mentioned. In the original Kochen-Specker paper [56] the triplets (edges with 3 vertices) were depicted as triangles and doublets (triplets from which one vertex was dropped) as straight lines—all together 117 vertices of 192 ones contained in 118 triplets. Their triangles are shown in Reference ([35], Figure 6(ii)). The same triangles are used in the Yu-Oh's set and are shown in Figure 2d. This triangle notation is a source of some confusion in the literature and research, though. For instance, in Reference [52] on p. 4, Figure 1b, where one line from one of the triangles from Yu-Oh's set is deleted, we read: "(b) G_{YO} minus one edge." However, the lines in the triangle are not edges. The whole triangle is an edge (triplet) as shown in Figure 2d. The lines within a triangle are orthogonalities and a removal of one of them means splitting the triplet into two doublets, that is, increasing the number of edges in the set. So, the set in Figure 1a of Reference [52] has 16 edges, while the set in Figure 1b has 17 edges. In any case the set (b) is not a subgraph of (a) nor is (a) a subgraph of (b). Of course, a removal of one of the orthogonalities must also be accompanied by a switch to a new coordinatization of the whole set.

In *The Kochen-Specker Theorem* article in the *Stanford Encyclopedia of Philosophy* only 117 vertices were considered. "[W]hat KS have shown is that a set of 117 yes-no observables cannot consistently be assigned 0-1 values" [70]. Jeffrey Bub writes, "This yields a total of 49 rays and 36 orthogonal triples. Now the only rays that occur in only one orthogonal triple are the 16 rays with a 5 as component. Removing these 16 rays from the 49 rays yields the following set of 33 rays that cannot be colored" [71]. However, 49 rays also cannot be colored and the 49–36 is critical, while 33–36 is not.

These facts offer the following approach, though. The aforementioned conditions 1 and 2 are also contained in the Definition 2 of an MMPH non-binary set and Peres' 33–40, Yu-Oh's 13–16, Bub's 33–36, Conway-Kochen's 31–37 and Kochen-Specker's 117–118 sets all violate conditions 1 and 2, thus confirming that these sets are MMPH non-binary sets. Moreover, they actually enable us to get many smaller MMPH non-binary sets from them because none of these sets is critical and they are all equipped with at least the coordinatization they inherit from their full scaled versions 57–40, 25–16, 49–36, 51–37, and 192–118, respectively, but often with even simpler ones.

The MMPH strings of the last three sets are:
Bub's 33–36 (derived from the full scale 49–36 ([46], Figure 19)): 12,134,156,67,48,9AB,CDE,6B,4E, 2FG,2HI,EG,GB,8I,I7,AJ,AK,C7L,MN9,HON,N3P,PL,MFQ,QL,M5R,RD,DO,STC,JHT,T5U,S3K,SFV,VW, 98W,WU,X9C.
Conway-Kochen's 31–37 (derived from the full scale 51–37 ([46], Figure 19)): 123,245,26,57,89A,BCD,5D, 3EF,3G,DF,FA,9H,87I,9J,CK,CL,LM,HN,M1N,KO,1OP,Q6R,QGH,BQS, PR,PJ, S4J,SET,NT,TI,RI,UV8, VGK,U6L,4V,UE,18B.
and the Kochen-Specker's 117–118 (derived from the original full scale 192–118 ([46], Figure 19)): 12,234,45, 56,678,81,9A,ABC,CD,DE,EFG,G9,HI,IJK,KL,LM,MNO,OH,PQ,QRS,ST,TU,UVW,WP,1X, XYZ,Za,ab, bcd,d1,ef,fgh,hi,ij,jkl,le,mn,nop,pq,qr,rst,tm,uv,vwx,xy,yz,z!","u,#$,$%&, &',’(,()*, *#,e-,-/:,:;,;<,<=>,>e,?@,@[\,\],]^,^_‘,‘?,{|,|}~,~+1,+1+2,+2+3+4,+4{, +5+6,+6+7+8, +8+9,+9+A,+A+B+C,+C+5,+D+E,+E+F+G,+G+H,+H+I,+I+J+K,+K+D,?+L,+L+M+N,+N+O, +O+P,+P+Q+R, +R?,37,BF,JN,RV,Yc,gk,os,w!,%),/=,[_,}+3,+7+B,+F+J,+M+Q,95e,HDe,PLe,aTe, mi?,uq?,y'?, ;#?,{]1,+5+11,+D+91,+O+H1,1e?.

All of them have coordinatizations and none of them is critical. They will be our MMPH non-binary *master sets* that we shall get smaller MMPH non-binary critical sets from in Section 2.4. Here, we want to stress that we have chosen the above sets to be our masters for historical reasons. But any set we obtain by stripping the original four KS sets from some other number of vertices being contained in only one edge can serve us as a master set. For instance, by stripping not 24 but 17 such vertices from Peres' 57–40 KS set, we obtain the following set which we can also use as our master set:
Peres' 40–40 (derived from the full scale 57–40 ([46], Figure 19)): 123,345,467,78,829,9A,A4,9B,5B,CD,CE, C7F,GAH,HIJ,JKE,KLM,MND,DIO,OPQ,QRK,OST,ET,UVW, UM1,UX,BQF,FYZ,ZaX,ab,bW,WYc,cd,da,cG, XG,e8,ebH,e5,Td1,1FH.

We present two smaller critical MMPH non-binary sets, 35–27 and 38–30, obtained from this 40–40 set, in Appendix A.3 because they are bigger than Peres' 33–40 and they are critical, while Peres' 33–40 is not. Also, criticals with 33 or less vertices we obtained from Peres' 33–40 and from Peres' 40–40 coincide. The difference is only in criticals with 34 to 38 vertices which we, of course, cannot obtain from Peres' 33–40 set.

2.4. Classes of MMPH Non-Binary Sets, Their Implementation, and Their Inequalities

From the MMPH non-binary master sets given in Section 2.3, we obtain smaller MMPH non-binary critical sets via STATES01. There is a principal difference in the feature of criticality between these sets and the full scale KS sets, though.

If we removed any of the edges of a full scale KS critical set, the remaining set would not be a KS set any more (see Definition 3). If we then continued to strip further edges from the remaining set, we would never arrive at a KS set again. This is not so with an MMPH non-binary critical set. When we remove any of its edges it does stop being an MMPH non-binary set, but if we removed further edges from the obtained set, it would often turn into a smaller MMPH non-binary critical set. Therefore we introduce:

Definition 4. *An MMPH non-binary set is called an MMPH non-binary critical set if a removal of any of its edges would turn the remaining set into an MMPH binary set. MMPH non-binary critical sets might properly contain smaller MMPH non-binary critical sets whose number of edges is smaller than the original critical set for at least 2 edges.*

Bub and Conway-Kochen's master sets share the coordinatization while Peres and Kochen-Specker's ones have different ones mutually and with respect to the former two sets. Therefore, also the classes of smaller MMPH non-binary critical sets we obtain from them will be structurally different.

From these master sets we generated classes of smaller MMPH non-binary critical sets by means of our programs [35,47], although the algorithms and programs should be redesigned and rewritten for an automated generation. MMPH sets generated from a master set we call a class of MMPH sets. So, we shall talk about Bub, Conway-Kochen, Peres and Kochen-Specker's classes. Distributions of their criticals are shown in Figure 4. The criticals are mostly the standard subgraphs of their masters obtained via our automated algorithms and programs, except for a limited number of smaller subgraphs we obtained via new algorithms which are still under development. Most subgraphs have a parity proof unlike most of the standard subgraphs of which only a very few have a parity proof.

Notice that the biggest critical sets in Figure 4a,c have the same number of vertices as their master sets, but 9,12 edges less, respectively.

A possible experimental implementation of MMPH non-binary sets might be made in analogy to the experimental implementation of KS sets carried out in Reference [12]. The difference is that the latter sets contain only triplets, while the former ones contain triplets and doublets, similarly to the Yu-Oh's 13–16 set, or even only doublets as in the 5–5 set. To carry out measurements on KS sets means that we have to verify that the probability of detecting a particle or a photon at each out-port of a gate representing an edge (triplet) is $1/3$. Yu-Oh's implementation rely on gates defined via Equations (2) and (4) by means of 13 vertices/vectors/rays/states and the gates representing 12 dropped vertices are not considered. Measurements on MMPH non-binary sets might be carried out as for KS sets with triplets (with the $1/3$ probability of detection at each out-port) and via calibrated detections of a particle or a photon at out-ports of a gate representing a doublet with the $1/2$ probability of detecting a particle at each of the two considered ports. When a vertex share a mixture of triplet and doublet edges the probability of detection is $1/p$, where $1/3 \leq p \leq 1/2$. The data obtained at the out-ports corresponding to the dropped third vertices are discarded or we simply do not measure them at all as in Yu-Oh's experiments [16,21,66]. To assure an equal distribution of outcomes at each port, the inputs to doublet gates should be scaled up with respect to the full triplet ones by $3/2$ and this is why we call them *calibrated*.

The inequalities to be experimentally verified for the MMPH non-binary sets can be defined as for the other two kinds of sets. For instance, for Yu-Oh's 13–16 set we verify their inequality given by Equation (6): $8.3 > 8$. Let us consider the set as shown in Figure 1b (excluding the gray dots). This set contains 4 triplets and 12 doublets. Vertices `A,K,L` share only triplets, so the probability of having a click along them is $1/3$. Vertices `3,7,D,H` share only doublets and the probability of getting clicks along them is $1/2$. Vertices `1,5,9,B,F,J` share a triplet and two doublets, each, what yields the probability $(1/2+1/2+1/3)/3 = 4/9$. Altogether, the probabilities for 13 vertices sum up to $3 \times 1/3 + 4 \times 1/2 + 6 \times 4/9 = 17/3$. This sum a quantum hypergraph index of an MMPH set and we denote it as HI_q. On the other hand, the set 13–16 allows at most four 1s. This is a classical upper bound for getting classical detection clicks, i.e., the maximal number of 1s we can assign to vertices of an MMPH non-binary set so as to satisfy the two conditions from Definition 2, i.e., a classical hypergraph index which we denote as HI_c. Hence, we obtain the inequality $HI_q[13\text{-}16] = 17/3 = 5.\dot{6} > HI_c[13\text{-}16] = 4$. Notice that even *uncalibrated* probabilities give us $HI_{q-unc}[13\text{-}16] = 13/3 = 4.\dot{3} > HI_c[13\text{-}16] = 4$. We obtain uncalibrated probabilities by measuring all vertices in all edges in Figure 1b, meaning with gray dots included. With each vertex in every edge we have a probability of getting a click, that is, of assigning 1 to it, being equal to 1/3. If we now selected the 13 red-dot vertices, we would get $13/3 = 4.\dot{3}$ which is also greater than $HI_c[13\text{-}16] = 4$. Notice also that the maximal number of 1s we can assign to vertices in the full scale 25–16 set is 11 and that gives us the inequality $HI_q[25\text{-}16] = 25/3 = 8.\dot{3} < HI_c[25\text{-}16] = 11$ which is yet another proof that 25–16 is not a KS set.

It is interesting that three of four considered masters also satisfy the uncalibrated inequality $HI_{q-unc} > HI_c$. Bub's 33–36: $HI_{q-unc}[33\text{-}36] = 11 > HI_c[33\text{-}36] = 10$, Conway-Kochen's 31–37: $HI_{q-unc}[31\text{-}37] = 10.\dot{3} > HI_c[31\text{-}37] = 8$, and Peres' 33–40 $HI_{q-unc}[33\text{-}40] = 11 > HI_c[33\text{-}40] = 6$.

Let us now present several smaller MMPH criticals from each class, consider their properties, and calculate Yu-Oh-like expressions and values for some of them.

The smallest Bub's critical subgraph with coordinatization we found is the pentagon 5–5 `12,23,34,45,51` (with the gray dots excluded) shown in Figure 5a. The full scale hypergraph 10–5 `162,273,384,495,5A1` is also shown Figure 5a (with the gray dots included).

As we proved in Reference [35], the smallest loop edges can form in a 3-dim space with vertices endowed with a real coordinatization is a pentagon. We could not find (with Mathematica) a complex coordinatization of any smaller MMPH, either. We conjecture that the filled pentagon MMPH 10–5 is the smallest MMPH with a coordinatization in the 3-dim Hilbert space. Its coordinatization is, for example, 1 = {0,0,1}, 2 = {0,1,0}, 3 = {1,0,1}, 4 = {1,1,−1}, 5 = {1,−1,0}, 6 = {1,0,0}, 7 = {1,0,-1}, 8 = {−1,2,1}, 9 = {1,1,2}, A = {1,1,0}. It, of course, includes the coordinatization of 5–5. As we can easily check, the maximal number of 1s assignable to vertices of 5–5, satisfying the two aforementioned condition, is 2. Thus we have the following contextual inequality $HI_q[5\text{-}5] = 5 \times 1/2 = 2.5 > HI_c[5\text{-}5] = 2$. Yu-Oh's approach does not offer us such a contextual distinguisher since for $\hat{L}$ and C of Equations (4)–(6) we get $\hat{L}_{10} = 2.5I$ and $C_{10} \leq 2.5$. Hence, $\langle\hat{L}_{10}\rangle = Max[C_{10}]$. MMPH non-binary subgraph 5–5 can actually be generated in all four MMPH classes, but we have not shown them for Conway-Kochen and Peres' classes in Figure 4. The pentagon 5–5 has a parity proof.

Subsequent small Bub's critical subgraphs we obtained, are 9–9 and 10–9. The latter is shown in Figure 5b. Its MMPH string can be easily read from the figure: `12,23,34,456,67,78,89,9A1,A5`. Its possible coordinatization is: 1 = {0,0,1}, 2 = {1,1,0}, 3 = {1,−1,1}, 4 = {0,1,1}, 5 = {2,−1,1}, 6 = {1,1,−1}, 7 = {1,0,1}, 8 = {1,2,−1}, 9 = {2,−1,0}, A = {1,2,0}. Vector component '2' is here because the set of `1-A` vertex coordinates is a subset of the `1-H` set of coordinates of the filled set 17–9. As for the contextuality verification, we have $HI_q[10\text{-}9] = 6 \times (1/2 + 1/3)/2 + 4 \times 1/2 = 9/2 = 4.5\dot{3} > HI_c[10\text{-}9] = 4$. On the other hand, we have $\hat{L}_{10} = 5.5I$ and $C_{10} \leq 5.5$. Hence, $\langle\hat{L}_{10}\rangle = Max[C_{10}]$. The set has a parity proof.

The first standard subgraph in the Bub's class we found is 14-11 shown in Figure 5c. Its coordinatization is 1 = {2,0,1}, 2 = {−1,−1,2}, 3 = {1,−1,0}, 4 = {1,1,1}, 5 = {2,−1,−1}, 6 = {0,1,−1}, 7 = {2,1,1}, 8 = {−1,1,1}, 9 = {1,1,0}, A = {1,−1,2}, B = {2,0,-1}, C = {1,0,2}, D = {0,1,0}, E = {−1,0,2}. $HI_q[14\text{-}11] = 4 \times 1/3 + 10 \times (1/2 + 1/3)/2 = 11/2 = 5.5\dot{3} > HI_c[14\text{-}11] = 5$. The Yu-Oh approach gives: $\hat{L}_{10} = 8.5I$ and $C_{10} \leq 8.75$. Hence, $\langle\hat{L}_{14}\rangle < Max[C_{14}]$. The set is one of the few standard subgraphs that have a parity proof. The only other Bub's criticals with a parity proof we found are 14–13, 18–15, 24–19, and 28–23.

Another critical with $\hat{L} = cI$ (c is a constant) we found is 14–13: `12,23,34,45,56,67,789,9A,AB,BC,CD,DE1,E8`. $\langle\hat{L}_{14}\rangle = 7.5 < Max[C_{14}] = 7.75$,

Yu-Oh's 13–16 is from the Peres' class but the only other critical with $\hat{L} = cI$ we found in Peres' class is the subgraph 13–11 shown in Figure 4d: `12,234,56,678,89,9A,ABC,CD,D1,35,4B7`. $\langle\hat{L}_{13}\rangle = 7.5 < Max[C_{13}] = 7.75$, The coordinatization is 1 = $\{1,1,\sqrt{2}\}$, 2 = $\{0,\sqrt{2},-1\}$, 3 = $\{0,1,\sqrt{2}\}$, 4 = {1,0,0}, 5 = $\{1,\sqrt{2},-1\}$, 6 = $\{\sqrt{2},-1,0\}$, 7 = {0,0,1}, 8 = $\{1,\sqrt{2},0\}$, 9 = $\{\sqrt{2},-1,1\}$, A = $\{1,0,-\sqrt{2}\}$, B = {0,1,0}, C = $\{\sqrt{2},0,1\}$, D = $\{1,1,-\sqrt{2}\}$. The components $\pm\sqrt{2}$ come from the coordinatization of the filled set 20–11 which requires the components $\pm\sqrt{2}, 3$, that is, more than Peres' master set itself. This is because 13–11 is a subgraph and not a standard subgraph of the master set. $HI_q[13\text{-}11] = 5.5\dot{3} > HI_c[13\text{-}11] = 5$. The critical 13–11 has a parity proof. We found no standard subgraph of Peres' master with a parity proof, though.

In Figure 4b, only critical standard subgraphs obtained via automated generation are shown. Hence, they are all subgraphs of Conway-Kochen's master but we shall explain how one can generate subgraphs from them.

Let us consider Conway-Kochen's critical 13–10 shown in Figure 6a: `12,234,45,56,678,89,9A1,ABC,3B7,CD5`. Its coordinatization is: 1 = {1,1,0}, 2 = {−1,1,1}, 3 = {1,0,1}, 4 = {1,2,−1}, 5 = {0,1,2}, 6 = {1,−2,1}, 7 = {1,0,−1}, 8 = {1,1,1}, 9 = {1,−1,0}, A = {0,0,1}, B = {0,1,0}, C = {1,0,0}, D = {0,2,−1}., after taking into account the filled 17–10 set. Similarly to Yu-Oh's set, the 13–10 set exhibits both contextual indices: $HI_q[\text{13-10}] = 4.9\dot{4} > HI_c[\text{13-10}] = 4$ and $HI_{q-unc}[\text{13-10}] = 13/3 = 4.\dot{3} > HI_c[\text{13-10}] = 4$. If we take out the vertex D (the gray dot in Figure 6a) the resulting subgraph 12–10 is critical too, which also shows that vertex-criticality is not consistent. Unlike Yu-Oh's set, neither 13–10 nor 12–10 have $\hat{L} = cI$ satisfied. $\hat{L}_{13}$ is not diagonal and $\hat{L}_{12}$ is diagonal but it is not a multiple of the unit matrix. The set 12–10 does not exhibit both contextual distinguishers: $HI_q[\text{12-10}] = 4.75\dot{4} > HI_c[\text{12-10}] = 4$ but $HI_{q-unc}[\text{12-10}] = 12/3 = 4 = HI_c[\text{12-10}] = 4$. It is, of course, due to the lower number of vertices, since the geometrical structure of the MMPHs, yielding the classical index 4, remains the same.

We find similar features within Kochen-Specker's MMPH class. Let us take two MMPH criticals from the middle of the distribution shown in Figure 4d. 32–25a: `45,5P7,76,6Q9,98,8V2,2UI,IHA,AB,BC,CG,GDK,KLJ,JYF,F3,3E,EWN,NMO,OR4,123,DE,STL,UTC,XMF,ZHG.` and 35–25b: `12,2TJ,JK,KQM,ML,LDF,FG,GZ3,34,4U6,65,5X7,78,8W9,9A,AV1,BC,DE,HI,NO,PO,RPI,SNH,YEC,OLB.` Their coordinatization is too long to be given here. Neither of them nor any other standard subgraph in the Kochen-Specker's class we obtained in Figure 4d has a parity proof.

Their different geometrical structure yield different classical hypergraph indices: $HI_c[\text{35-25a}] = 11$ and $HI_c[\text{35-25b}] = 12$. However, the number of vertices and therefore the quantum uncalibrated hypergraph indices of both MMPHs are the same: $HI_{q-unc}[\text{35-25}] = 35/3 = 11.\dot{6}$. That means that 35–25a exhibits contextuality even for uncalibrated measurement outcomes, while 35–25a does not. Their calibrated indices are: $HI_q[\text{35-25a}] = 12.\dot{4} > HI_c[\text{35-25a}] = 11$ and $HI_q[\text{35-25b}] = 13.75 > HI_c[\text{35-25b}] = 12$. Pentagons in 35–25b in Figure 6c are subgraphs of Kochen-Specker's master unlike the pentagon 5–5 (without gray dots) in Figure 5, which is a subgraph. If we removed all gray dots, the resulting set 25–25 will not be critical any more, but if we leave `S` and `R` in the red pentagon, the resulting 27–25 set will be critical. This cannot be achieved with the green pentagon—leaving `Y` as the only gray dot in the 26–25 set will not make it critical. $\hat{L}$ of the double pentagon is not diagonal.

In Appendix A we give chosen MMPH non-binary critical sets which are standard subgraphs of the four MMPH master sets.

3. Discussion

In the last half a century, a vast number of constructive proofs of quantum contextuality were obtained in even dimensional Hilbert spaces, but only a very few in odd dimensional ones. In particular, in the 3-dim space—Bub, Conway-Kochen, Peres, and Kochen-Specker's KS sets, Yu-Oh contextual set and Klyachko-Can-Binicioğlu-Shumovsky's pentagram/pentagon state-dependent set; in total, 6 sets.

In this paper, we present n-dim hypergraph contextuality which consists of generating sets which preclude binary assignments of values 0 and 1 to vertices of a hypergraph, such that 1 is assigned to only one of the vertices in each edge of the hypergraph, where an edge can contain less than n mutually orthogonal vertices. Such a set which we call an n-dim MMPH non-binary set, is defined by Definition 2. We stay with $n = 3$, that is, we deal with qutrits only, although the method can be extrapolated to any dimension. The method serves us to distinguish classical models with predetermined binary values, which can be assigned to measurement outcomes of classical observables underlying classical computation, from quantum models that do not allow for such values and that underlie quantum computation.

Let us make use of a graphical representation of an n-dim MMPH to describe the method. Vertices within an MMP hypergraph are drawn as dots and edges containing mutually orthogonal vertices are drawn with the help of straight or curved lines connecting these "orthogonal dots" as shown in

Figures 1–3, 5 and 6. There can be a different number of vertices/dots in edges. Our program then verifies whether a chosen MMPH k-l violates or obeys the 0,1 assignment rules from Definition 2. Edges in MMPH k-l might contain 3 or 2 vertices. We then consider a filled MMPH k'-l in which we add a vertex to each edge which contains only 2 vertices and try to find a coordinatization for it. If successful, we make a one-to-one correspondence between vertices and vectors in the n-dim Hilbert space, that is, for the MMPH k'-l set. The MMPH k-l set inherits the coordinatization from from the MMPH k'-l set. If we implemented the MMPH k'-l, each edge would be a gate with n outcomes and the probability of detecting an outcome would be $1/n$.

Now, our approach consists of discarding the outcomes corresponding to chosen vertices which share (are contained in) only one edge from chosen edges and considering outcomes only of the remaining vertices. In the 3-dim Hilbert space, that means that some of the edges/gates should be taken as doublets and the others as triplets. Our programs can handle such MMPHs because they are written for edges of mixed sizes. Measurements on MMPH non-binary sets might then be carried for triplets in a standard manner, that is, with the probability of 1/3 of obtaining a click (value 1) at each of the three ports, at a gate corresponding to an edge/triple and via a *calibrated* detection at out-ports of a gate representing a doublet with the probability of 1/2. For vertices that share triplet and doublet edges, the probability would be equal to $1/p$, where $1/3 \leq p \leq 1/2$. Calibration consists of sending three input particles to a doublet gate for each two sent to a triplet gate, that is, the ratio of doublet to triplet inputs should be 3/2.

To obtain a measure of quantum contextuality of an MMPH non-binary set we define hypergraph indices. A classical hypergraph index HI_c is the maximal number of 1s we can assign to vertices within edges of an MMPH so as to obey the 0,1 assignment rules from Definition 2. A (calibrated) quantum hypergraph index HI_q is the sum of calibrated probabilities for all k vertices of the aforementioned k-l MMPH. An uncalibrated quantum hypergraph index HI_{q-unc} is the sum of 1/3-probabilities for all k' vertices of the aforementioned k'-l MMPH. A basic measure of quantum contextuality of an MMPH non-binary set is the inequality $HI_c < HI_q$. If it were satisfied, the MMPH would be contextual. If not, it would not. A stronger measure of quantum contextuality of an MMPH non-binary set is the inequality $HI_c < HI_{q-unc}$. Some of the considered MMPHs do satisfy both inequalities. For instance, Yu-Oh's set 13–16, MMPH 13–10 shown in Figure 6a, MMPH 35–25a shown in Figure 6b and the MMPH master sets considered in Section 2.4. Other considered critical non-binary MMPHs satisfy only calibrated inequalities but that is sufficient for experimental verification of contextuality and possible applications.

We get thousands of MMPH non-binary sets as follows. For the time being, we start with the previously found KS sets—Bub 49–36, Conway-Kochen 51–37, Peres 57–40, and Kochen-Specker's 192–118 which are all critical, that is, if we took out any edge from any of them they would stop being KS ([46], Definition 3). However, when we strip all the vertices contained in only one edge we obtain Bub 33–36, Conway-Kochen 32–37, Peres 33–40 and Kochen-Specker's 117–118 master sets, none of which are critical. This enables us to generate thousands of new smaller MMPH critical sets from them via our programs. Their distributions are shown in Figure 4. Chosen MMPHs critical sets are given in Section 2.4 and Appendix A and shown in Figures 5 and 6. They can be easily implemented, in particular the smaller ones.

The large number of obtained sets can also be used for an automated testing of Yu-Oh's operators and inequalities along the examples we gave in Sections 2.2 and 2.4. For that we are developing new algorithms and programs. This is a work in progress.

Next, one can make use of the obtained MMPHs to formulate new entropic tests of contextualities following Kurzyński, Ramanathan and Kaszlikowski [54]. In 2012, they only had one pentagram/pentagon set [53] at their disposal. The pentagon 5–5 set is the simplest MMPH set we obtained (see Figure 4) and many other generated small sets can now serve the purpose.

Also, the methods for evaluating conditions for being a SIC set developed in References [51,52] and the methods of Cabello-Severini-Winter graph-theoretic approach to quantum correlations [72] require samples of hypergraphs and that is what our method offers—a constructive probabilistic generation of arbitrary MMPH sets when coupled with automated vector generation algorithms we developed in Reference [47].

Finally, we stress that the MMPH constructive generation of non-binary quantum sets from operationally chosen vectors out of all possible ones within such sets contribute to our understanding of the physical origin of quantum correlations since they represent a new MMPH *scenario* for getting "quantum correlations from simple assumptions" presented in Reference [73].

4. Methods

The methods we use to handle quantum contextual sets rely on algorithms and programs within the MMP language—VECFIND, STATES01, MMPSTRIP, MMPSHUFFLE, SUBGRAPH, LOOP and SHORTD developed in References [6,35,38,39,57,58,74,75]. They are freely available at https://www.irb.hr/users/mpavicic/programs/. MMPHs can be visualized via hypergraph figures consisting of dots and lines and represented as a string of ASCII characters. The latter representation enables the processing of billions of MMPHs simultaneously via supercomputers and clusters. For the latter elaboration, we developed other dynamical programs specifically to handle and parallelize jobs with arbitrary number of MMP hypergraph vertices and edges.

Acknowledgments: Technical supports of Emir Imamagić and Daniel Vrčić from Isabella and CRO-NGI are gratefully acknowledged.

Abbreviations

The following abbreviations are used in this manuscript:

KS	Kochen-Specker
MMPH	McKay-Megill-Pavičić hypergraph

Appendix A. ASCII Strings from MMPH Non-Binary Classes

Below we give several chosen standard subgraphs from the four classes of critical MMPH sets shown in Figure 4. The first number in each line is m of the biggest m-gon loop for the MMPH in the line. The second and third numbers are the numbers of the MMPH vertices and edges, respectively. Three commas ",,," denote the end of a loop and * behind an ASCII symbol means that the symbol belongs to the loop.

Appendix A.1. Bub's Class

10-v18-e13 `213,36,6GC,CDB,BH8,89,9I4,45,5EA,A2,,,73*,9*2*,FD*7.`

11-v21-e16 `213,3A,AHG,GFE,E57,76,6KL,LD8,89,9IC,C2,,,45*,B3*,D*2*,JF*B,H*8*4.`

14-v24-e18 `12,2L3,34,4KG,GHI,I85,56,6B,BC,CA,A9,9FE,ED,DO1,,,78*,JH*F*,MN7,ND*C*.`

13-v27-e20 `213,3L4,45,5B,BC,CMN,NOE,E6F,FD9,9A,AJI,IHG,GP2,,,6*7,87,D*3*,KL*H*,QRO*,O*82*,RD*B*.`

17-v30-e23 `543,3PC,CB,BA,AON,NJ6,67,7KL,L2S,SRI,IH,HTM,MGD,DE,E9,98,8Q5,,,12*3*,FG*,J*5*,P*O*M*,US*Q*,N*F2*.`

17-v33-e26 `45,5CL,L7E,EF,FG,GBH,HIJ,JN,NRS,SWO,O6P,PMQ,QA2,2V8,89,9TU,U34,,,12*3*,6*7*,A*B*,C*D,KL*H*,G*D,M*J*,O*H*3*,XU*R*.`

Appendix A.2. Conway-Kochen's Class

8-v15-e11 12,2E7,78,8D3,34,4C6,65,5F1,,,9AB,B7*6*,A3*1*.

12-v22-e16 67,7GF,FB,B5D,D3,3ME,EC,CK8,89,9HI,I2A,AL6,,,12*3*,45*,A*B*C*,J42*.

14-v26-e19 312,2F,FMN,NL5,596,67,7OJ,JIE,EB,BA,APD,DC,CQH,H3,,,45*3*,89*,G2*,KL*G,I*H*8.

15-v29-e22 12,2RH,HQ3,34,47M,MT9,9A,AJE,ED,DIF,FG,GC,CB,B5N,NS1,,,5*6,7*8,H*I*,KL8,LD*6,OPG*,PN*M*.

17-v30-e24 12,2TD,DH,HRO,O87,76,65,5P4,43,3SJ,JK,K9L,LIM,MQN,NCE,EB,BU1,,,8*9*,AB*,C*D*,FG,I*G,P*GA,Q*J*F.

Appendix A.3. Peres' Class

10-v15-e12 12,2A,AC8,87,7D5,56,6B9,94,43,3E1,,,E*C*B*,FE*D*.

14-v19-e16 12,23,34,4E,EGA,A9,9HB,BC,CFD,D8,87,7I6,65,51,,,I*G*F*,JI*H*.

14-v27-e19 12,2QD,DE,E3I,IJK,KM5,56,6L8,87,7PG,GHF,FAB,BC,CR1,,,3*4,9A*,E*C*,NJ*9,OH*4.

20-v35-e27 213,3G,GLM,MNE,EF,FVX,XYU,UP5,5I,IT7,78,89,9S6,6J,JZQ,QHA,AB,BKD,DC,CR2,,,45*6*,H*I*,F*3*,OP*K*,V*Q*L*,T*S*2*,WX*R*.

22-v38-e30 345,5SU,UTH,HI,IR2,2cZ,ZFa,aJW,WVX,XQG,G7,76,6LC,CB,BMD,DE,EYA,A9,98,8ON,NbK,KP3,,,12*3*,F*G*,J*5*,J*I*,K*E*,P*Q*L*,R*S*M*,a*Y*O*.

Appendix A.4. Kochen-Specker' Class

7-v12-e9 12,23,34,456,6A9,987,7C1,,,5*1*,B8*3*.

12-v19-e14 12,2IA,AB,BC8,87,7E5,56,6D4,43,3FG,G9H,HJ1,,,9*A*,H*D*C*.

16-v30-e21 312,2E,EMN,NL8,8RC,CD,D7,76,6GH,HP9,9SA,AB,BQ4,45,5IJ,JO3,,,8*9*3*,F2*,KL*F,TO*B*,UP*D*.

18-v38-e27 34,4VD,DE,ETG,GF,FcO,ON,NHJ,JK,KSC,CB,BZ5,56,6Y9,9A,AX7,78,8W3,,,12,H*I,LM,PQ,RQ,UMI,aR2,bP1,QN*L.

18-v46-e33 56,6a8,87,7e9,9A,AcC,CB,BdD,DE,EbG,GF,FiP,PQ,QYX,XT,THJ,JK,Kh5,,,12,34,H*I,LM,NO,RS,T*US,VU,WU,ZRO,fI4,gM3,jW2,kV1,T*NL.

12-v54-e39 78,8oV,VW,WgX,XY,YZ,ZfU,UT,TpA,A9,9Ps,sN7,,,12,34,56,BC,DE,FG,HI,JK,LM,N*O,P*Q,RS,ab,cb,dcS,eaR,hK2,iI1,jM6,kOG,lQ5,mC4,nE3,qY*D,rbB,s*LH,s*JF.

References

1. Cabello, A.; D'Ambrosio, V.; Nagali, E.; Sciarrino, F. Hybrid Ququart-Encoded Quantum Cryptography Protected by Kochen-Specker Contextuality. *Phys. Rev. A* **2011**, *84*, 030302. [CrossRef]
2. Nagata, K. Kochen-Specker Theorem as a Precondition for Secure Quantum Key Distribution. *Phys. Rev. A* **2005**, *72*, 012325. [CrossRef]
3. Howard, M.; Wallman, J.; Veitech, V.; Emerson, J. Contextuality Supplies the 'Magic' for Quantum Computation. *Nature* **2014**, *510*, 351–355. [CrossRef] [PubMed]
4. Bartlett, S.D. Powered by Magic. *Nature* **2014**, *510*, 345–346. [CrossRef] [PubMed]
5. Kurzyński, P.; Cabello, A.; Kaszlikowski, D. Fundamental Monogamy Relation between Contextuality and Nonlocality. *Phys. Rev. Lett.* **2014**, *112*, 100401. [CrossRef] [PubMed]
6. Pavičić, M.; McKay, B.D.; Megill, N.D.; Fresl, K. Graph Approach to Quantum Systems. *J. Math. Phys.* **2010**, *51*, 102103. [CrossRef]
7. Megill, N.D.; Pavičić, M. Kochen-Specker Sets and Generalized Orthoarguesian Equations. *Ann. Henri Poinc.* **2011**, *12*, 1417–1429. [CrossRef]
8. Simon, C.; Żukowski, M.; Weinfurter, H.; Zeilinger, A. Feasible Kochen-Specker Experiment with Single Particles. *Phys. Rev. Lett.* **2000**, *85*, 1783–1786. [CrossRef] [PubMed]
9. Michler, M.; Weinfurter, H.; Żukowski, M. Experiments towards Falsification of Noncontextual Hidden Variables. *Phys. Rev. Lett.* **2000**, *84*, 5457–5461. [CrossRef] [PubMed]

10. Amselem, E.; Rådmark, M.; Bourennane, M.; Cabello, A. State-Independent Quantum Contextuality with Single Photons. *Phys. Rev. Lett.* **2009**, *103*, 160405. [CrossRef] [PubMed]
11. Liu, B.H.; Huang, Y.F.; Gong, Y.X.; Sun, F.W.; Zhang, Y.S.; Li, C.F.; Guo, G.C. Experimental Demonstration of Quantum Contextuality with Nonentangled Photons. *Phys. Rev. A* **2009**, *80*, 044101-1-4. [CrossRef]
12. D'Ambrosio, V.; Herbauts, I.; Amselem, E.; Nagali, E.; Bourennane, M.; Sciarrino, F.; Cabello, A. Experimental Implementation of a Kochen-Specker Set of Quantum Tests. *Phys. Rev. X* **2013**, *3*, 011012. [CrossRef]
13. Huang, Y.F.; Li, C.F.; Yong-Sheng Zhang, J.W.P.; Guo, G.C. Realization of All-or-nothing-type Kochen-Specker Experiment with Single Photons. *Phys. Rev. Lett.* **2002**, *88*, 240402.
14. Huang, Y.F.; Li, C.F.; Zhang, Y.S.; Pan, J.W.; Guo, G.C. Experimental Test of the Kochen-Specker Theorem with Single Photons. *Phys. Rev. Lett.* **2003**, *90*, 250401. [CrossRef] [PubMed]
15. Lapkiewicz, R.; Li, P.; Schaeff, C.; Langford, N.K.; Ramelow, S.; Wieśniak, M.; Zeilinger, A. Experimental Non-Classicality of an Indivisible Quantum System. *Nature* **2011**, *474*, 490–493. [CrossRef] [PubMed]
16. Zu, C.; Wang, Y.X.; Deng, D.L.; Chang, X.Y.; Liu, K.; Hou, P.Y.; Yang, H.X.; Duan, L.M. State-Independent Experimental Test of Quantum Contextuality in an Indivisible System. *Phys. Rev. Lett.* **2012**, *109*, 150401. [CrossRef] [PubMed]
17. Cañas, G.; Etcheverry, S.; Gómez, E.S.; Saavedra, C.; Xavier, G.B.; Lima, G.; Cabello, A. Experimental Implementation of an Eight-Dimensional Kochen-Specker Set and Observation of Its Connection with the Greenberger-Horne-Zeilinger Theorem. *Phys. Rev. A* **2014**, *90*, 012119. [CrossRef]
18. Cañas, G.; Arias, M.; Etcheverry, S.; Gómez, E.S.; Cabello, A.; Saavedra, C.; Xavier, G.B.; Lima, G. Applying the Simplest Kochen-Specker Set for Quantum Information Processing. *Phys. Rev. Lett.* **2014**, *113*, 090404. [CrossRef] [PubMed]
19. Zhan, X.; Zhang, X.; Li, J.; Zhang, Y.; Sanders, B.C.; Xue, P. Realization of the Contextuality-Nonlocality Tradeoff with a Qubit-Qutrit Photon Pair. *Phys. Rev. Lett.* **2016**, *116*, 090401. [CrossRef] [PubMed]
20. Li, T.; Zeng1, Q.; Song, X.; Zhang, X. Experimental Contextuality in Classical Light. *Sci. Rep.* **2017**, *7*, 44467. [CrossRef] [PubMed]
21. Li, T.; Zeng, Q.; Zhang, X.; Chen, T.; Zhang, X. State-Independent Contextuality in Classical Light. **2019**, in press.
22. Frustaglia, D.; Baltanás, J.P.; Velázquez-Ahumada, M.C.; Fernández-Prieto, A.; Lujambio, A.; Losada, V.; Freire, M.J.; Cabello, A. Classical Physics and the Bounds of Quantum Correlations. *Phys. Rev. Lett.* **2016**, *116*, 250404. [CrossRef] [PubMed]
23. Zhang, A.; Xu, H.; Xie, J.; Zhang, H.; Smith, B.J.; Kim, M.S.; Zhang, L. Experimental Test of Contextuality in Quantum and Classical Systems. *Phys. Rev. Lett.* **2004**, *122*, 080401. [CrossRef] [PubMed]
24. Hasegawa, Y.; Loidl, R.; Badurek, G.; Baron, M.; Rauch, H. Quantum Contextuality in a Single-Neutron Optical Experiment. *Phys. Rev. Lett.* **2006**, *97*, 230401. [CrossRef] [PubMed]
25. Cabello, A.; Filipp, S.; Rauch, H.; Hasegawa, Y. Proposed Experiment for Testing Quantum Contextuality with Neutrons. *Phys. Rev. Lett.* **2008**, *100*, 130404. [CrossRef] [PubMed]
26. Bartosik, H.; Klepp, J.; Schmitzer, C.; Sponar, S.; Cabello, A.; Rauch, H.; Hasegawa, Y. Experimental Test of Quantum Contextuality in Neutron Interferometry. *Phys. Rev. Lett.* **2009**, *103*, 040403. [CrossRef] [PubMed]
27. Kirchmair, G.; Zähringer, F.; Gerritsma, R.; Kleinmann, M.; Gühne, O.; Cabello, A.; Blatt, R.; Roos, C.F. State-Independent Experimental Test of Quantum Contextuality. *Nature* **2009**, *460*, 494–497. [CrossRef]
28. Moussa, O.; Ryan, C.A.; Cory, D.G.; Laflamme, R. Testing Contextuality on Quantum Ensembles with One Clean Qubit. *Phys. Rev. Lett.* **2010**, *104*, 160501. [CrossRef] [PubMed]
29. Jerger, M.; Reshitnyk, Y.; Oppliger, M.; Potočnik, A.; Mondal, M.; Wallraff, A.; Goodenough, K.; Wehner, S.; Juliusson, K.; Langford, N.K.; Fedorov, A. Contextuality without Nonlocality in a Superconducting Quantum System. *Nat. Commun.* **2016**, *7*, 12930. [CrossRef] [PubMed]
30. Barrett, J.; Kent, A. Noncontextuality, Finite Precision Measurement and the Kochen-Specker. *Stud. Hist. Philos. Mod. Phys.* **2004**, *35*, 151–176. [CrossRef]
31. Kunjwal, R.; Spekkens, R.W. From the Kochen-Specker Theorem to Noncontextuality Inequalities without Assuming Determinism. *Phys. Rev. Lett.* **2015**, *115*, 110403. [CrossRef] [PubMed]
32. Kunjwal, R. Hypergraph Framework for Irreducible Noncontextuality Inequalities from Logical Proofs of the Kochen-Specker Theorem. *arXiv* **2018**, arXiv:1805.02083. Available online: https://arxiv.org/abs/1805.02083 (accessed on 12 November 2019).

33. Bengtsson, I.; Blanchfield, K.; Cabello, A. A Kochen–Specker Inequality from a SIC. *Phys. Lett. A* **2012**, *376*, 374–376. [CrossRef]
34. Cabello, A.; Estebaranz, J.M.; García-Alcaine, G. Bell-Kochen-Specker Theorem: A Proof with 18 Vectors. *Phys. Lett. A* **1996**, *212*, 183–187. [CrossRef]
35. Pavičić, M.; Merlet, J.P.; McKay, B.D.; Megill, N.D. Kochen-Specker Vectors. *J. Phys. A* **2005**, *38*, 1577–1592. [CrossRef]
36. Waegell, M.; Aravind, P.K. Critical Noncolorings of the 600-Cell Proving the Bell-Kochen-Specker Theorem. *J. Phys. A* **2010**, *43*, 105304. [CrossRef]
37. Waegell, M.; Aravind, P.K. Parity Proofs of the Kochen-Specker Theorem Based on 60 Complex Rays in Four Dimensions. *J. Phys. A* **2011**, *44*, 505303. [CrossRef]
38. Megill, N.D.; Fresl, K.; Waegell, M.; Aravind, P.K.; Pavičić, M. Probabilistic Generation of Quantum Contextual Sets. *Phys. Lett. A* **2011**, *375*, 3419–3424. [CrossRef]
39. Pavičić, M.; Megill, N.D.; Aravind, P.K.; Waegell, M. New Class of 4-Dim Kochen-Specker Sets. *J. Math. Phys.* **2011**, *52*, 022104. [CrossRef]
40. Waegell, M.; Aravind, P.K.; Megill, N.D.; Pavičić, M. Parity Proofs of the Bell-Kochen-Specker Theorem Based on the 600-cell. *Found. Phys.* **2011**, *41*, 883–904. [CrossRef]
41. Waegell, M.; Aravind, P.K. Proofs of Kochen-Specker Theorem Based on a System of Three Qubits. *J. Phys. A* **2012**, *45*, 405301. [CrossRef]
42. Waegell, M.; Aravind, P.K. Proofs of the Kochen-Specker Theorem Based on the N-Qubit Pauli Group. *Phys. Rev. A* **2013**, *88*, 012102. [CrossRef]
43. Waegell, M.; Aravind, P.K. Parity Proofs of the Kochen-Specker Theorem Based on 120-Cell. *Found. Phys.* **2014**, *44*, 1085–1095. [CrossRef]
44. Waegell, M.; Aravind, P.K. Parity Proofs of the Kochen-Specker Theorem Based on the Lie Algebra E8. *J. Phys. A* **2015**, *48*, 225301. [CrossRef]
45. Waegell, M.; Aravind, P.K. The Penrose Dodecahedron and the Witting Polytope Are Identical in $\mathbb{CP}^3$. *Phys. Lett. A* **2017**, *381*, 1853–1857. [CrossRef]
46. Pavičić, M. Arbitrarily Exhaustive Hypergraph Generation of 4-, 6-, 8-, 16-, and 32-Dimensional Quantum Contextual Sets. *Phys. Rev. A* **2017**, *95*, 062121. [CrossRef]
47. Pavičić, M.; Megill, N.D. Vector Generation of Quantum Contextual Sets in Even Dimensional Hilbert Spaces. *Entropy* **2018**, *20*, 928. [CrossRef]
48. Pavičić, M.; Waegel, M.; Megill, N.D.; Aravind, P. Automated Generation of Kochen-Specker Sets. *Sci. Rep.* **2019**, *9*, 6765. [CrossRef] [PubMed]
49. Yu, S.; Oh, C.H. State-Independent Proof of Kochen-Specker Theorem with 13 Rays. *Phys. Rev. Lett.* **2012**, *108*, 030402. [CrossRef] [PubMed]
50. Xu, Z.P.; Chen, J.L.; Su, H.Y. State-independent contextuality sets for a qutrit. *Phys. Lett. A* **2015**, *379*, 1868–1870. [CrossRef]
51. Ramanathan, R.; Horodecki, P. Necessary and Sufficient Condition for State-Independent Contextual Measurement Scenarios. *Phys. Rev. Lett.* **2014**, *112*, 040404. [CrossRef] [PubMed]
52. Cabello, A.; Kleinmann, M.; Budroni, C. Necessary and Sufficient Condition for Quantum State-Independent Contextuality. *Phys. Rev. Lett.* **2014**, *114*, 250402. [CrossRef] [PubMed]
53. Klyachko, A.A.; Can, M.A.; Binicioğlu, S.; Shumovsky, A.S. Simple Test for Hidden Variables in Spin-1 Systems. *Phys. Rev. A* **2008**, *101*, 020403. [CrossRef] [PubMed]
54. Kurzyński, P.; Ramanathan, R.; Kaszlikowski, D. Entropic Test of Quantum Contextuality. *Phys. Rev. Lett.* **2012**, *109*, 020404. [CrossRef] [PubMed]
55. Braunstein, S.L.; Caves, C.M. Information- Theoretic Bell Inequalities. *Phys. Rev. Lett.* **1988**, *61*, 662–665. [CrossRef] [PubMed]
56. Kochen, S.; Specker, E.P. The problem of hidden variables in quantum mechanics. *J. Math. Mech.* **1967**, *17*, 59–87. [CrossRef]
57. McKay, B.D.; Megill, N.D.; Pavičić, M. Algorithms for Greechie Diagrams. *Int. J. Theor. Phys.* **2000**, *39*, 2381–2406. [CrossRef]
58. Pavičić, M.; Megill, N.D.; Merlet, J.P. New Kochen-Specker Sets in Four Dimensions. *Phys. Lett. A* **2010**, *374*, 2122–2128. [CrossRef]

59. Gleason, A.M. Measures on the closed subspaces of a Hilbert space. *J. Math. Mech.* **1957**, *6*, 885–893. [CrossRef]
60. Zimba, J.; Penrose, R. On Bell Non-Locality without Probabilities: More Curious Geometry. *Stud. Hist. Phil. Sci.* **1993**, *24*, 697–720. [CrossRef]
61. Peres, A. Two Simple Proofs of the Bell-Kochen-Specker Theorem. *J. Phys. A* **1991**, *24*, L175–L178. [CrossRef]
62. Bengtsson, I. Gleason, Kochen-Specker, and a Competition that Never Was. In Proceedings of the AIP, Vaxjo, Sweden, 11–14 June 2012; Volume 1508, pp. 125–135.
63. Larsson, J.Å. A Kochen-Specker Inequality. *Europhys. Lett.* **2002**, *58*, 799–805. [CrossRef]
64. Held, C. Kochen-Specker Theorem. In *Compendium of Quantum Physics*; Greenberger, D., Hentschel, K., Weinert, F., Eds.; Springer: New York, NY, USA, 2009; pp. 331–335.
65. Ruuge, A.E. New Examples of Kochen–Specker-Type Configurations on Three Qubits. *J. Phys. A* **2012**, *45*, 465304. [CrossRef]
66. Cabello, A.; Amselem, E.; Blanchfield, K.; Bourennane, M.; Bengtsson, I. Proposed Experiments of Qutrit State-Independent Contextuality and Two-Qutrit Contextuality-Based Nonlocality. *Phys. Rev. A* **2012**, *85*, 032108. [CrossRef]
67. Kleinmann, M.; Budroni, C.; Larsson, J.Å.; Gühne, O.; Cabello, A. Optimal Inequalities for State-Independent Contextuality. *Phys. Rev. Lett.* **2012**, *109*, 250402. [CrossRef] [PubMed]
68. Clifton, R. Getting Contextual and Nonlocal Elements-of-Reality the Easy Way. *Am. J. Phys.* **1993**, *61*, 443–447. [CrossRef]
69. Svozil, K. *Quantum Logic*; Discrete Mathematics and Theoretical Computer Science; Springer-Verlag: New York, NY, USA, 1998.
70. Held, C. The Kochen-Specker Theorem. In *The Stanford Encyclopedia of Philosophy*; Zalta, E.N., Ed.; Stanford University: Stanford, CA, USA, 2018. Available online: https://plato.stanford.edu/archives/spr2018/entries/kochen-specker (accessed on 12 November 2019).
71. Bub, J. Schütte's Tautology and the Kochen-Specker Theorem. *Found. Phys.* **1996**, *26*, 787–806. [CrossRef]
72. Cabello, A.; Severini, S.; Winter, A. Graph-Theoretic Approach to Quantum Correlations. *Phys. Rev. Lett.* **2014**, *112*, 040401. [CrossRef] [PubMed]
73. Cabello, A. Quantum Correlations from Simple Assumptions. *Phys. Rev. A* **2019**, *100*, 032120. [CrossRef]
74. Pavičić, M.; Megill, N.D. Quantum Logic and Quantum Computation. In *Handbook of Quantum Logic and Quantum Structures*; Engesser, K., Gabbay, D., Lehmann, D., Eds.; Elsevier: Amsterdam, The Netherlands, 2007; Chapter Quantum Structures, pp. 751–787.
75. Megill, N.D.; Pavičić, M. New Classes of Kochen-Specker Contextual Sets (Invited Talk). In Proceedings of the 40th MIPRO Convention, Opatija, Croatia, 22–26 May 2017; IEEE: Piscataway, NJ, USA, 2017; pp. 195–200, ISBN 9781509049691.

13

Uniqueness of Minimax Strategy in View of Minimum Error Discrimination of Two Quantum States

Jihwan Kim, Donghoon Ha and Younghun Kwon *

Department of Applied Physics, Hanyang University, Ansan, Kyunggi-Do 425-791, Korea
* Correspondence: yyhkwon@hanyang.ac.kr

Abstract: This study considers the minimum error discrimination of two quantum states in terms of a two-party zero-sum game, whose optimal strategy is a minimax strategy. A minimax strategy is one in which a sender chooses a strategy for a receiver so that the receiver may obtain the minimum information about quantum states, but the receiver performs an optimal measurement to obtain guessing probability for the quantum ensemble prepared by the sender. Therefore, knowing whether the optimal strategy of the game is unique is essential. This is because there is no alternative if the optimal strategy is unique. This paper proposes the necessary and sufficient condition for an optimal strategy of the sender to be unique. Also, we investigate the quantum states that exhibit the minimum guessing probability when a sender's minimax strategy is unique. Furthermore, we show that a sender's minimax strategy and a receiver's minimum error strategy cannot be unique if one can simultaneously diagonalize two quantum states, with the optimal measurement of the minimax strategy. This implies that a sender can confirm that the optimal strategy of only a single side (a sender or a receiver but not both of them) is unique by preparing specific quantum states.

Keywords: quantum state discrimination; quantum minimax; uniqueness of strategy; guessing probability

---

1. Introduction

Quantum information processing can be achieved by discriminating quantum states, where classical information is encoded. Quantum states which are orthogonal to each other can be perfectly distinguishable. However, non-orthogonal quantum states cannot be perfectly discriminated. Therefore, one needs to have a discrimination strategy for non-orthogonal quantum states, and there are various strategies [1–4] such as minimum error discrimination (MD) [4–7], unambiguous discrimination [8–12], maximum confidence discrimination [13], and discrimination of fixed rate inconclusive result [14–18]. Unambiguous discrimination is a strategy where there is no error in the conclusive result by allowing an inconclusive result. Maximum confidence is a strategy where one maximizes the confidence of a conclusive result. Discrimination of fixed rate inconclusive result is a strategy where one may fix the rate of an inconclusive result. Among these strategies, the MD strategy can conclusively discriminate quantum states with a prior probability.

The MD strategy is employed for quantum states with a given prior probability, and the quantum states are optimally measured. MD strategy is that one maximizes the probability that the result of measurement of a receiver correctly points out the quantum state that a sender transmitted when only a conclusive result is permitted. The maximum probability is called guessing probability. One can investigate the behavior of MD in terms of a prior probability when quantum states are given.

Because the guessing probability is obtained based on prior probability, a change in prior probability results in different guessing probabilities, which implies that prior probabilities can be considered as a strategy of a sender. Even though one has discussed the uniqueness of measurement strategy in discrimination of two quantum states, the strategy of preparation such as a prior probability, which can be a strategy of a sender, has not been discussed in terms of identical guessing probability and optimal measurement strategy.

Quantum minimax approach is obtained by applying the minimax approach of a statistical decision to quantum state discrimination. Von Neumann, the inventor of game theory, showed that there exists a solution to the minimax problem when sender and receiver can choose a finite number of strategies in a two-person zero-sum game. Wald proved that the necessary and sufficient condition to the existence of a solution to the minimax problem is that the set of strategy for sender and receiver is countable [19]. Hirota and Ikehara discussed quantum minimax theorem, using the fact that the set of measurement strategy satisfies compactness [20]. They suggested the necessary and sufficient condition for minimax strategy in quantum state discrimination.

Further, by mean value theorem D'Ariano showed that there exists a quantum minimax strategy for two quantum state discrimination and provided a sufficient condition for the strategy [21]. However, in spite of these studies, the necessary and sufficient condition for uniqueness of minimax strategy in two quantum state discrimination is not known yet. Even more, the uniqueness of minimax strategy in two quantum state discrimination is not understood in terms of sender's strategy, which is a selection of prior probability.

This study investigates a two-person zero-sum game where the payoff is defined by the correct probability of two quantum states [19–22]. The optimal strategy of the game is a minimax strategy, where the minimax strategy of a receiver is to select the optimal measurement providing MD and the minimax strategy of a sender is to choose the prior probability providing the minimum of guessing probability, which is displayed in Figure 1.

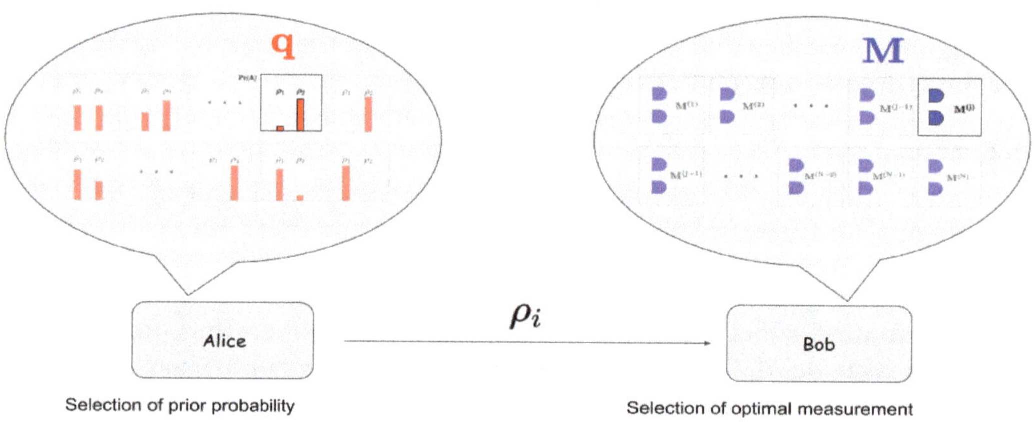

Figure 1. The strategy of the sender(Alice) and the receiver(Bob) in two-person zero-sum quantum game. The strategy of Alice is to choose the optimal prior probability **q**, which is the probability of quantum states prepared in the quantum system, to minimize the payoff. The strategy of Bob is to choose the optimal measurement to maximize payoff.

In this scenario, the prior probability and the measurement in MD are constructed as the strategy of a sender Alice and a receiver Bob [20,21]. First, Alice sends the quantum states, where classical information ($x = 1, 2$) is encoded, to Bob. Because the quantum states are not orthogonal to each other, a single measurement of Bob cannot perfectly discriminate the quantum states. Therefore, a suitable strategy is needed. Here Bob should choose a measurement strategy that can perform MD.

Meanwhile, a suitable selection of prior probability can be obtained by Alice, as a sender's strategy. Alice's strategy is to interfere with the minimum error strategy of Bob to minimize the guessing probability. Because Bob should perform MD without noticing Alice's strategy, Bob tries to find an optimal strategy to obtain a payoff. Therefore, the minimum of guessing probability implies that a suitable selection of prior probability lets Bob obtain the minimum of guessing probability when Bob performs an optimal measurement. Furthermore, if Bob cannot perform an optimal measurement, he obtains a probability less than the guessing probability.

The quantum minimax theorem [20,21] can be used to prove that Alice and Bob can set up an optimal strategy on both sides. However, it is not known whether the minimax strategy is unique or not. The uniqueness of the optimal strategy of the game is important in performing the game. There is no alternative to a unique optimal strategy. Therefore, a strategy cannot be optimal if an error occurs when performing the strategy. However, a strategy can still be optimal if it is not unique, even though an error occurs in the strategy. In this light, it is crucial to know whether the minimax strategy is unique, when the strategy is optimal in the game. Here, we investigate the condition for uniqueness of the optimal strategy of a sender. The condition is described by the quantum states and the minimax strategy of a receiver. More explicitly, we study the necessary and sufficient condition for the uniqueness of a sender's strategy. Using the condition, we investigate the quantum states that exhibit the minimum of guessing probability when a sender's minimax strategy is unique.

Also, we show that a sender's minimax strategy and a receiver's minimum error strategy cannot be unique if two quantum states are simultaneously diagonalized with the optimal measurement of minimax strategy. Therefore, a sender can make the optimal strategy of only a single side unique by preparing specific quantum states. Our investigation can be applied to various fields. As the first example of our investigation, we explain how the BB84 protocol [23] with equal prior probability is optimal in terms of the minimax strategy. We also discuss how the results of this study can be applied to building a quantum random number generator(QRNG) [24–26].

This paper is organized as follows. In Section 2, we explain the necessary background of our investigation. In Section 3, for the minimax strategy of a sender, we provide the necessary and sufficient condition for uniqueness of the optimal strategy. We investigate the uniqueness of the strategy of the sender for some quantum states by using this condition. Furthermore, we obtain the condition under which both the sender's minimax strategy and the receiver's optimal minimum error strategy cannot be unique. Finally, we discuss the results and conclusions in Section 4.

2. Preliminaries

For two quantum states ρ_1 and ρ_2, the minimal subspace $\mathcal{H}$ for discriminating ρ_1 and ρ_2 should satisfy $\mathcal{H} = \mathrm{Supp}(\rho_1 + \rho_2)$. In this study, we assume that the rank of quantum state is finite. Then, by the relation $\dim \mathcal{H} \leq \mathrm{rank}(\rho_1) + \mathrm{rank}(\rho_2)$, a quantum state or an optimal measurement can be represented as an operator on finite dimensional Hilbert space.

The MD of two quantum states ρ_1 and ρ_2 is a strategy to determine the maximum value of correct probability $\mathrm{P}_{\mathrm{corr}} = q\mathrm{tr}(\rho_1 M_1) + (1-q)\mathrm{tr}(\rho_2 M_2)$, which is called guessing probability, by performing an optimal measurement. The maximum value of the correct probability is known as Helstrom bound [27].

Assuming that the prior probabilities of two quantum states ρ_1 and ρ_2 are q and $1-q$, respectively, one can obtain the following lemma in the MD of the two quantum states. (The proof can be found in the Appendix A).

Lemma 1 (Optimal condition of MD for two quantum states [27,28]). *The necessary and sufficient condition for optimal measurement $\{M_x\}_{x=1}^2$ is given by*

$$(-1)^x \left((1-q)\rho_2 - q\rho_1\right) M_x \geq 0 \quad \forall x \in \{1,2\}. \tag{1}$$

In general, the optimal measurement in MD is not unique. If the nullity of operator $\Lambda \equiv (1-q)\rho_2 - q\rho_1$ is d, there exist at least 2^d number of optimal extreme POVMs. A convex combination of these POVM also provides an optimal measurement of MD. When Λ has full rank, the optimal measurement is unique. Quantum minimax theorem tells that among optimal MD strategies there is at least a POVM of minimax strategy in a prior probability providing the minimum of guessing probability [20,21].

Theorem 1 (Quantum minimax theorem [20,21]). *There exists an a priori probability* $\mathbf{q}^\star = (q^\star, 1 - q^\star)$ *for the states* ρ_1 *and* ρ_2*, and a measurement* $\mathbf{M}^\star = (M_1^\star, M_2^\star)$ *such that*

$$\min_{\mathbf{q}}\max_{\mathbf{M}} \mathrm{P}_{\mathrm{corr}}(\mathbf{q}, \mathbf{M}) = \mathrm{P}_{\mathrm{corr}}(\mathbf{q}^\star, \mathbf{M}^\star) = \max_{\mathbf{M}}\min_{\mathbf{q}} \mathrm{P}_{\mathrm{corr}}(\mathbf{q}, \mathbf{M}) \tag{2}$$

where $q^\star \in (0,1)$, $P_{\mathrm{corr}}(\mathbf{q}, \mathbf{M}) = \sum_{x=1}^{2} q_x \mathrm{tr}(\rho_x M_x)$.

Note that when quantum states are prepared in a prior probability $\mathbf{q}$, $\max_{\mathbf{M}}\min_{\mathbf{q}} \mathrm{P}_{\mathrm{corr}}(\mathbf{q}, \mathbf{M}) = \mathrm{P}_{\mathrm{corr}}(\mathbf{q}^\star, \mathbf{M}^\star)$ implies that the measurement of $\mathbf{M}^\star$ is optimal and $\min_{\mathbf{q}}\max_{\mathbf{M}} \mathrm{P}_{\mathrm{corr}}(\mathbf{q}, \mathbf{M}) = \mathrm{P}_{\mathrm{corr}}(\mathbf{q}^\star, \mathbf{M}^\star)$ implies that the prior probability of $\mathbf{q}^\star$ provides the minimum of guessing probability. However, every optimal MD in the prior probability of $\mathbf{q}^\star$ is not a minimax strategy of Bob. Suppose that a measurement of $\mathbf{N} = (N_1, N_2)$ in the prior probability of $\mathbf{q}^\star$ is an optimal strategy of MD, satisfying $\mathrm{tr}(\rho_1 N_1) > \mathrm{tr}(\rho_2 N_2) > 0$. Then, the strategy of Alice in $\mathbf{q}^\star$ cannot be a prior probability for the minimax strategy, as $\tilde{\mathbf{q}} = (0,1)$ of Alice's strategy provides a lower guessing probability than that of $\mathbf{q}^\star$:

$$P_{\mathrm{corr}}(\mathbf{q}^\star, \mathbf{N}) = q^\star \mathrm{tr}(\rho_1 N_1) + (1 - q^\star)\mathrm{tr}(\rho_2 N_2) > \mathrm{tr}(\rho_2 N_2) = P_{\mathrm{corr}}(\tilde{\mathbf{q}}, \mathbf{N}). \tag{3}$$

Therefore, the first condition that the minimax strategy $\mathbf{M}^\star$ of Bob should satisfy is $\mathrm{tr}(\rho_1 M_1^\star) = \mathrm{tr}(\rho_2 M_2^\star)$. Because the measurement of $\mathbf{M}^\star$ is an optimal strategy for the prior probability of $\mathbf{q}^\star$, it satisfies the optimal condition of MD, which is the second condition. Inversely, the fulfillment of the two conditions is the sufficient condition for the minimax strategy.

Here, the conditions can be explained as follows. Suppose that a measurement $M^\circ = (M_1^\circ, M_2^\circ)$ satisfies $\mathrm{tr}(\rho_1 M_1^\circ) = \mathrm{tr}(\rho_2 M_2^\circ)$ and is optimal for the prior probability of $\mathbf{q}^\circ$. Then, we find the following relation:

$$\min_{\mathbf{q}}\max_{\mathbf{M}} \mathrm{P}_{\mathrm{corr}}(\mathbf{q}, \mathbf{M}) \leq \max_{\mathbf{M}} \mathrm{P}_{\mathrm{corr}}(\mathbf{q}^\circ, \mathbf{M}) = \mathrm{P}_{\mathrm{corr}}(\mathbf{q}^\circ, \mathbf{M}^\circ) = \min_{\mathbf{q}} \mathrm{P}_{\mathrm{corr}}(\mathbf{q}, \mathbf{M}^\circ). \tag{4}$$

The last equality holds by $\mathrm{tr}(\rho_1 M_1^\circ) = \mathrm{tr}(\rho_2 M_2^\circ)$. Because of $\min_{\mathbf{q}}\max_{\mathbf{M}} \mathrm{P}_{\mathrm{corr}}(\mathbf{q}, \mathbf{M}) \geq \min_{\mathbf{q}} \mathrm{P}_{\mathrm{corr}}(\mathbf{q}, \mathbf{M}^\circ)$, we find $\min_{\mathbf{q}}\max_{\mathbf{M}} \mathrm{P}_{\mathrm{corr}}(\mathbf{q}, \mathbf{M}) = \min_{\mathbf{q}} \mathrm{P}_{\mathrm{corr}}(\mathbf{q}^\circ, \mathbf{M}^\circ)$. And the following relation holds:

$$\max_{\mathbf{M}}\min_{\mathbf{q}} \mathrm{P}_{\mathrm{corr}}(\mathbf{q}, \mathbf{M}) \leq \min_{\mathbf{q}} \mathrm{P}_{\mathrm{corr}}(\mathbf{q}, \mathbf{M}^\circ) = \mathrm{P}_{\mathrm{corr}}(\mathbf{q}^\circ, \mathbf{M}^\circ) = \max_{\mathbf{M}} \mathrm{P}_{\mathrm{corr}}(\mathbf{q}^\circ, \mathbf{M}) \tag{5}$$

The first equality is obtained by $\mathrm{tr}(\rho_1 M_1^\circ) = \mathrm{tr}(\rho_2 M_2^\circ)$. Because of $\max_{\mathbf{M}}\min_{\mathbf{q}} \mathrm{P}_{\mathrm{corr}}(\mathbf{q}, \mathbf{M}) \geq \max_{\mathbf{M}} \mathrm{P}_{\mathrm{corr}}(\mathbf{q}^\circ, \mathbf{M})$, we obtain $\max_{\mathbf{M}}\min_{\mathbf{q}} \mathrm{P}_{\mathrm{corr}}(\mathbf{q}, \mathbf{M}) = \max_{\mathbf{M}} \mathrm{P}_{\mathrm{corr}}(\mathbf{q}^\circ, \mathbf{M}^\circ)$ and $\min_{\mathbf{q}}\max_{\mathbf{M}} \mathrm{P}_{\mathrm{corr}}(\mathbf{q}, \mathbf{M}) = \mathrm{P}_{\mathrm{corr}}(\mathbf{q}^\circ, \mathbf{M}^\circ) = \max_{\mathbf{M}}\min_{\mathbf{q}} \mathrm{P}_{\mathrm{corr}}(\mathbf{q}, \mathbf{M})$. It implies that $(\mathbf{q}^\circ, \mathbf{M}^\circ)$ is a minimax strategy. Then, one can obtain the following lemma.

Lemma 2. *When MD is performed for a given prior probability, the minimum of guessing probability is obtained iff an optimal measurement* $\{M_x\}_{x=1}^2$ *satisfies* $\mathrm{tr}(\rho_1 M_1) = \mathrm{tr}(\rho_2 M_2)$.

3. Results

This section presents the necessary and sufficient condition that ensures the uniqueness of the minimax strategy of a sender. Because there always exists a minimax strategy for the quantum minimax theorem, when one finds a minimax strategy, one can obtain the condition by which the strategy is unique. When MDs with different prior probabilities can provide the same guessing probability, the following lemma provides the condition by which the MDs with different prior probabilities can have the same optimal measurement (The proof of this lemma can be found in the Appendix A).

Lemma 3. *The quantum ensembles of S_1 and S_2 are given as $\{p_x, \rho_x\}_{x=1}^2$ and $\{q_x, \rho_x\}_{x=1}^2$, respectively, where $p_1 \neq q_1$. Suppose that in the MD of a quantum ensemble S_x, the guessing probability is $p_{\text{guess}}^{(x)}$ and the minimum value of guessing probability is $p_{\text{guess}}^{\star}$. Then, when $p_{\text{guess}}^{(1)} = p_{\text{guess}}^{(2)}$, if there exists an measurement that can simultaneously perform MD on two quantum ensembles S_1 and S_2, one can obtain $p_{\text{guess}}^{(1)} = p_{\text{guess}}^{\star}$.*

Note that the optimal measurement performing simultaneous MD on two quantum ensemble S_1 and S_2 satisfies the equal probabilities of correct detection. It is the minimax strategy of the receiver. Here, the set of prior probability providing the minimum of guessing probability is a convex set. It can be shown in the following way. Suppose that the prior probabilities of $\mathbf{q}$ and $\mathbf{p}$ provide the minimum of guessing probability $p_{\text{guess}}^{\star}$. Then, by Lemma 3 there exists a measurement $\mathbf{M}$ that can perform MD on both the quantum states, satisfying $\sum_{x=1}^{2} q_x \text{tr}(\rho_x M_x) = p_{\text{guess}}^{\star} = \sum_{x=1}^{2} p_x \text{tr}(\rho_x M_x)$. Now, one can see that the relation of $\sum_{x=1}^{2}(\theta q_x + (1-\theta)p_x)\text{tr}(\rho_x M_x) = p_{\text{guess}}^{\star}$ holds for $\theta \in [0,1]$. If one assumes that the minimax strategy $(\mathbf{q}, \mathbf{M})$ is not unique and there is another strategy $\mathbf{p}$ for a sender, then the minimax strategy of the sender forms a convex set, and one can find the prior probability where $\mathbf{M}$ is optimal in the ϵ-neighborhood of $\mathbf{q}$ for an arbitrary positive number of ϵ. Therefore, one can check the uniqueness of the prior probability of $\mathbf{q}$ providing the minimum of guessing probability, by deciding whether there exists a prior probability exhibiting optimal $\mathbf{M}$ in the ϵ neighborhood of $\mathbf{q}$ after finding the optimal POVM $\mathbf{M}$ for minimax strategy in the prior probability of $\mathbf{q}$ providing the minimum of guessing probability. Proposition 1 shows the necessary and sufficient condition for the non-uniqueness of a prior probability $\mathbf{q}$ of which $\mathbf{M}$ is optimal in the ϵ neighborhood (The proof of this proposition can be found in the Appendix A).

Proposition 1. *The prior probability providing the minimum of guessing probability is not unique if and only if $\{M_x\}_{x=1}^2$ satisfies the following conditions.*

1. $[\rho_x, M_1] = 0 \quad \forall x \in \{1,2\}$,
2. *For some $x \in \{1,2\}$, every $|v\rangle \in \text{Supp}(M_x)$ satisfies $\langle v|\rho_1|v\rangle : \langle v|\rho_2|v\rangle \neq 1-q:q$.*

where $[A,B] = AB - BA$.

Lemma 2 and Proposition 1 can be applied to check whether the strategy under a situation is unique. By applying Lemma 2, one can explain why the identical prior probability in the BB84 protocol is the best strategy of a sender. The quantum states used in the BB84 protocol are $\{|0\rangle, |1\rangle, |+\rangle, |-\rangle\}$ [23]. Alice encodes (a_0, a_1) into quantum states and sends them to Bob. In general, a_0 is selected by Alice, but a_1 is randomly chosen. Suppose that Table 1 is used for encoding bit. Here, encoding means that a quantum state corresponding to a_0a_1 is prepared for communication.

Table 1. Encoding table for Alice.

a_0a_1	**Quantum States**
00	$\|0\rangle$
01	$\|1\rangle$
10	$\|-\rangle$
11	$\|+\rangle$

When Alice chooses 0 as the value of a_0, the quantum state is determined by a_1. If the value of a_1 is 0, the quantum state becomes $|0\rangle$. However, when a_1 is 1, $|1\rangle$ is prepared for the quantum state. If the quantum state does not interact with the environment, Bob receives the quantum state prepared by Alice. Then, Bob performs the following measurements:

$$M_0 = \{|0\rangle\langle 0|, |1\rangle\langle 1|\}, \quad M_1 = \{|+\rangle\langle +|, |-\rangle\langle -|\} \tag{6}$$

When for the quantum states $\{|0\rangle, |1\rangle, |+\rangle, |-\rangle\}$ the prior probability of the quantum states is identical, the optimal measurement becomes

$$M = \{\frac{1}{2}|0\rangle\langle 0|, \frac{1}{2}|1\rangle\langle 1|, \frac{1}{2}|+\rangle\langle +|, \frac{1}{2}|-\rangle\langle -|\}. \tag{7}$$

The optimal measurement satisfies the following relation:

$$\mathrm{tr}(|0\rangle\langle 0| \frac{1}{2}|0\rangle\langle 0|) = \mathrm{tr}(|1\rangle\langle 1| \frac{1}{2}|1\rangle\langle 1|) = \mathrm{tr}(|+\rangle\langle +| \frac{1}{2}|+\rangle\langle +|) = \mathrm{tr}(|-\rangle\langle -| \frac{1}{2}|-\rangle\langle -|) = 0.5 \tag{8}$$

It implies that the identical prior probability provides the minimum of guessing probability. It is because if there exists an optimal measurement satisfying the above condition, the prior probability provides the minimum of guessing probability. It should be noted that Lemma 3 implies that the measurement of M is optimal for every prior probability providing the minimum of guessing probability. Meanwhile, if the prior probability is not identical, all the quantum states in the BB84 protocol does not have M as an optimal measurement. It can be shown in the following way. Let us assume that the probability of a_1 to become 0 or 1 is equal, and the probability of a_0 to become 0 or 1 is q. Then, the prior probability of each quantum state becomes $q/2, q/2, (1-q)/2, (1-q)/2$. We can show that if the measurement M is optimal at $q \neq 0.5$, the following inequalities should be satisfied [2,28]:

$$\frac{q}{4}|0\rangle\langle 0| + \frac{q}{4}|1\rangle\langle 1| + \frac{1-q}{4}|+\rangle\langle +| + \frac{1-q}{4}|-\rangle\langle -| - \frac{q}{2}|0\rangle\langle 0| \geq 0 \tag{9}$$

$$\frac{q}{4}|0\rangle\langle 0| + \frac{q}{4}|1\rangle\langle 1| + \frac{1-q}{4}|+\rangle\langle +| + \frac{1-q}{4}|-\rangle\langle -| - \frac{1-q}{2}|+\rangle\langle +| \geq 0 \tag{10}$$

However, when $q \neq 0.5$, one of these inequalities cannot be satisfied. Therefore, the prior probability providing the minimum of the guessing probability is only the case of $q = 1/2$.

Using Proposition 1, we can investigate the quantum states of the unique prior probability, which provides the minimum of the guessing probability. Here we consider the MD of the following two quantum states:

$$\rho_1 = \frac{2}{3}|\phi^-\rangle\langle\phi^-| + \frac{I}{12}, \tag{11}$$

$$\rho_2 = \frac{1}{3}|\phi^-\rangle\langle\phi^-| + \frac{I}{6}. \tag{12}$$

From Figure 2, we can check whether the prior probability providing the minimum of guessing probability is unique. We can see that the prior probabilities providing the minimum of guessing probability are $q_1 = \frac{2}{5}$ and $q_2 = \frac{3}{5}$. The optimal measurement for the quantum ensemble is $\{M_1 = \frac{4}{5}|\phi^-\rangle\langle\phi^-|, M_2 = I - \frac{4}{5}|\phi^-\rangle\langle\phi^-|\}$, since the measurement satisfies Lemma 1 as follows:

$$(q\rho_1 - (1-q)\rho_2)M_1 = \left(\frac{2}{5}\rho_1 - \frac{3}{5}\rho_2\right)M_1 = \left(\frac{1}{15}|\phi^-\rangle\langle\phi^-| - \frac{1}{15}I\right)\frac{4}{5}|\phi^-\rangle\langle\phi^-| = 0 \tag{13}$$

$$\begin{aligned}((1-q)\rho_2 - q\rho_1)M_2 &= \left(\frac{3}{5}\rho_2 - \frac{2}{5}\rho_1\right)M_2 \\ &= \left(\frac{1}{15}I - \frac{1}{15}|\phi^-\rangle\langle\phi^-|\right)\left(I - \frac{4}{5}|\phi^-\rangle\langle\phi^-|\right) = \frac{1}{15}\left(I - |\phi^-\rangle\langle\phi^-|\right) \geq 0\end{aligned} \tag{14}$$

In addition, $\{M_x\}_{x=1}^2$ satisfies the relation of $\mathrm{tr}(\rho_1 M_1) = \mathrm{tr}(\rho_1 - \frac{4}{5}|\phi^-\rangle\langle\phi^-|) = 0.6 = \mathrm{tr}(\rho_2\left(I - \frac{4}{5}|\phi^-\rangle\langle\phi^-|\right)) = \mathrm{tr}(\rho_2 M_2)$. From Lemma 2, the prior probability of $q_1 = \frac{2}{5}$ and $q_2 = \frac{3}{5}$ provides the minimum of guessing probability. Now, we verify the uniqueness of the prior probability which provides the minimum of the guessing probability for the given quantum states. The following relations show $[\rho_1 + \rho_2, M_1] = [\rho_2, M_1] = 0$, which is the first condition of Proposition 1:

$$(\rho_1 + \rho_2)M_1 = \left(|\phi^-\rangle \langle\phi^-| + \frac{I}{4} \right) |\phi^-\rangle \langle\phi^-| = |\phi^-\rangle \langle\phi^-| \left(|\phi^-\rangle \langle\phi^-| + \frac{I}{4} \right) = M_1(\rho_1 + \rho_2) \quad (15)$$

$$\rho_2 M_1 = \left(\frac{1}{3} |\phi^-\rangle \langle\phi^-| + \frac{2}{3}\frac{I}{4} \right) |\phi^-\rangle \langle\phi^-| = |\phi^-\rangle \langle\phi^-| \left(\frac{1}{3} |\phi^-\rangle \langle\phi^-| + \frac{2}{3}\frac{I}{4} \right) = M_1\rho_2 \quad (16)$$

However, $|\phi^-\rangle$ which is the support of M_1 and M_2, satisfies the following relation:

$$\langle\phi^-| \rho_1 |\phi^-\rangle : \langle\phi^-| \rho_2 |\phi^-\rangle = \frac{9}{12} : \frac{6}{12} = \frac{3}{5} : \frac{2}{5} = 1 - q : q \quad (17)$$

Because the second condition of Proposition 1 cannot be satisfied, the prior probability providing the minimum of the guessing probability is unique.

Now, to investigate the case of non-unique prior probability, which provides the minimum of the guessing probability, we consider the following quantum states:

$$\rho_1 = \begin{pmatrix} 0.3 & 0 \\ 0 & 0.7 \end{pmatrix}, \quad \rho_2 = \begin{pmatrix} 0.7 & 0 \\ 0 & 0.3 \end{pmatrix}$$

From Figure 2, we can see non-uniqueness of the prior probability, which can provide the minimum of guessing probability.

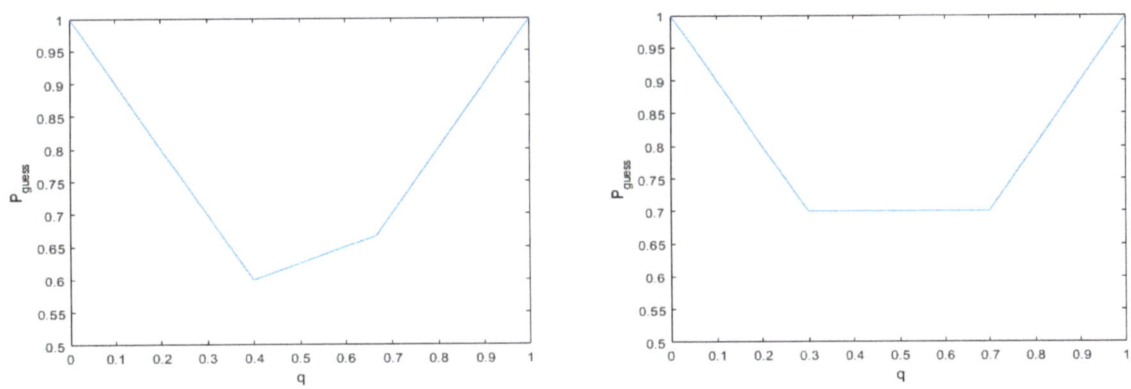

Figure 2. **(Left:)** Example of unique prior probability providing the minimum of guessing probability. The guessing probability of two quantum states $\rho_1 = \frac{2}{3} |\phi^-\rangle \langle\phi^-| + \frac{I}{12}$ and $\rho_2 = \frac{1}{3} |\phi^-\rangle \langle\phi^-| + \frac{I}{6}$ is shown in terms of prior probability $(q, 1-q)$. **(Right:)** Example of non-unique prior probability providing the minimum of guessing probability. The guessing probability of two quantum states $\rho_1 = \text{diag}[0.3, 0.7]$ and $\rho_2 = \text{diag}[0.7, 0.3]$ is shown in terms of prior probability $(q, 1-q)$.

For ρ_1 and ρ_2 with the prior probability of $q = 0.5$, we can obtain the minimum of the guessing probability, which is 0.7. Then, the optimal measurements are $M_1 = \begin{pmatrix} 0 & 0 \\ 0 & 1 \end{pmatrix}$ *and* $M_2 = \begin{pmatrix} 1 & 0 \\ 0 & 0 \end{pmatrix}$. Because of $\text{tr}(\rho_1 M_1) = \text{tr}(\rho_2 M_2) = 0.7$, the minimum of the guessing probability becomes 0.7 at $q = 0.5$. And because of $(\rho_1 + \rho_2)M_1 = M_1$, the support of M_1 is $|e_2\rangle = (0, 1)^T$, which is unique. Then, one has

$$\langle e_2| q\rho_1 - (1-q)\rho_2 |e_2\rangle = \langle e_2| \frac{1}{2} \begin{pmatrix} -0.4 & 0 \\ 0 & 0.4 \end{pmatrix} |e_2\rangle = 0.2 > 0.$$

Further, because of $(\rho_1 + \rho_2)M_2 = M_2$, the support of M_2 is $|e_1\rangle = (1,0)^T$, which is unique. We have

$$\langle e_1| q\rho_1 - (1-q)\rho_2 |e_1\rangle = \langle e_1| \frac{1}{2}\begin{pmatrix} -0.4 & 0 \\ 0 & 0.4 \end{pmatrix} |e_1\rangle = -0.2 < 0.$$

Then, the prior probability providing the minimum of the guessing probability is not unique.

From Proposition 1, the unique prior probability providing the minimum of the guessing probability has two cases. The interesting case of the two cases is one where the prior probability providing the minimum of the guessing probability is unique, with the condition that the second inequality of Proposition 1 does not hold. This is because, in this case, Bob's optimal MD strategy is not unique. When the second condition of Proposition 1 is satisfied, an element $|v_1\rangle$ in the support of M_1 satisfies the relation of $\langle v_1| \rho_1 |v_1\rangle : \langle v_1| \rho_2 |v_1\rangle = 1-q : q$. Then, from Lemma A1 in Appendix B, there exists $\epsilon > 0$ providing $M_1 - \epsilon |v_1\rangle \langle v_1| \geq 0$. Now, we define M_1' and M_2' as $M_1 - \epsilon |v_1\rangle \langle v_1|$ and $M_2 + \epsilon |v_1\rangle \langle v_1|$, respectively. Then, M_1' and M_2' are positive semidefinite operators. Because of $M_1' + M_2' = (M_1 - \epsilon |v_1\rangle \langle v_1|) + (M_2 + \epsilon |v_1\rangle \langle v_1|) = I$, $\mathbf{M}' = (M_1', M_2')$ is a POVM. We can verify whether $\mathbf{M}'$ is an optimal measurement at $\mathbf{q}$. First, from the relation of $\langle v_1| \rho_1 |v_1\rangle : \langle v_1| \rho_2 |v_1\rangle = 1-q : q$, we have $\langle v_1| (1-q)\rho_2 - q\rho_1 |v_1\rangle = 0$. For $|v_1\rangle \in \mathrm{Supp}(M_1)$, by Lemma A2, we can obtain $((1-q)\rho_2 - q\rho_1) |v_1\rangle \langle v_1| = 0$. Then, we have the following relations that show that $\mathbf{M}'$ is an optimal measurement at $\mathbf{q}$:

$$\begin{aligned} ((1-q)\rho_2 - q\rho_1) M_1' &= ((1-q)\rho_2 - q\rho - 1) (M_1 - \epsilon |v_1\rangle \langle v_1|) = ((1-q)\rho_2 - q\rho_1) M_1 \geq 0 \\ (q\rho_1 - (1-q)\rho_2) M_2' &= (q\rho_1 - (1-q)\rho_2) (M_2 + \epsilon |v_1\rangle \langle v_1|) = (q\rho_1 - (1-q)\rho_2) M_2 \geq 0 \end{aligned}$$

According to Lemma A3 in the Appendix B, the necessary and sufficient condition that two conditions of $[\rho_1 + \rho_2, M_1] = 0$ and $[\rho_2, M_1] = 0$ are satisfied is that the optimal measurement $\mathbf{M}$ of a receiver can be simultaneously diagonalized with two quantum states ρ_1 and ρ_2. Therefore, if the optimal measurement $\mathbf{M}$ of a receiver is simultaneously diagonalized with two quantum states ρ_1 and ρ_2, the uniqueness of the sender's minimax strategy cannot be compatible with the uniqueness of the receiver's MD strategy. The following Corollary summarizes this result.

Corollary 1. *If the optimal measurement* $\mathbf{M}$ *can be simultaneously diagonalized with two quantum states* ρ_1 *and* ρ_2*, the uniqueness of minimax strategy of a sender and the uniqueness of MD of a receiver cannot be compatible.*

The above result can be applied to cases of building quantum random number generator(QRNG). Suppose that only one side's strategy is unique. Therefore, either the minimax strategy of a sender or the minimum error strategy is unique. The randomness in QRNG is defined as the min-entropy to the classical bit in the quantum-classical state and depends on the prior probability [25,29]. If the prior probability providing minimum guessing probability is not unique, we can build QRNG that is not sensitive to the prior probability. When QRNG is built such that the receiver's strategy is unique, even a slight error in the measurement leads to the loss of the optimality of the receiver's strategy. The quantum states with a unique receiver's strategy in QRNG can be found by using Corollary 1.

4. Conclusions

We studied the two person zero sum game where the payoff is defined by the correct probability of the two quantum states. Because it is known that the optimal strategy of the game is a minimax strategy, and it is important to verify its uniqueness of the minimax strategy, we focused on the uniqueness condition of the minimax strategy of a sender and the minimax strategy of a receiver. In this study, we obtained the necessary and sufficient condition for the uniqueness of the sender's strategy. Using this condition, we investigated the quantum states providing the minimum guessing probability when a sender's minimax strategy is unique. Further, we found the condition where both the sender's minimax strategy and the receiver's optimal minimum error strategy cannot be unique.

Our result helps to understand the fundamental aspect of minimax strategy. We studied the minimax strategy in the quantum state discrimination of two quantum states. The uniqueness of the minimax strategy in the quantum state discrimination of more than two quantum states is not known yet. In our future work, we hope to investigate this problem.

Author Contributions: Conceptualization, J.K. and D.H.; Investigation, J.K.; Methodology, J.K.; Supervision and Critical advice Y.K.; Writing—first draft, J.K.; Revision, Y.K. All authors have read and approved the final manuscript.

Abbreviations

The following abbreviations are used in this manuscript:

MD	Minimum error Discrimination
QRNG	Quantum Random Number Generator

Appendix A. Proofs

Proof of Lemma 1. ($\Rightarrow$) Suppose that measurement $\{M_x\}_{x=1}^2$ satisfies the above condition. We define the operator K to be $q\rho_1 M_1 + (1-q)\rho_2 M_2$. Then, we obtain the following relations:

$$K - q\rho_1 = q\rho_1 M_1 + (1-q)\rho_2 M_2 - q\rho_1 = ((1-q)\rho_2 - q\rho_1)\, M_2 \geq 0$$
$$K - (1-q)\rho_2 = q\rho_1 M_1 + (1-q)\rho_2 M_2 - (1-q)\rho_2 = (q\rho_1 - (1-q)\rho_2)\, M_1 \geq 0.$$

For an arbitrary measurement $\{N_x\}_{x=1}^2$, we obtain:

$$\mathrm{tr}(K) - \mathrm{tr}(q\rho_1 N_1 + (1-q)\rho_2 N_2) = \mathrm{tr}((K - q\rho_1)\, N_1) + \mathrm{tr}((K-(1-q)\rho_2)\, N_2) \geq 0.$$

This implies that the measurement $\{M_x\}_{x=1}^2$ is optimal.
($\Leftarrow$) Let us assume that measurement $\{M_x\}_{x=1}^2$ is optimal in the MD of two quantum state. This implies that the measurement provides the guessing probability:

$$\begin{aligned} p_{\text{guess}} &= q\mathrm{tr}(\rho_1 M_1) + (1-q)\mathrm{tr}(\rho_2 M_2) \\ &= \frac{1}{2}\left(1 + \mathrm{tr}(((1-q)\rho_2 - q\rho_1)\,(M_2 - M_1))\right) \\ &= \frac{1}{2}\left(1 + \mathrm{tr}(\Lambda(M_2 - M_1))\right), \end{aligned}$$

where Λ is $(1-q)\rho_2 - q\rho_1$. Because Λ is a Hermitian operator, from the spectrum theorem, we know that there is a projection operator onto the eigenspace:

$$\Lambda = \sum_{i\in\Omega} \lambda_i P_i = \sum_{i\in\Omega_>} \lambda_i P_i + \sum_{i\in\Omega_<} \lambda_i P_i$$

Here, λ_i are eigenvalues and $P_i P_j = P_i \delta_{ij}$ is satisfied for every $i, j \in \Omega$. Further, $\Omega_> = \{i \in \Omega : \lambda_i > 0\}$ and $\Omega_< = \{i \in \Omega : \lambda_i < 0\}$. Then, $I - \sum_{i\in\Omega_>\cup\Omega_<} P_i$ is a projection onto the kernel of Λ. Because measurement $\{M_x\}_{x=1}^2$ is optimal, the general form is given as:

$$M_1 = \sum_{i\in\Omega_<} P_i + N_1, \quad M_2 = \sum_{i\in\Omega_>} P_i + N_2$$

Here, we have $N_1 \geq 0$, $N_2 \geq 0$ and $N_1 + N_2 = I - \sum_{i\in\Omega_>\cup\Omega_<} P_i$. First, M_1 is optimal and Λ contains the projector $\sum_{i\in\Omega_<} P_i$ onto the eigenspace of negative eigenvalues. And M_2 is optimal and Λ includes projector $\sum_{i\in\Omega_>} P_i$ onto eigenspace of positive eigenvalues. However, for $\Omega \neq \Omega_> \cup \Omega_<$, $\sum_{i\in\Omega_>} P_i + \sum_{i\in\Omega_<} P_i = I$ is not generally satisfied. Meanwhile, $I - \sum_{i\in\Omega_>\cup\Omega_<} P_i$ is a projector onto the null space of Λ, which does not affect optimization. Therefore, for $N_1 \geq 0$ and $N_2 \geq 0$, one can find $N_1 + N_2 = I - \sum_{i\in\Omega_>\cup\Omega_<} P_i$. Because $\Lambda N_x = 0 (x = 1, 2)$, for $x \in \{1, 2\}$, we have

$$(-1)^x \Lambda M_x = (-1)^x \Lambda (M_x - N_x) = (-1)^x (M_x - N_x) \Lambda (M_x - N_x) \geq 0.$$

Therefore, when measurement $\{M_x\}_{x=1}^2$ is optimal in the MD of two quantum states, the relation of $(-1)^x ((1-q)\rho_2 - q\rho_1) M_x \geq 0$ $(x = 1, 2)$ is satisfied. □

Proof of Lemma 3. ($\Rightarrow$) Suppose that a measurement $\{M_x\}_{x=1}^2$ can simultaneously perform MD on $\{p_x, \rho_x\}_{x=1}^2$ and $\{q_x, \rho_x\}_{x=1}^2$. This implies that $\sum_{x=1}^2 p_x \mathrm{tr}(\rho_x M_x) = \sum_{x=1}^2 q_x \mathrm{tr}(\rho_x M_x)$ and one has $\sum_{x=1}^2 (p_x - q_x)\mathrm{tr}(\rho_x M_x) = 0$. Therefore, $(p_1 - q_1)(\mathrm{tr}(\rho_1 M_1) - \mathrm{tr}(\rho_2 M_2)) = 0$. Because of $p_1 \neq q_1$, we obtain $\mathrm{tr}(\rho_1 M_1) = \mathrm{tr}(\rho_2 M_2)$. Note that only in the prior probability providing the minimum guessing probability, there exists an optimal measurement that satisfies $\mathrm{tr}(\rho_1 M_1) = \mathrm{tr}(\rho_2 M_2)$. Therefore, when an optimal measurement of the simultaneous MD on $\{p_x, \rho_x\}_{x=1}^2$ and $\{q_x, \rho_x\}_{x=1}^2$ exists, we have $p_{\mathrm{guess}}^{(1)} = p_{\mathrm{guess}}^{\star}$.

($\Leftarrow$) When a prior probability can provide the minimum guessing probability, there exists at least one optimal measurement $\{M_x\}_{x=1}^2$ satisfying $\mathrm{tr}(\rho_1 M_1) = \mathrm{tr}(\rho_2 M_2)$. Because of $p_{\mathrm{guess}}^{(1)} = p_{\mathrm{guess}}^{\star}$, an optimal measurement $\{M_x\}_{x=1}^2$ satisfies $\mathrm{tr}(\rho_1 M_1) = \mathrm{tr}(\rho_2 M_2)$ and we have $\sum_{x=1}^2 q_x \mathrm{tr}(\rho_x M_x) = \sum_{x=1}^2 p_x \mathrm{tr}(\rho_x M_x) = p_{\mathrm{guess}}^{(1)} = p_{\mathrm{guess}}^{(2)}$. This implies that the optimal measurement $\{M_x\}_{x=1}^2$ which performs the minimum error discrimination on $\{p_x, \rho_x\}_{x=1}^2$ can discriminate $\{q_x, \rho_x\}_{x=1}^2$ with minimum error. Therefore, when $p_{\mathrm{guess}}^{(1)} = p_{\mathrm{guess}}^{\star}$, there exists an optimal measurement that can simultaneously perform MD on $\{p_x, \rho_x\}_{x=1}^2$ and $\{q_x, \rho_x\}_{x=1}^2$. □

Proof of Proposition 1. ($\Rightarrow$) Suppose that the prior probability providing the minimum of guessing probability is not unique. For example, prior prbability $(p, 1-p)$ or $(q, 1-q)$ exhibits the minimum of the guessing probability. In this case, we will show $[\rho_1 + \rho_2, M_1] = 0$ and $[\rho_2, M_1] = 0$. By Lemmas 2 and 3, measurement $\{M_x\}_{x=1}^2$ is optimal in both prior probabilities p and q to ρ_1. Therefore, one can have the following relations:

$$(q\rho_1 - (1-q)\rho_2)M_1 \geq 0 \tag{A1}$$
$$(p\rho_1 - (1-p)\rho_2)M_1 \geq 0. \tag{A2}$$

Note that the two operators are Hermitian and their difference is $(p-q)(\rho_1 + \rho_2)M_1$, which is also Hermitian. Because of $p \neq q$, $(\rho_1 + \rho_2)M_1$ is a Hermitian operator and the relation of $[\rho_1 + \rho_2, M_1] = 0$ holds. This is because $(\rho_1 + \rho_2)M_1 = ((\rho_1 + \rho_2)M_1)^\dagger = M_1^\dagger(\rho_1 + \rho_2)^\dagger = M_1(\rho_1 + \rho_2)$ when $(\rho_1 + \rho_2)M_1$ is Hermitian. Further, $(p-q)\rho_2 M_1$ is also Hermitian, and because of $p \neq q$ and $\rho_2 M_1$ is Hermitian. Because of $\rho_2 M_1 = (\rho_2 M_1)^\dagger = M_1^\dagger \rho_2^\dagger = M_1 \rho_2$, we have $[\rho_2, M_1] = 0$. This implies that both $(\rho_1 + \rho_2)M_1$ and $\rho_2 M_1$ are positive semidefinite operator. $\rho_1 + \rho_2$ and M_1 (ρ_2 and M_1) commute each other and are simultaneously diagonalizable. Furthermore, the positive semidefinite operator $\sqrt{M_1}$

can be diagonalized by any basis that can diagonalize M_1. Therefore, we can find $[\rho_1+\rho_2,\sqrt{M_1}]=0$ and $[\rho_2,\sqrt{M_1}]=0$. Then, we can obtain the following relations for a vector $|v\rangle$:

$$\langle v|(\rho_1+\rho_2)M_1|v\rangle = \langle v|\sqrt{M_1}(\rho_1+\rho_2)\sqrt{M_1}|v\rangle \geq 0 \tag{A3}$$
$$\langle v|\rho_2 M_1|v\rangle = \langle v|\sqrt{M_1}\rho_2\sqrt{M_1}|v\rangle \geq 0 \tag{A4}$$

Hence, $(\rho_1+\rho_2)M_1 \geq 0$ and $\rho_2 M_1 \geq 0$. In the prior probability $(p, 1-p)$ because $\{M_x\}_{x=1}^2$ is optimal, $(p\rho_1-(1-p)\rho_2)M_1$ and $((1-p)\rho_2-p\rho_1)M_2$ are positive semidefinite. Therefore we obtain the following relations:

$$(q\rho_1-(1-q)\rho_2)M_1 \geq (q-p)(\rho_1+\rho_2)M_1 \tag{A5}$$
$$((1-q)\rho_2-q\rho_1)M_2 \geq (p-q)(\rho_1+\rho_2)M_2 \tag{A6}$$

We will show that for $x\in\{1,2\}$ an element $|v\rangle$ of support of M_x does not satisfy $\langle v|\rho_1|v\rangle : \langle v|\rho_2|v\rangle = 1-q:q$. First, we consider the case of $p<q$. Then, $q\rho_1-(1-q)\rho_2$ and M_1 commute each other and a vector $|v\rangle$ satisfies the following relation:

$$\begin{aligned}\langle v|\sqrt{M_1}(q\rho_1-(1-q)\rho_2)\sqrt{M_1}|v\rangle &= \langle v|(q\rho_1-(1-q)\rho_2)M_1|v\rangle\\ &\geq (q-p)\langle v|(\rho_1+\rho_2)M_1|v\rangle\\ &= (p-q)\langle v|\sqrt{M_1}(\rho_1+\rho_2)\sqrt{M_1}|v\rangle \geq 0\end{aligned} \tag{A7}$$

Here, the first and the last equalities are satisfied by Corollary A1. $\sqrt{M_1}$ and M_1 have the same support and $\sqrt{M_1}|v\rangle$ is a element of M_1's support. Every element of the support of M_1 can be expressed by $\sqrt{M_1}|v\rangle$ for a vector $|v\rangle$. Moreover, $\rho_1+\rho_2$ is full rank and we obtain $\langle v|\rho_1+\rho_2|v\rangle>0$ for a non-zero vector $|v\rangle$. Therefore, the condition for equality in the last inequality becomes $\sqrt{M_1}|v\rangle=0$, which implies that $|v\rangle$ is a kernel of M_1. Therefore, a non-zero element $|v_1\rangle$ in the support of M_1 satisfies the inequality $\langle v_1|(q\rho_1-(1-q)\rho_2)|v_1\rangle>0$.

Now, we consider the case of $p>q$. By the completeness condition of POVM, $q\rho_1-(1-q)\rho_2$ and M_2 commute each other and for a vector $|v\rangle$ we have the following relation:

$$\begin{aligned}\langle v|\sqrt{M_2}((1-q)\rho_2-q\rho_1)\sqrt{M_2}|v\rangle &= \langle v|((1-q)\rho_2-q\rho_1)M_2|v\rangle\\ &\geq (p-q)\langle v|(\rho_1+\rho_2)M_2|v\rangle\\ &= (p-q)\langle v|\sqrt{M_2}(\rho_1+\rho_2)\sqrt{M_2}|v\rangle \geq 0\end{aligned} \tag{A8}$$

The first and the last equalities are obtained by Corollary A1. $\sqrt{M_2}$ and M_2 have the same support and $\sqrt{M_2}|v\rangle$ is an element of M_2's support. Every element of the support of M_2 can be expressed by $\sqrt{M_2}|v\rangle$ for an element of $|v\rangle$. The condition for the equality in the last inequality is $\sqrt{M_2}|v\rangle$, which implies that $|v\rangle$ is a kernel of M_2. Then, a non-zero element $|v_2\rangle$ in the support of M_2 satisfies $\langle v_2|(1-q)\rho_2-q\rho_1|v_2\rangle>0$. Therefore, when $p<q$, any vector $|v_1\rangle$ in the support of M_1 does not satisfy $\langle v_1|\rho_1|v_1\rangle : \langle v_1|\rho_2|v_1\rangle = 1-q:q$. When $p>q$, any vector $|v_2\rangle$ in the support of M_2 does not satisfy $\langle v_2|\rho_1|v_2\rangle : \langle v_2|\rho_2|v_2\rangle = 1-q:q$.

In summary, if the prior probability providing the minimum of guessing probability is not unique, the relations of $[\rho_1+\rho_2,M_1]=0$ and $[\rho_2,M_1]=0$ hold and for $x\in\{1,2\}$, any vector $|v_x\rangle$ in the support of M_x does not satisfy $\langle v_x|\rho_1|v_x\rangle : \langle v_x|\rho_2|v_x\rangle = 1-q:q$. This contradicts the assumption that condition 1 and 2 hold. Therefore, when condition 1 and 2 are satisfied, the prior probability providing the minimum of guessing probability is unique.

($\Leftarrow$) Assume that the measurement $\{M_x\}_{x=1}^2$ satisfies $[\rho_1+\rho_2,M_1]=0$, $[\rho_2,M_1]=0$ and for some $x'\in\{1,2\}$ there is no $|v_{x'}\rangle$ of the support of $M_{x'}$ that satisfies the relation $\langle v_{x'}|\rho_1|v_{x'}\rangle : \langle v_{x'}|\rho_2|v_{x'}\rangle =$

$1-q:q$. For every $x \in \{1,2\}$ the support of M_x is a subspace of the direct sum of the non-negative eigenspace of $(-1)^x((1-q)\rho_2 - q\rho_1)$. Therefore, for an element $|v_x\rangle$ in the support of M_x the relation of $(-1)^x((1-q)\rho_2 - q\rho_1)$ holds. However, when $x = x'$, because of $\langle v_{x'}|(1-q)\rho_2 - q\rho_1|v_{x'}\rangle \neq 0$, we can have $(-1)^{x'}\langle v_{x'}|(1-q)\rho_2 - q\rho_1|v_{x'}\rangle > 0$.

Now, let us find the other prior probability which can share the optimal measurement. We define p as $q + \frac{(-1)^{x'}}{2}\min_{|v\rangle\in\mathrm{Supp}(M_{x'})}(-1)^{x'}\langle v|(1-q)\rho_2 - q\rho_1|v\rangle$. When $x' = 1$, we have $p < q$. By $\min_{|v\rangle\in\mathrm{Supp}(M_1)}\langle v|q\rho_1 - (1-q)\rho_2|v\rangle \leq q$ we have $p \geq 0$. When $x' = 2$, one has $p > q$ and by $\min_{|v\rangle\in\mathrm{Supp}(M_2)}\langle v|(1-q)\rho_2 - q\rho_1|v\rangle \leq 1-q$ we obtain $p \leq 1$.

Then, we will show that $\{M_x\}_{x=1}^2$ is optimal in $(q, 1-q)$. Note that the following two relations hold:

$$\langle v_1|q\rho_1 - (1-q)\rho_2|v_1\rangle \geq -(p-q)\langle v_1|\rho_1+\rho_2|v_1\rangle \quad \text{for all } |v_1\rangle \in \mathrm{Supp}(M_1) \tag{A9}$$

$$\langle v_2|(1-q)\rho_2 - q\rho_1|v_2\rangle \geq (p-q)\langle v_2|\rho_1+\rho_2|v_2\rangle \quad \text{for all } |v_2\rangle \in \mathrm{Supp}(M_2). \tag{A10}$$

Here, $\rho_1+\rho_2$ is full rank and for every vector $|v\rangle$ one has $\langle v|\rho_1+\rho_2|v\rangle > 0$. When $x' = 1$, $p < q$ and because of $\langle v_2|(1-q)\rho_2 - q\rho_1|v_2\rangle \geq 0$ the second condition holds. By the following relation the first inequality (A13) holds:

$$\begin{aligned}\langle v_1|q\rho_1 - (1-q)\rho_2|v_1\rangle &\geq \min_{|v\rangle\in\mathrm{Supp}(M_1)}\langle v|q\rho_1 - (1-q)\rho_2|v\rangle \\ &= -2(p-q) \geq -(p-q)\langle v_1|\rho_1+\rho_2|v_1\rangle\end{aligned} \tag{A11}$$

Let us consider the case of $x' = 2$. Because $p > q$ and $\langle v_1|q\rho_1 - (1-q)\rho_2|v_1\rangle \geq 0$, the first condition holds. By the following relation, the second inequality holds:

$$\begin{aligned}\langle v_2|(1-q)\rho_2 - q\rho_1|v_2\rangle &\geq \min_{|v\rangle\in\mathrm{Supp}(M_2)}\langle v|(1-q)\rho_2 - q\rho_1|v\rangle \\ &= 2(p-q) \geq (p-q)\langle v_2|\rho_1+\rho_2|v_2\rangle\end{aligned} \tag{A12}$$

Because $[\rho_1+\rho_2, M_1] = 0$ and $[\rho_2, M_1] = 0$, have the relation $[p\rho_1 - (1-p)\rho_2, M_1] = [p(\rho_1+\rho_2) - \rho_2, M_1] = 0$. This implies that $p\rho_1 - (1-p)\rho_2$ and M_1 are simultaneously diagonalizable. Then, $p\rho_1 - (1-p)\rho_2$ and the positive semidefinite operator satisfying $N_1^2 = M_1$ are simultaneously diagonalizable, which has the same support as to that of M_1. Therefore, for every vector $|v\rangle$ the following relation holds:

$$\langle v|(p\rho_1 - (1-p)\rho_2)M_1|v\rangle = \langle v|N_1(p\rho_1 - (1-p)\rho_2)N_1|v\rangle \geq 0 \tag{A13}$$

Note that $N_1|v\rangle$ is an element of M_1. Therefore, we have $(p\rho_1 - (1-p)\rho_2)M_1 \geq 0$. In the same manner, by the completeness relation of POVM, $p\rho_1 - (1-p)\rho_2$ and M_2 are simultaneously diagonalizable. $p\rho_1 - (1-p)\rho_2$ and positive semidefinite operator N_2 satisfying $N_2^2 = M_2$ are simultaneously diagonalizable, which has the same support as that of M_2 by Corollary A1. Therefore, for every vector $|v\rangle$, the following relation holds:

$$\langle v|((1-p)\rho_2 - p\rho_1)M_2|v\rangle = \langle v|N_2((1-p)\rho_2 - p\rho_1)N_2|v\rangle \geq 0 \tag{A14}$$

Note that $N_2|v\rangle$ is support of M_2 and one has $((1-p)\rho_2 - p\rho_1)M_2 \geq 0$.

By Lemma 1, for every $x \in \{1,2\}$, one finds $(-1)^x((1-p)\rho_2 - p\rho_1)M_x \geq 0$ and $\{M_x\}_{x=1}^2$ is optimal at the prior probability $(p, 1-p)$. This contradicts the assumption that the identical measurement cannot be shared in the different prior probabilities. Therefore, when the prior probability providing the minimum guessing probability is unique, the condition 1(or 2) holds. □

Appendix B. Lemmas

Let $\mathcal{H}$ be a finite dimensional Hilbert space.

Lemma A1. *Let A be a positive semidefinite operator. Let $|v\rangle$ be an element of support of A. Then there exists $\epsilon > 0$ such that $A - \epsilon |v\rangle \langle v| \geq 0$.*

Proof of Lemma A1. Let us assume that there exists no $\epsilon > 0$ satisfying $A - \epsilon |v\rangle \langle v| \geq 0$. This implies that for any $\epsilon > 0$, there exists $|w\rangle \in \mathcal{H}$ such that $\langle w| (A - \epsilon |v\rangle \langle v|) |w\rangle < 0$. Thus $\langle w| A |w\rangle < \epsilon |\langle w|v\rangle|^2$. Because $A \geq 0$, it follows that $0 \leq \langle w| A |w\rangle < \epsilon |\langle w|v\rangle|^2$.

Note that $\langle w| A |w\rangle$ cannot be zero. Because when we assume $\langle w| A |w\rangle = 0$, $|w\rangle$ is an element of $\ker(A)$. Since $|v\rangle$ is orthogonal to $\ker(A)$; this directly implies $\langle w|v\rangle = 0$. Hence, $0 < 0$, which is a contradiction. Therefore $\langle w| A |w\rangle > 0$.

Any vector in $\mathcal{H}$ can be decomposed as a linear combination of elements of $\mathrm{Supp}(A)$ and $\ker(A)$. Furthermore $|w\rangle$ cannot be an element of kernel. Thus there exists $|s\rangle \in \mathrm{Supp}(A)$ and $|k\rangle \in \ker(A)$ such that $|w\rangle = c_1 |s\rangle + c_2 |k\rangle$ and where $c_1 \neq 0$.

Because A is an operator on the finite dimensional Hilbert space, there exists $\gamma > 0$ such that $\gamma \equiv \inf_{|s\rangle \in \mathrm{Supp}(A)} \langle s| A |s\rangle$. When $\epsilon = \gamma$,

$$\langle w| A |w\rangle = |c_1|^2 \langle s| A |s\rangle \geq |c_1|^2 \epsilon \geq |c_1|^2 \epsilon |\langle s|v\rangle|^2 \geq \epsilon |\langle w|v\rangle|^2.$$

This contradicts the initial assumption. Therefore there exists $\epsilon > 0$ such that $A - \epsilon |v\rangle \langle v| \geq 0$. □

Lemma A2. *Let A be a Hermitian operator on $\mathcal{H}$. Let B be a positive semidefinite operator on $\mathcal{H}$. Suppose that A and B are commutable. If $|v\rangle \in \mathrm{Supp}(B)$ satisfies $\langle v| A |v\rangle = 0$, then $|v\rangle \in \ker(A)$.*

Proof of Lemma A2. Because A and B are commutable, there exists an orthonormal basis $\{|\phi_i\rangle\}_i$ such that $A = \sum_{i \in \chi_A} a_i |\phi_i\rangle \langle \phi_i|$, $B = \sum_{i \in \chi_B} b_i |\phi_i\rangle \langle \phi_i|$, where $a_i \neq 0$ for all $i \in \chi_A$ and $b_i > 0$ for all $i \in \chi_B$. Then $AB = \sum_{i \in \chi_A \cap \chi_B} a_i b_i |\phi_i\rangle \langle \phi_i|$. Because $AB \geq 0$, $a_i > 0$ for all $i \in \chi_A \cap \chi_B$. As $|v\rangle \in \mathrm{Supp}(B)$, we can express $|v\rangle$ as $\sum_{i \in \chi_B} c_i |\phi_i\rangle$. Then

$$\begin{aligned}\sum_{i \in \chi_A \cap \chi_B} |c_i|^2 a_i &= \sum_{i,j \in \chi_B} c_i^* \langle \phi_i| \sum_{k \in \chi_A} a_k |\phi_k\rangle \langle \phi_k| |\phi_j\rangle c_j \\ &= \langle v| A |v\rangle = 0.\end{aligned}$$

Because $a_i > 0$, it follows that $|c_i|^2 = 0$ for all $i \in \chi_A \cap \chi_B$. This implies that $c_i = 0$. Thus

$$A |v\rangle = \sum_{k \in \chi_A} a_k |\phi_k\rangle \langle \phi_k| \sum_{i \in \chi_B} c_i |\phi_i\rangle = \sum_{i \in \chi_A \cap \chi_B} a_i c_i |\phi_i\rangle = 0$$

Therefore if $|v\rangle \in \mathrm{Supp}(B)$ satisfies $\langle v| A |v\rangle = 0$, then $|v\rangle \in \ker(A)$ □

Let $\mathcal{H}$ be a Hilbert space with dimension d. Let A, B be Hermitian operators on $\mathcal{H}$.

Lemma A3. *If $[A, B] = 0$, then A, B can be simultaneously diagonalizable.*

Proof of lemma A3. Because A and B are Hermitian operators and $[A, B] = 0$, it follows that $(AB)^\dagger = B^\dagger A^\dagger = BA = AB$. This implies that AB is a Hermitian operator. Let $\sum_{i=1}^{d} a_i |a_i\rangle \langle a_i|$ be a spectral decomposition of A. Then

$$AB = \sum_{i,j=1}^{d} a_i |a_i\rangle \langle a_i| B |a_j\rangle \langle a_j| = \sum_{i,j=1}^{d} a_i \langle a_i| B |a_j\rangle |a_i\rangle \langle a_j|.$$

Because AB is a Hermitian operator, it follows that

$$a_l \langle a_k| B |a_l\rangle = (AB)^*_{lk} = (AB)_{kl} = a_k \langle a_k| B |a_l\rangle \quad \text{for all } k, l \in \{1, 2, \cdots, d\}$$

This implies that $(a_l - a_k) \langle a_k| B |a_l\rangle = 0$. Thus $a_l = a_k$ or $\langle a_k| B |a_l\rangle = 0$.

Let us define a set of indices $I \subset \{1,2,\cdots,d\}$ such that for every $i \in I$, if $j \in \{1,2,\cdots,d\}\backslash\{i\}$ satisfies $\langle a_i| B |a_j\rangle \neq 0$, then $j \in I$ and there is no non-empty subset $J \subset I$ such that for every $i \in I\backslash J$ and for every $j \in J$, $\langle a_i| B |a_j\rangle = 0$. Using the result above, $a_i = a_j$ for all $i,j \in I$. This implies that $\sum_{i,j\in I} \langle a_i| B |a_j\rangle |a_i\rangle \langle a_j|$ can be represented with a block matrix with a basis of $\{|a_i\rangle\}_{i\in I}$ by rearranging the indices. Define a' as the positive number satisfying $a_i = a'$ for all $i \in I$. Then $\sum_{i\in I} |a_i\rangle \langle a_i|$ is a projection operator on the eigenspace of A providing eigenvalue a'. Note that the eigenspace has a degree of freedom in choosing the orthonormal basis.

Now let us explain how to choose a basis that can diagonalize A and B simultaneously. Because $\sum_{i,j\in I} \langle a_i| B |a_i\rangle |a_i\rangle \langle a_i|$ is a Hermitian operator, it can be diagonalized with some orthonormal basis. Suppose that $\{|c_i\rangle\}_{i\in I}$ is the basis. Then $\sum_{i,j\in I} \langle a_i| B |a_j\rangle |a_i\rangle \langle a_j| = \sum_{i\in I} c_i |c_i\rangle \langle c_i|$ holds, where c_i is a eigenvalue of B. Furthermore, because $\mathrm{span}\{|a_i\rangle\}_{i\in I}$ is an eigenspace of A, $\sum_{i\in I} a' |a_i\rangle \langle a_i|$ can be rewritten as $\sum_{i\in I} a' |c_i\rangle \langle c_i|$. Similarly, we can find a basis that diagonalizes each block matrix of B. A can be diagonalized using this basis. Therefore if $[A,B] = 0$, then A and B can be simultaneously diagonalizable. □

Let $\sum_{i=1}^{r} \lambda_i P_i$ be a spectral decomposition of B. Then $C = \sum_{i=1}^{r} \sqrt{\lambda_i} P_i$ satisfies $C^2 = B$. We show that the positive semidefinite operator satisfying $C^2 = B$ is unique.

Suppose that C is not unique. Then there exists another positive semidefinite operator C' such that $C'^2 = B$. Because $[B, C'] = 0$, from Lemma A3, B, C' are simultaneously diagonalizable. That is, there exists an orthornormal basis $\{|v_i^{(j)}\rangle\}_{i,j=1}^{r,d_i}$ such that

$$B = \sum_{i=1}^{r} \lambda_i \left(\sum_{j=1}^{d_i} |v_j^{(i)}\rangle \langle v_j^{(i)}| \right), \quad C' = \sum_{i=1}^{r} \left(\sum_{j=1}^{d_i} v_{ij} |v_i^{(j)}\rangle \langle v_i^{(j)}| \right)$$

for some non-negative $\lambda_i, v_{i,j} \in \mathbb{R}$. Because $C'^2 = B$, it follows that

$$B - C'^2 = \sum_{i=1}^{r} \sum_{j=1}^{d_i} (\lambda_i - v_{ij}^2) |v_i^{(j)}\rangle \langle v_i^{(j)}| = 0.$$

Thus $v_{ij} = \sqrt{\lambda_i}$ for all $i,j \in \{1,2,\cdots,d\}$. Further, because $\sum_{j=1}^{d} |v_i^{(j)}\rangle \langle v_i^{(j)}| = P_i$, it follows that $C' = \sum_{i=1}^{r} \sqrt{\lambda_i} \left(\sum_{j=1}^{d_i} |v_i^{(j)}\rangle \langle v_i^{(j)}| \right) = \sum_{i=1}^{r} \sqrt{\lambda_i} P_i = C$. This contradicts the initial assumption. Therefore, the positive semidefinite operator satisfying $C^2 = B$ is unique.

Let A, B be positive semidefinite operators on $\mathcal{H}$. Let C be a positive semidefinite operator on $\mathcal{H}$ satisfying $C^2 = B$.

Corollary A1. *If* $[A,B] = 0$, *then* $[A,C] = 0$.

Proof of Corollary A1. According to Lemma A3, $[A,B] = 0$ implies that A and B are simultaneously diagonalizable. That is there exists an orthonormal basis $\{|\lambda_i\rangle\}_{i=1}^{d}$ such that

$$A = \sum_{i=1}^{d} a_i |\lambda_i\rangle \langle \lambda_i|, \quad B = \sum_{i=1}^{d} b_i |\lambda_i\rangle \langle \lambda_i|,$$

for some $a_i \geq 0$ and $b_i \geq 0$. Then C is uniquely defined as $C = \sum_{i=1}^{d} \sqrt{b_i} |\lambda_i\rangle \langle \lambda_i|$ by the statement above. Note that there exists an orthornormal basis $\{|\lambda_i\rangle\}_{i=1}^{d}$ diagonalizing A, C simultaneously. Therefore if $[A,B] = 0$, then $[A,C] = 0$. □

References

1. Chefles, A. Quantum state discrimination. *Contemp. Phys.* **2000**, *41*, 401–424. [CrossRef]
2. Barnett, S.M.; Croke, S. Quantum state discrimination. *Adv. Opt. Photon.* **2009**, *1*, 238–278. [CrossRef]
3. Bergou, J.A. Discrimination of quantum states. *J. Mod. Opt.* **2010**, *57*, 160–180. [CrossRef]

4. Bae, J.; Kwek, L. Quantum state discrimination and its applications. *J. Phys. A Math Theor.* **2015**, *48*, 083001. [CrossRef]
5. Ha, D.; Kwon, Y. Complete analysis for three-qubit mixed-state discrimination. *Phys. Rev. A* **2013**, *87*, 062302. [CrossRef]
6. Ha, D.; Kwon, Y. Discriminating N-qudit states using geometric structure. *Phys. Rev. A* **2014**, *90*, 022330. [CrossRef]
7. Herzog, U. Minimum-error discrimination between a pure and a mixed two-qubit state. *J. Opt. B Quantum Semiclass. Opt.* **2004**, *6*, S24–S28. [CrossRef]
8. Ivanovic, I. How to differentiate between non-orthogonal states. *Phys. Lett. A* **1987**, *123*, 257–259. [CrossRef]
9. Dieks, D. Overlap and distinguishability of quantum states. *Phys. Lett. A* **1988**, *126*, 303–306. [CrossRef]
10. Peres, A. How to differentiate between non-orthogonal states. *Phys. Lett. A* **1988**, *128*, 19. [CrossRef]
11. Jaeger, G.; Shimony, A. Optimal distinction between two non-orthogonal quantum states. *Phys. Lett. A* **1995**, *197*, 83–87. [CrossRef]
12. Chefles, A. Unambiguous discrimination between linearly independent quantum states. *Phys. Lett. A* **1998**, *239*, 339–347. [CrossRef]
13. Croke, S.; Andersson, E.; Barnett, S.M.; Gilson, C.R.; Jeffers, J. Maximum Confidence Quantum Measurements. *Phys. Rev. Lett.* **2006**, *96*, 070401. [CrossRef] [PubMed]
14. Chefles, A.; Barnett, S.M. Strategies for discriminating between non-orthogonal quantum states. *J. Mod. Opt* **1998**, *45*, 1295–1302. [CrossRef]
15. Zhang, C.W.; Li, C.F.; Guo, G.C. General strategies for discrimination of quantum states. *Phys. Lett. A* **1999**, *261*, 25–29. [CrossRef]
16. Fiurášek, J.; Ježek, M. Optimal discrimination of mixed quantum states involving inconclusive results. *Phys. Rev. A* **2003**, *67*, 012321. [CrossRef]
17. Eldar, Y.C. Mixed-quantum-state detection with inconclusive results. *Phys. Rev. A* **2003**, *67*, 042309. [CrossRef]
18. Ha, D.; Kwon, Y. An optimal discrimination of two mixed qubit states with a fixed rate of inconclusive results. *Quantum Inf. Process.* **2017**, *16*, 273. [CrossRef]
19. Wald, A. Generalization of a Theorem By v. Neumann Concerning Zero Sum Two Person Games. *Ann. Math.* **1945**, *46*, 281–286. [CrossRef]
20. Hirota, O.; Ikehara, S. Minimax Strategy in the Quantum Detection Theory and Its Application to Optical Communications. *Trans. IECE Jpn.* **1982**, *E65*, 627–633.
21. D'Ariano, G.M.; Sacchi, M.F.; Kahn, J. Minimax quantum-state discrimination. *Phys. Rev. A* **2005**, *72*, 032310. [CrossRef]
22. Nakahira, K.; Kato, K.; Usuda, T.S. Minimax strategy in quantum signal detection with inconclusive results. *Phys. Rev. A* **2013**, *88*, 032314. [CrossRef]
23. Bennett, C.H.; Brassard, G. Quantum cryptography: Public key distribution and coin tossing. *Theor. Comput. Sci.* **2014**, *560*, 7–11. [CrossRef]
24. Cao, Z.; Zhou, H.; Ma, X. Loss-tolerant measurement-device-independent quantum random number generation. *New J. Phys.* **2015**, *17*, 125011. [CrossRef]
25. Bischof, F.; Kampermann, H.; Bruß, D. Measurement-device-independent randomness generation with arbitrary quantum states. *Phys. Rev. A* **2017**, *95*, 062305. [CrossRef]
26. Ma, X.; Yuan, X.; Cao, Z.; Qi, B.; Zhang, Z. Quantum random number generation. *npj Quantum Inf.* **2016**, *2*, 16021. [CrossRef]
27. Helstrom, C.W. *Quantum Detection and Estimation Thoery*; Academic Press: New York, NY, USA, 1976.
28. Eldar, Y.C.; Megretski, A.; Verghese, G.C. Designing optimal quantum detectors via semidefinite programming. *IEEE Trans. Inf. Theory* **2003**, *49*, 1007–1012. [CrossRef]
29. Konig, R.; Renner, R.; Schaffner, C. The Operational Meaning of Min- and Max-Entropy. *IEEE Trans. Inf. Theory* **2009**, *55*, 4337–4347. [CrossRef]

14

Representation Lost: The Case for a Relational Interpretation of Quantum Mechanics

Raffael Krismer

Department of Philosophy, University of Vienna, 1090 Vienna, Austria; raffael.krismer@univie.ac.at

Abstract: Contemporary non-representationalist interpretations of the quantum state (especially *QBism*, *neo-Copenhagen views*, and the *relational interpretation*) maintain that quantum states codify observer-relative information. This paper provides an extensive defense of such views, while emphasizing the advantages of, specifically, the relational interpretation. The argument proceeds in three steps: (1) I present a classical example (which exemplifies the spirit of the relational interpretation) to illustrate why some of the most persistent charges against non-representationalism have been misguided. (2) The special focus is placed on dynamical evolution. Non-representationalists often motivate their views by interpreting the collapse postulate as the quantum mechanical analogue of Bayesian probability updating. However, it is not clear whether one can also interpret the Schrödinger equation as a form of rational opinion updating. Using results due to Hughes & van Fraassen as well as Lisi, I argue that unitary evolution has a counterpart in classical probability theory: in both cases (quantum and classical) probabilities relative to a *non-participating* observer evolve according to an entropy maximizing principle (and can be interpreted as rational opinion updating). (3) Relying on a thought-experiment by Frauchiger and Renner, I discuss the differences between quantum and classical probability models.

Keywords: foundations of quantum theory; entropy maximizing principle; information-theoretic approaches to the quantum state; relational interpretation; toy-models; realism debate

---

1. Introduction

The idea that quantum states do not *represent* (or *correspond to*) physical reality is as old as quantum theory itself. Niels Bohr, e.g., has famously been alleged to assert that "There is no quantum world ... It is wrong to think that the task of physics is to find out how nature *is*. Physics concerns what we can say about nature" (attributed to Bohr by Petersen [1] (p. 12)). However perplexing such claims may appear, there still exists a colorful variety of contemporary views that have carried the idea of non-representational quantum states into the 21st century. This paper develops a broad line of defense on behalf of non-representationalist interpretations of the quantum state. In no particular order, the ones I shall focus on are: *QBism* [2–7], *neo-Copenhagen approaches* [8–10], and the *relational interpretation* [11–13]. But even though parts of the argument below might be adopted to suit the purposes of any of these interpretations, I will emphasize the advantages of what appears to me to be the most promising one: a slightly modified version of Rovelli's relational interpretation. To this end, I present a classical example, which exemplifies how defenders of the relational approach think about quantum theory. Relying on this example, I will be able to discharge several worries that have been levelled against non-representationalists more generally.

Let me start by providing the motivation for this project. The crux of the aforementioned interpretations lies in their commitment to the claim that quantum theory's probabilistic predictions should be accounted for by information-theoretic means, where the information in question is thought to be relative to some observer. Hence, according to these views, the quantum state is regarded as an irredeemably *relational* concept. While the different interpretations differ substantively over what the quantum state is allegedly relative *to* (i.e., what they mean by the term "observer" -QBism/neo-Copenhagn views: a decision-making agent or subject [2–6], [10] (p. 4); relational interpretation: a physical reference system [11] (pp. 1–2)), this shared commonality runs deep–both from a philosophical but also from a technical viewpoint.

The first set of issues that have plagued non-representationalist views are rooted in the (legitimate) fear that if quantum state ascriptions are observer-relative, objective reality will escape our theoretical clutch. This has both philosophical but also technical dimensions. On the philosophical side, non-representationalists, by virtue of stripping quantum theory of the ability to offer third-person descriptions of the world, have been charged with solipsism or skepticism. Relatedly, the introduction of a subjective element into science has been the source of significant unease. On the technical side, the "psi-ontology" theorems that have emerged in recent years, especially the theorem due to Pusey, Barrett and Rudolph [14] (for an extended review, see [15]), have been interpreted as pulling towards realist interpretations of the quantum state. My first goal will be to ease the pressure that derives from these types of worries. This will be achieved by presenting a classical example of a blatantly non-representational modelling practice which: (1) portrays striking similarities to quantum theory (and the relational way of thinking about quantum theory in particular), and which (2) allows us to demonstrate, using purely classical intuitions, why these arguments against non-representationalism (although *prima facie* plausible) are ultimately guilty of what Dennett once called the "Philosophers' Syndrome: Mistaking a failure of the imagination for an insight into necessity." [16] (p. 406).

The second core issue that this paper addresses is more specific, and concerns the question of dynamical evolution. To see why dynamics would play an important role for non-representationalists, notice, first, that an important consequence of their shared commitment to observer-relative states is that the textbook dynamical postulates—von Neumann's [17] *collapse postulate* and the *unitary evolution* of the Schrödinger equation—are not to be understood in terms of a mechanical/substance-type story of some entity collapsing or evolving. Instead, those changes in the quantum state are to be understood as the process in which the observer rationally updates her opinion—either literally (QBism, neo-Copenhagen), or at least "on the model of" (relational interpretation).

What is thus often cited as a motivation for these views is the analogy between *Lüder's rule* [18] and its classical counterpart, i.e., *Bayes' theorem*. [2,19] Lüder's rule can be taken to justify the idea that one may view von Neumann's *collapse postulate* as a form of probabilistic conditionalization: the projected (i.e., "collapsed") state agrees, in its probability assignments to quantum mechanical observables, with the (canonical) generalization of the notion of conditional probability to quantum mechanics. (Recall that in classical probability theory, conditional probability is defined as follows: let P be a probability measure on a Borel field. Then, define a derivative probability measure $P'(-) = P(-|A)$ for each Borel set A, which is the unique probability measure on the Borel subsets of A such that (1) $P'(A) = 1$ and (2) the probability ratios (i.e., the "odds") are preserved: $\frac{P(A\&X)}{P(A\&Y)} = \frac{P'(A\&X)}{P'(A\&Y)}$. From this, Bayes' theorem can be derived, and this way of thinking about conditional probability can be generalized to the case in which the underlying domain isn't a Boolean algebra but has the particular structure of an ortho-modular lattice. [20] (pp. 171–175), [21], [22] (pp. 170–173)).

If this analogy between quantum and classical probability models indeed lends credence to non-representationalist approaches, however, we immediately run into a problem: *what is the classical analogue of unitary evolution?* Brown [23], e.g., raises this point in his discussion of QBism. If, for a realist conception of a unitarily evolving wavefunction, the measurement problem was "mysteriously solved" by the projection postulate, then, according to Brown, on the QBist framework it now seems "as if von Neumann's two motions in quantum mechanics have reappeared in a different guise! The difference now is that the mystery lies with the unitary evolution." (p. 17) Insofar as unitary evolution appears mysterious on the observer-relative interpretations, this certainly presents challenge: if these interpretations are (at least in part) inspired by the analogy between classical and quantum probability models, is there a classical counterpart of unitary evolution? And if so, can we make sense of it as a form of rational opinion updating?

A central goal of the discussion below will be to show that unitary evolution indeed has a counterpart in classical probability theory. Hence, in *both* cases, classical and quantum, *probabilistic information evolves differently relative to different observers* (such that, as will be shown, the probabilities relative to an external—i.e., "non-participating"—observer are the solutions to an *entropy maximizing problem*). My own relational biases notwithstanding, this is good news for non-representationalists more generally. The presentation of this theorem will follow Hughes and van Fraassen in Ref. [24] (cf. [20]), but since that theorem doesn't seem to be all that well-known, it is a worthwhile task to repeat it. And while the mathematical details will mirror those of Hughes and van Fraassen, I will: (1) put their theorem in the context of the previously developed example, which will engender certain specific advantages; (2) make explicit in what sense the classical theorem can indeed be viewed as the counterpart of unitary evolution (here I will rely on a result due to Lisi [25]).

The structure of this paper is as follows. Section 2 introduces the terms and intuitions by means of a classical example of a modelling practice that closely resembles the way information is encoded in quantum theory. Section 3 generalizes the initial example to derive, following Hughes and van Fraassen [24], a version of the Schrödinger equation for classical probability models. While Sections 2 and 3 are intelligible from a purely mathematical perspective, my interpretational motives will be laid bare in Section 4. I will make explicit how the set-up presented in Sections 2 and 3 instantiates how defenders of the relational interpretation think about quantum theory, and I will also point out several advantages of such an approach. Section 5 makes the transition to quantum theory, where the first goal will be to illustrate how quantum mechanics mirrors the classical way of reasoning presented in Sections 2–4. Using a result due to Lisi [25], the analogy between the classical and the quantum case (and the relational approach in particular) will become strikingly clear: by an analogous argument as in the classical case, unitary evolution of probabilities relative to an external observer can be shown to be the result of an entropy maximizing principle (subject to analogous constraints as in the classical case). Having discussed the similarities between classical and quantum probability models, Section 6 will discuss their differences. I argue that a key difference between quantum and classical probability models is that the latter *can*, but the former *cannot* (in general), be supplemented with an ontological story about what the world is like. This will be achieved by placing the previous discussion in the context of a recent thought-experiment proposed by Frauchiger and Renner [26], which can be interpreted to show that, in general, quantum theory has no room for the notion of observer-independent facts.

2. The Basic Set-Up

To set up the classical example, consider the model of a (presumably) familiar reality that is given in Figure 1:

	Team	Games Played	Won	Drawn	Lost	Points
1	Manchester United	38	28	7	3	91
2	FC Arsenal	38	22	7	9	73
3	Leeds United	38	21	6	11	69
4	FC Liverpool	38	19	10	9	67
5	FC Chelsea	38	18	11	9	65
6	Aston Villa	38	15	13	10	58

Figure 1. The English Premier League 1999.

The specifics of Figure 1 will not turn out to be important, and below I will only discuss the general strategy for how such tables as Figure 1 are produced. Proceeding by means of an example, however, has the advantage of allowing me to introduce the central terminology in intuitive terms.

2.1. Measurements

The aim of this modelling practice is to characterize how successful a football team is—we want to create an ordering of the teams on the scale of the natural numbers. This goal will be achieved by letting the teams compete against one another. The games are to be regarded as real processes, real *interactions*, between existing entities (the teams), such that each game has a determinate outcome, which is either *win*, *draw*, or *loss*. Notice that not all combinations of outcomes are possible: although both teams can draw in a single game, it is not the case that both teams can win (or lose). Thus: (1) games are *events*, (2) the outcomes of these events are *definitive* of "what the world is like" (i.e., how the teams will be ranked), and (3) there will be correlations in the descriptions of teams (if one team wins, its opponent must have lost). The set of possible outcomes {win, draw, lose} will be referred to as the "measurement context"; the individual games are called "measurements". Due to the central importance of these measurement interactions, I will refer to the probabilistic model that results from the considerations below as an "interactional probability model".

2.2. States

Once we have determined a measurement context, we can collect information about the teams. This information will be called a "state." However, the need will arise to distinguish between different kinds of states, and to introduce different kinds of mathematical structures.

2.2.1. Betting-States

One way to encode information about the teams would be to provide a list of the outcomes of all the individual games. But any attempt to define a relation " ... is better than ... " by virtue of, e.g., "K is better than L if and only if K has won against L" (information that would be provided by our list) might lead to inconsistencies (if K wins against L, L wins against M, and M wins against K). To achieve our initial aim, of creating a ranking of the teams, we will do much better if we begin by characterizing each team by its total number of wins, draws, and losses. Such a triple of numbers will be called a *betting-state*:

- *Betting-states*. The *betting-state* ascribed to a team is a triple $\langle w, d, l \rangle$ where w, d, and l denote the number of wins, draws, and losses respectively.

Clearly, betting-states represent only the outcomes of the games, but not the underlying *mechanisms* by which these outcomes are produced. There may or may not be any systematic way of modelling these mechanism—the point is that we may choose not to worry about such *vastly* complicated things. Hence, we wisely trade descriptive accuracy for predictive success.

Given this definition, it is natural to inquire into the structure of the set of betting-states. Suppose a team is characterized by $\langle w_1, d_1, l_1 \rangle$ during the first m games, and by $\langle w_2, d_2, l_2 \rangle$ during the subsequent n games. Then, the overall betting-state is given by $\langle w_1 + w_2, d_1 + d_2, l_1 + l_2 \rangle$, and hence we can define a component-wise addition for betting-states associated with a single team: $\langle w_1, d_1, l_1 \rangle + \langle w_2, d_2, l_2 \rangle := \langle w_1 + w_2, d_1 + d_2, l_1 + l_2 \rangle$ (clearly, it makes no sense to add betting-states that are associated with *different* teams). Similarly, we can define a (component-wise) multiplication by a scalar λ: $\langle \lambda w, d, l \rangle = \langle \lambda w, \lambda d, \lambda l \rangle$ (where λ might represent the number of rounds in which the same result was obtained, such that component-wise multiplication yields the overall betting-state after λ rounds). Equipped with these operations, the set of betting-states is now a "vector space" (informally speaking, of course; most notably, we are lacking an additive inverse and a multiplicative inverse for multiplication by a scalar).

2.2.2. Odds Comparison

Let's try and put our betting-states to use. Suppose there is a game coming up—can we use the betting-states to inform our betting behavior? One might, first, propose that betting-states can be used to make *probability assignments* for future events: if a team's betting-state is given by $\langle w,\ d,\ l\rangle$ (such that $w + d + l = n$, where n denotes the total number of games), then the respective probabilities are given as: $p_\alpha = \alpha/n$ ($\alpha = w, d, l$). The suggestion is that these probabilities should guide our betting behavior.

However, there is something a little naive about using the probabilities generated from the betting-states like $p_\alpha = \alpha/n$ ($\alpha = w, d, l$). Here is why: think about a team near the bottom of the league. This team might have a low probability of winning ($p_w = \frac{w}{n} < \frac{1}{2}$). But while this may be so, we should be wary not to take this "absolute" description (of a seemingly *intrinsic* property) of the team all that seriously. *Winning*, after all, is inherently relational: you can only win *against* some other team. Suppose that the lowly ranked team plays against a team that is located even further down the table. Its chances of winning, in this case, may actually not be all that low. A sophisticated bettor won't read too much into the probabilities $p_\alpha = \alpha/n$, but will acknowledge that, for each team, the probability of winning, losing, and drawing is correctly specified *relative* to its opponent. Betting behavior, in other words, must be informed by mathematical structures that are sensitive to the measurement that is being performed, rather than those that aim for an "absolute" description of each team.

These ideas can be modelled via what Hughes and van Fraassen [24] (cf. [20] (p. 70)) call an "*odds comparison*." If two teams are assigned betting-states $\langle w_1,\ d_1,\ l_1\rangle$ and $\langle w_2,\ d_2,\ l_2\rangle$ (after both teams have played the same number of games), then we may define their *odds comparison* like this: $\langle w_1/w_2,\ d_1/d_2,\ l_1/l_2\rangle$. Determining these *relative* odds of two teams (recall that odds, by definition, are probability ratios [20] (p. 69)) will certainly be a most valuable piece of information if we wish to be even moderately sophisticated about our predictions for the outcomes of specific games.

2.2.3. Number-States

Even the odds comparisons, however, aren't sufficient to unambiguously determine, for any arbitrary pair of teams, which team is better. It is still unclear, e.g., which of the two betting-states (associated with different teams) is better: $\langle 5,\ 1,\ 5\rangle$ or $\langle 3,\ 7,\ 1\rangle$?

A successful way of ordering the teams proceeds by defining a function *s*, which takes as its input a betting-state and assigns to that state a *number*. Depending on whether this number is higher, lower, or equal for the betting-states $\langle w_1,\ d_1,\ l_1\rangle$ and $\langle w_2,\ d_2,\ l_2\rangle$ that are associated with teams K and L respectively, we will, by definition, know whether K is ranked above, below, or equal to L. These functions will be called *number-state functions*. Let me reemphasize that the primary role of these number-state functions is to assign numerical values to betting-states, which in turn *generates a relative ordering of the teams* on the scale of the natural numbers.

The current number-state function is $\langle s(w,\ d,\ l)\rangle = 3w + d$, where a win is assigned 3 points, a draw is assigned 1 point, and a loss is assigned 0 points. Certainly, there are other sensible options: we could, e.g., define number-state functions "projectively," such that $p(\langle w,\ d,\ l\rangle) = w$. This would result in a model that would deem only the number of wins to be relevant. Clearly, there are many potential choices of number-state functions, and any such choice is going to be conventional.

This immediately invites the question of how arbitrary our convention is going to be. A natural constraint is that number-state functions should pay tribute to the vector-space structure of the betting-states. Hence, we demand that $s(\sum\langle w_i,\ d_i,\ l_i\rangle) = \sum s(\langle w_i,\ d_i,\ l_i\rangle)$. This is "natural" because we have previously decided to collect, in the betting-states, only information about the outcome of each game, but no information about the *order* of the results.

Before proceeding, let me summarize these remarks by formally introducing two closely related concepts:

- *Number-states.* A *number-state ascription* is an ascription of a numerical value to a betting-state.
- *Number-state functions. Number-state functions* are linear functions from betting-states to the natural numbers.

The reason to insist on this distinction between the number-state functions and the number-states themselves is because one could, in principle, assign numbers to betting-states in an arbitrary way. Thus, it is not trivial to require (as I will) that all number-states derive from a choice of number-state function, but a condition that must be put in by hand.

From the linearity constraint (and the additional constraint that $s(\langle 0, 0, 0\rangle) = 0$) we conclude that number-state functions must be of the form $s(\langle w, d, l\rangle) = s_1 w + s_2 d + s_3 l$. Hence, they can *also* be written as triples $\bar{s} = \langle s_1, s_2, s_3\rangle$, and thus we can (informally, again) consider them to be vectors as well. Number-state functions are therefore of the same mathematical type as the betting-states. The identity of mathematical representations of both types of states, however, should not distract from the fact that they should be interpreted differently.

These observations have two noteworthy consequences. First, notice that if the number-state functions are of the form $s(\langle w, d, l\rangle) = s_1 w + s_2 d + s_3 l$, we can interpret the resulting number-states (somewhat informally) as being *expectation values* for the overall number of points a team will receive after n games. (Observe that, strictly speaking, this yields an expectation value only if we divide this expression by the total number of games n (the sum of the components): $s(\langle w, d, l\rangle)/n = s_1 p_w + s_2 p_d + s_3 p_l$, where $p_\alpha = \alpha/n$ is the relative frequency of each of the outcomes as specified in Section 2.2.2. Even though these probabilities were previously argued to not be particularly useful (and actively misleading) they are, of course, still probabilities in a mathematical sense, since the frequencies calculated via $p_\alpha = \alpha/n$ satisfy the probability axioms.) Secondly, the similarity of mathematical representations of betting-states and number-state functions can be exploited in the following way. The number-states, i.e., the function values $s(\langle w, d, l\rangle)$, can be written in the form of a *dot product* between vectors: $s(\langle w, d, l\rangle) = \bar{s}\cdot\bar{x} = s_1 w + s_2 d + s_3 l$. [24] (p. 72).

Before moving on to the main point, which will concern the dynamical evolution of the betting-states, let me add three important remarks:

(1) The correspondence between betting-states and number-states is many-to-one. If, e.g., a team is assigned 15 points after 10 games (by the standard number-state function $\bar{s} = \langle 3, 1, 0\rangle$), this is compatible with the team being in betting-states $\langle 5, 0, 5\rangle$ or $\langle 4, 3, 3\rangle$ or $\langle 3, 6, 1\rangle$. In general, therefore, knowledge of a team's number-state only restricts, but does not determine, which betting-state the team can be said to be in. Although it is natural to say that number-states "encode information about the betting-states," that information is not fully recoverable from the number-states. (There is common ground here between the football example and the toy-model developed by Spekkens in Ref. [27]: since the number-states put a limit on what can be known about the betting-states, they echo what Spekkens' refers to as the "knowledge balance principle", which he introduces as a postulate; cf. [27] (p. 3).) Introducing a further piece of terminology, I will say that a number-state "declares possible" all the betting-states that are compatible with it (so that, e.g., the number-state ascription "K has 15 points after 10 games" declares possible the betting-states $\langle 5, 0, 5\rangle$, $\langle 4, 3, 3\rangle$ and $\langle 3, 6, 1\rangle$ *relative* to the choice of number-state function $\bar{s} = \langle 3, 1, 0\rangle$).

(2) The specific number-state assigned to a betting-state has no objective significance, in the sense that (a) the choice of number-state function (from which it derives) is conventional, and (b) the relative ordering of the teams is not preserved under a general change of number-state function. In other words, since different choices of number-state functions (generically) produce different tables, the *relevant* relations— " ... is better than ... ," " ... is worse than ... ," and " ... is equal to ... "—are inherently relative to the choice of number-state function. Thus, these relations cannot be said to *reflect* (or *represent*) objective states of affairs.

(3) Using a piece of terminology familiar from foundational studies on the reality of the quantum state, we would say that the number-states are "ontic" rather than "epistemic" [14,15]. Since this terminology will prove useful again below, it is worth outlining the main idea behind the *ontological models framework*, from which this terminology derives (cf. [15] esp. (pp. 82–88) for a comprehensive overview of the relevant issues). A model is called *ontological*, if each state

in the model's state-space, which will be denoted by Π, corresponds to a classical probability distribution over some measurable space (Λ, Σ) (where Λ is called the *ontic state-space*, and Σ is a Borel (σ-) algebra on Λ). An ontological model is called *ontic* if, for any two states n and m in the state-space Π, every element in the ontic state-space Λ which n declares possible, m declares impossible (this is a somewhat loose, though I hope appropriate, way of paraphrasing the definition given in Ref. [15] for the case in which the elements of the state-space Π don't ascribe concrete probabilities to the elements of the ontic state-space Λ that they declare possible).

To see why our football example can be regarded as an instance of the ontological models framework, we reason as follows. First, take the betting-states to be elements of the ontic state-space Λ (and let Σ be the standard Borel field obtained from the set of—jointly exhaustive and mutually exclusive—*elementary propositions* "K has won/drawn/lost against L"). The number-states are elements of the state-space Π (which will thus be a subset of the natural numbers). Although the number-states fall short of providing a probability distribution over the betting-states, they declare possible a set of betting-states. However, each of these betting-states is declared possible by exactly one number-state (the last two points follow because, as was noted above, the correspondence between betting-states and number-states is many-to-one). Therefore, number-states are ontic. Despite being ontic, however, there is clearly nothing in the world that "corresponds" to a choice of number-state (since, as was already noted, different choices of number-state functions induce different orderings of the teams). This establishes that number-states are counterexamples to the argument that infers claims about "objective existence in reality" from a state's *logical* property of "being ontic." *Number-states are both ontic and non-representational*, and this impairs on the logical validity of the argument that aims to ground representationalist interpretations of the *quantum* state in the recent theorem by Pusey, Barrett & Rudolph [14] (which shows that pure quantum states are ontic).

2.3. Dynamics

What is still lacking from our analysis is an account of *dynamical evolution*: we would like to know how betting-states change over time. Consider, thus, how the situation looks *from the point of view of different observers*, who both know that the initial betting-states of some team is $\langle w, d, l \rangle$. Suppose that observer A was lucky enough to have acquired a ticket for the ensuing games. Observer B, however, has been less fortunate, and she doesn't know the outcomes the subsequent games. Since B, unlike A, isn't collecting any new descriptive information, I will refer to B as an *external* or *non-participating* observer. The question now arises: *how does each observer describe the change in betting-state for a given team?* (Notice that this set-up is the classical counterpart of the Wigner's friend thought-experiment [28] that will also be discussed below).

2.3.1. How the Situation Looks from A's Perspective

From A's perspective, the situation is clear. The correct state to assign after the next game is either $\langle w+1, d, l \rangle$ or $\langle w, d+1, l \rangle$ or $\langle w, d, l+1 \rangle$.

2.3.2. How the Situation Looks from B's Perspective

For B, the situation is more complicated. If B only knows *that* a certain number of games has taken place (i.e., if she agrees with A on the total number of games that were played), but not what the *outcomes* of these games has been, she will have to hedge her bets more carefully. Lacking the relevant descriptive information, she will be left to guesswork and speculation. However, not all speculation is equally good, and in the next section I will outline in what sense B can make a *best* guess (subject to certain constraints) as to what the final betting-state (into which the initial betting-states will have evolved) will be. From here onwards, since B's best guess is no longer defined as a partial description of the actual outcomes of the games, I will refer to this best guess as the "betting-state relative to

B" (or: "B's betting-state"). This is intended to indicate that the concept of a betting-state functions differently relative to B than it does relative to A.

In formal epistemology, the question of what B's best guess consists in has been discussed under the heading: "How do probabilities evolve if we do not assume an underlying determinism?" [20] (p. 68) In the next section, this problem will, following the presentation in Ref. [24], be addressed in its most general form. This will lead to a classical version of the Schrödinger equation as the correct equation governing the evolution of B's betting-state.

3. A Classical Version of the Schrödinger Equation for Optimal Opinion Updating Relative to Non-Participating Observers

Before we can address B's dynamics problem, it will be useful to first generalize the situation described thus far. Box 1 summarizes the kinematics of interactional probability models, for which Section 2 gave a specific example.

Box 1. Kinematics of Interactional Probability Models.

Kinematics of Interactional Probability Models

1. **Measurement Context:** In the general case, our *measurement context* consists of n distinguishable outcomes of the interactions between the entities within the model's scope.
2. **Betting-states relative to B**: The *betting-state relative to B* is an n-tuple of numbers $\overline{x} = x_1, \ldots, x_n$, the components of which specify B's best guess for how many times a particular outcome was observed (this best guess will be a *true* guess, if B happens to know the outcomes of the games). The set of betting-states has the following structure:
    a. *Betting-states are vectors.* Betting-states can be added (component-wise) and multiplied by a scalar (component-wise).
    b. *Odds-Comparison:* The *odds comparison* of two betting-states $\overline{x} = \langle x_1, \ldots, x_n \rangle$ and $\overline{y} = \langle y_1, \ldots, y_n \rangle$ is defined (for $\sum x_i = \sum y_i$) as $\overline{x}/\overline{y} := \langle x_1/y_1, \ldots, x_n/y_n \rangle$ (if well-defined).
3. **Number-states**: A *number-state ascription* is an ascription of a numerical value to a betting-state. All number-states that will be considered are required to arise from a choice of number-state function.
    a. *Number-state functions. Number-state functions* are linear functions from betting-states to some choice of number field. Thus, number-state functions are characterized by n numbers, $s = \langle s_1, \ldots, s_n \rangle$, such that: $s(\overline{x}) = \sum s_i x_i$. Hence, they are vectors of the same mathematical type as the betting-states.
    b. *Number-states are expectation values.* From the point of view of interpretation, the equation $\langle \overline{x} \rangle = s(\overline{x}) = \sum s_i x_i$ yields an *expectation value* (for the number of points associated with a team after a certain number of games, for caveats, cf. Section 2.2.3).
    c. *Number-state functions are coordinatizations.* Since the primary role of number-state functions is that of generating an ordering of the teams, I will say that number-states are "coordinatizations" of the betting-states (cf. Section 4).

Let's return to the problem of dynamical evolution (from the point of view of an external or non-participating observer B). Suppose that B knows the initial betting-state $\overline{x}(0)$ of a team, at time $t = 0$, and wishes to update her state to a final time t_f, yielding a state $\overline{x}(t_f)$. The question we are facing is this: what is B's best guess for the final betting-state $\overline{x}(t_f)$?

Before she can even begin to address this question, B must make an initial assumption. This concerns that fact that so far, the components of the betting-states were integers. However, since B is now confronted with the problem of having to account for the *changes* that occur in the components of her betting-states, and since it can be computationally very difficult to model discrete changes, she will do well to transform her problem into one that can be handled more easily. This can be achieved by allowing the components of her betting-state to evolve *continuously* (with respect to a parameter t). Therefore, she will embed (in the sense of providing an injective structure-preserving map) the betting-states—which are of the form $\langle w, d, l \rangle$ (with the components elements of the natural numbers)—into the set of triples of the form $\langle w, d, l\rangle$ (where the components now lie in the *real* numbers). This embedding is simply the trivial embedding of the natural numbers into real numbers (i.e., the identity map: $\mathrm{id} : a \longmapsto a$). The reason why this assumption is justified, is because the embedding preserves the relevant algebraic structure (in particular: the vector-space structure and the odds comparison). Therefore, although the problem has now been transformed, it has not been significantly altered. [24] (p. 852) In a slight abuse of terminology, I will still refer to these new states as "betting-states relative to B".

In this more general scenario, we can ask what types of constraints should be respected, for the evolution of $\overline{x}(0)$ to $\overline{x}(t_f)$. One reasonable constraint is that dynamical evolution shouldn't mess up the odds comparison of the teams. Otherwise, the evolution would (unnaturally) privilege certain teams over others, which might distort—not the facts, mind you!—but what is rational to believe from the point of view of observer B. To implement this, we will need the concept of a symmetry.

- *Symmetry*. A *symmetry* of the space of betting-states is a linear map U that maps the space onto itself such that the odds comparison of betting-states is preserved: $U(\overline{x})/U(\overline{y}) = \overline{x}/\overline{y}$ (if well-defined). [24] (p. 857), [20] (p. 72).

Our first constraint, that odds comparisons should be preserved, therefore becomes: B's evolved state should be the result of a symmetry transformation on her initial state.

Before proceeding, let me add some brief comments. (1) As an important observation, notice that Bayes' theorem is also derived from a symmetry condition (this is a key element that Hughes and van Fraassen in Ref. [24] draw attention to). (2) Notice that the justification for the first condition relies on the *relational nature* of our description—what we want to be preserved is a relational quantity (the odds comparison) rather than the quantities $p_\alpha = \alpha/n$ $(\alpha = w, d, l)$, which characterize a *single* team. (3) We will have a chance to use the following theorem:

Theorem. *A transformation U is a symmetry if and only if there exist positive real numbers $u_1, \ldots, u_n$ such that $U(\overline{x}) = \langle u_1\, x_1, \ldots, u_n\, x_n\rangle$. [24] (p. 857), [20] (pp. 71–73).*

Proof. See Appendix A. □

The next condition is that evolving by a time t_1 and then by t_2 should be the same as evolving by $t_1 + t_2$. ("The set of evolution operators form a semi-group"). This gives rise to another definition:

- *Uniform motion.* A *uniform motion* on the space of betting-states is an element of the set $\{U(t),\, t \geq 0 : U(t_1) \circ U(t_2) = U(t_1 + t_2)\}$ of symmetries labelled by a continuous parameter t. [24] (p. 857), [20] (p. 72).

There is another straightforward theorem that will prove useful:

Theorem. *If two betting-states $\overline{x}(0)$ and $\overline{x}(t)$ are related via a uniform motion $\overline{x}(t) = U(t)(\overline{x}(0)\,) = u_1(t)\, x_1(0), \ldots, u_n(t)\, x_n(0)$, then there exist positive real numbers $k_1, \ldots, k_n$ such that $u_i(t) = e^{k_i t}$. [24] (p. 858), [20] (pp. 72–73).*

Proof. See Appendix A. □

The next constraint will put on the breaks. We still don't know anything about the coefficients k_i that figure in the previous theorem. If those are chosen at random, the final state $\overline{x}(t_f)$ might be arbitrarily distant from the initial state. To prevent this, which we should, if B's guess is to be taken as a "best" guess, we impose that the *overall* change induced by the evolution operators is minimal. Now, the total rate of change in B's betting-state is given by: $\frac{\partial}{\partial t} \sum x_i(t) = \sum k_i x_i(t)$. Furthermore, since the k_i's are positive real, we know that all derivatives $\frac{\partial^{(m)}}{\partial t^{(m)}} \sum x_i(t)$ are positive real. Thus, in particular, the second derivative will be greater than 0, which means the first derivative will be a monotonically increasing function. To minimize the overall change in B's betting-state, we can therefore require that the quantity $\frac{\partial}{\partial t}\sum x_i(t) = \sum k_i x_i(t)$ be minimal for $\overline{x}(t_f)$. Notice that, since $k_i \sim \ln\left(\frac{x_i(t_f)}{x_i(0)}\right)$, this has the form of an *entropy maximizing condition*. [24] (pp. 858–860).

There are two final conditions. First, we want the total number of games to be known: the betting-state relative to B should be *normalized*, in the sense that B knows the total number of games that have been played at the final time: $\sum x_i(t_f) = n(t_f)$. Secondly, we impose that the number-state (which has played no role so far) at the final time is fixed, i.e., that $\sum s_i x_i(t_f) = r(t_f)$ (thus $r(t_f)$ denotes the final number-state of the team). The interpretational spin we could put on this is that the number-states assigned to the teams should be *the same* for both observers. Hence, we allow that A communicates to B (after all games have been played) what the final number-state of the team is. Since there might be many different betting-states that give rise to the same number-state, this condition plays an important role: the allowed betting-states are those that the number-state declares possible (in the terminology from Section 2.2.3).

Summarizing the discussion, we can now present the problem of B's dynamical evolution as the following *optimization* problem:

Box 2. Non-participating observer B's dynamical problem.

Optimal Rational Opinion Updating Relative to Non-Participating Observers
B's dynamical problem: Find a set of evolution operators U that relate the betting-states $\overline{x}(0)$ and $\overline{x}(t_f)$ (at time 0 and t_f respectively) such that:

1. U is a uniform motion: $\overline{x}(t_f) = U(t_f)(\overline{x}(0))$.

According to the theorems mentioned above, this means that we already know that $\overline{x}(t_f) = \langle u_1(t_f)\, x_1(0), \ldots, u_n(t_f)\, x_n(0)\rangle$, and that there exist positive real numbers $k_1, \ldots, k_n$ such that $u_i(t_f) = e^{k_i t_f}$.

2 Find real numbers $k_1, \ldots, k_n$ such that $\sum k_i x_i(t_f) \to \min$. This is subject to the constraints that:

a. The final betting-state relative to B is normalized: $\sum x_i(t_f) = n(t_f)$.
b. The final number-state is agreed upon by both observers A and B: $\sum s_i x_i(t_f) = r(t_f)$.

It now can be proven that [24] (pp. 860–862):

Theorem. *There exist constants v & w such that the* $u_i\left(t_f\right)$ *'s are given by* $u_i\left(t_f\right) = e^{vt_f}\, e^{ws_i t_f}$.

Proof. See Appendix A. □

In other words, *B's optimal opinion change is given by something that looks a lot like a classical version of the Schrödinger equation:* the constant w plays the role of Planck's constant and the s_i's play the role of the eigenvalues of the Hamiltonian. w and v are Lagrange multipliers that are uniquely determined by the boundary-conditions 2a/b in Box 2. [24] (p. 861) Observe, also, that evolution depends on a choice of number-state function; hence different such choices induce different "shifts" in different "bases." Therefore, the final betting-state relative to B—which will be a different state than the betting-state relative to observer A—is uniquely fixed by the above conditions. Therefore, the evolution of B's betting-state is *deterministic*.

This concludes the mathematical discussion. The challenges ahead, of course, are still quite significant. For one, we haven't said anything about quantum mechanics yet. While the above theorem certainly resembles, rather closely, the form of the solutions to the Schrödinger equation, there are still important differences (most notably, the additional factor of e^{vt_f} and the absence of the imaginary unit *i*). Section 5 tries to substantiate the claim that the above theorem can indeed be viewed as a classical version of the Schrödinger equation. Specifically, I will argue that it is the form of the *problem*, as well as the form of the *solution*, that justifies viewing the result of the above theorem as the classical analogue of the Schrödinger equation. To pave the way for this discussion, Section 4 discusses some conceptual subtleties that, so far, haven't received the attention they deserve. (There is another worry, unrelated to the subsequent discussion, to which an anonymous referee has alerted me. This concerns the fact that that the theorem predicts that the total number of games n(t)—i.e., the sum of the components—evolves as a sum of exponential functions, which might appear counterintuitive. Why would B conclude that n(t) evolves in this fashion (as opposed to, say, making the more reasonable assumption that events occur at a constant rate)? To see that the theorem produces "reasonable" results, there is an explicit example worked out in Appendix B.).

4. Preliminary Discussion—Some Advantages of the Example

The setting I have chosen will likely have struck the reader as somewhat peculiar. At the very least, this *interactional probability model* is quite detached from the paradigm cases of probabilistic models, such as coin-flipping and dice-tossing, which still often guide our thinking about these matters. Here, I argue that this is a good thing, which will be achieved by locating the example relative to some familiar issues in the philosophy of probability. I will also expose the sense in which Rovelli's relational interpretation [11–13] (or: at least one potential version of it; cf. Section 5) is inspired by such interactional probability models as the one that was presented above. Using the classical example, I will try to illustrate why the relational view might enjoy some important advantages over its closest neighbors (such as QBism or neo-Copenhagen views).

Now, if we had chosen a more paradigmatic setting, such as a coin being tossed, the theorem that fixes the evolution of B's probabilities would still have gone through. After all, Hughes & van Fraassen's presentation in Ref. [24] can be interpreted in this way: how do the probabilities of a *single* coin, or a dice, evolve relative to different rationality constraints (which specify the epistemic situation of different observers: observer A has knowledge of *outcomes*, while observer B only knows *that* the coin was tossed)? For this reason, proponents of observer-dependent interpretations of the quantum state other than the relational view might wish to include the result of Section 3 in their argumentative toolbox. Nevertheless, the set-up from Sections 2 and 3 has some strong conceptual advantages over the more traditional examples, which merit closer inspection.

4.1. The Betting-States are not "Absolute Descriptions"

The claim that the betting-states are not "absolute" descriptions of the teams is intended in the following sense: had we chosen a more traditional setting of a coin being tossed, the possible outcomes *heads* and *tails* certainly characterize fully *intrinsic* properties of the coin. But while the concepts of *winning*, *losing*, and *drawing* may give away the impression of characterizing the teams "absolutely", this is only an appearance: you can only win, lose, or draw *against* another team. There are two noteworthy aspects to this.

4.1.1. Betting-States are Correlations

The first remark relates back to the observation that there will be *correlations* in the betting-states. After each round, as many teams will have one more win as there will be teams with one additional loss. Hence, if you ask a team how they played, and they tell you that they have won, you know—instantaneously and without hesitation—that the other team has lost. For that reason, the above example exemplifies a central commitment of Rovelli's relational view, namely that the ascription of any particular state (which is interpreted as codifying observer-dependent information) is equivalent to asserting that there exists a certain *correlation* between systems: "*The fact that the pointer variable in O has information about S (has measured q) is expressed by the existence of a correlation between the q variable of S and the pointer variable of O.*" [11] (p. 9).

This is to be contrasted, in particular, with neo-Copenhagen interpretations of especially Zeilinger and Brukner [8,9], according to which quantum theory is also grounded in information-theoretic considerations, but in a different sense. Zeilinger [8], e.g., postulates his "foundational principle", which states that each elementary system carries one bit of information. This, clearly, suggests that the amount of information somehow characterizes an intrinsic property of certain systems, leading to the vexing claim that "information is physical" (cf. Timpson in Ref. [29] (pp. 67–73; 152–158) for a pointed discussion of the problems associated with such claims). On the relational view, information is simply the existence of correlations, and hence, this view sidesteps these types of debates (of whether we can make coherent sense of the suggestion that information is physical). This advantage of the relational view over the neo-Copenhagen approach (one of its closest allies) carries over to the second point I would like to stress.

4.1.2. "How Things Are" vs. "How Things Affect One Another"

It is important to observe that the betting-state relative to B isn't useful because it characterizes "how physical systems are." On the contrary, it is useful primarily because it determines the odds for how a team will fare when playing against some other team (since we required the odds comparison to be preserved). To put it in a slogan: B's problem is all about "how things affect one another" rather than about "how things are." In that sense, this example perfectly instantiates what is perhaps *the* core commitment of Rovelli's relational view: "The core idea is to read the theory [of quantum mechanics] as a theoretical account of the way distinct physical systems *affect each other* when they interact (and not the way physical systems 'are')." [12] In other words, on the relational interpretation, quantum theory is interpreted along the following lines: from the previous point, we know that quantum state ascriptions are the result of preparation procedures, in which a correlation between measurement apparatus and system has been established. [11] (pp. 9–10) At the same time, the point of collecting information in this way is that this allows predictions to be made (from the point of view of different observers) for the outcomes of possible future interaction with other systems (as opposed to: preparation procedures playing the role of characterizing "how quantum mechanical systems are").

4.2. Completeness

The football example is "universal," in the sense that the interacting entities (the teams) are, albeit *physically* distinct (since they are characterized by different betting-/number-states), *metaphysically* equal. This is important for the following reason. It is uncontroversial, though easily overlooked in paradigmatic cases of probabilistic processes (such as a coin being tossed), that probabilities are a peculiar mixture of being both "absolute" descriptions of "how things are," but also descriptions of "how things affect one another," at least in the following sense: suppose, e.g., that we want to toss a coin, but that this coin is also a magnet, such that if it is tossed over a sufficiently strong magnetic field, it will always show *heads*. Since we might, e.g., turn the magnetic field on and off, the probabilities that are to be assigned for each toss should always be thought of as functions of: (1) of the coin itself, but *also* (2) the environment with which the coin, once tossed, will interact. In the traditional settings, however, the environment will generically obey *different* (perhaps non-*probabilistic*) laws than the coin itself (which is modelled probabilistically). In that sense, although both environment and system may obey well-defined sets of physical laws, they will have to be viewed as metaphysically distinct. This situation changes in the football example. The model universally applies to all entities within its scope, and this shows how a theory of "how things affect one another" can consistently held to be *complete*. We simply don't require any story about "how things are" in order to have a substantive and powerful account of "how things affect one another." And this, in turn, captures the intuition behind why Rovelli's relational interpretation aspires to be *complete:* on the relational view every system will be treated as a *quantum mechanical* system; all systems will be metaphysically equal, though physically distinct, partners in the interactions, and the description in terms of *transition probabilities* (from earlier to later states) will be held to be *complete*. [11] (p. 7), [11] Since completeness is a clear virtue of any interpretational strategy, the relational view will enjoy this advantage over its next-door neighbors such as QBism or the neo-Copenhagen view, which require further resources to describe reality (QBists in terms of direct experience [7]; neo-Copenhagens in terms of classical mechanics [10]).

4.3. *The Role of Number-States*

The football example reserves an explicit role for the number-states, which deepens the analogy to quantum states in at least two ways.

4.3.1. Number-States are Expectation Values

Notice, first, that we can construct a counterpart to the problem of "choosing a number-state function" for the traditional example of a coin being tossed. For example, since we could associate any arbitrary pair of numbers with either side, we could, given any such assignment, calculate a numerical expectation value (just as in the football example). And this expectation value is also going to be conventional (as the numbers we could choose to associate with each side might be selected at random). First of all, this would be somewhat artificial. Secondly, this is beset with some vexing philosophical problems. For suppose we have decided to label the different sides of the coin with the numbers 1 and 2. In the next step, we will have to ask ourselves whether there is a correct way of labelling the different sides of the coin. If the coin is slightly biased or asymmetrical, which must be assumed for any non-ideal case, then it makes a difference which number we assign to which side. But we can't know what that bias is, if we don't already know what the correct labelling consists in (arguing that there is a primitive labelling of *heads* and *tails* only pushes the problem one step back). In the football example, these problems, which echo some traditional problems from the philosophy of probability, simply disappear (which, of course, is due to the fact that in the football example, symmetry considerations don't concern any intrinsic property of a single team, but a relational quantity, i.e., the preservation of the odds comparison).

4.3.2. Number-States are Coordinatizations

The number-states have the peculiar role of generating a *relative ordering* of the teams (as, e.g., the relation "better than" was induced by a choice of number-state). This has an obvious counterpart in physics: the claim that a particle is *located at position x*, must always be relative to some reference system (which defines our origin). Hence, physics is in the same business of *coordinating systems* (a) relative to other systems and (b) relative to a choice of coordinates ("S is m meters to the left of O" maps onto "K is n points better than L"). Notice, in particular, that the condition that number-states were agreed upon by both A and B (which implies that the ordering of the teams is the same for both observers) played an important role in the derivation of the equation that governs the evolution of B's betting-states.

5. Quantum vs. Classical: The Similarities

Let's now make the transition to quantum theory. Here, I focus on the similarities between the classical example and quantum theory. In Section 6, I will discuss the differences between the classical and the quantum case.

Rather than immersing ourselves in the details of these arguments straight away, however, I would like to briefly comment on what such a discussion, of the similarities and differences between quantum theory and a classical example, could possibly establish. Now, we might observe that insofar as the dynamical evolution of betting-states relative to B was a *result*, a natural approach would be to place the previous discussion in the context of existing "reconstruction programs," and investigate which assumptions would have to be modified in order to obtain a *different* result than in the classical case (i.e., the *actual* Schrödinger equation). What this would mean, provided that some of the more promising reconstruction projects proceed in terms of imposing information-theoretic constraints [8,9,11,30–32] is that we would have to investigate in what sense quantum theory limits the amount of knowledge an observer could, in principle, obtain. And certainly, this would sharpen our sense for what is "quantum" about quantum mechanics. While there is a lot to say about the similarities between the classical example and various reconstruction programs, the focus here will not be to provide such a systematic comparison or analysis. The reason is because it is not clear what these reconstruction programs establish from the point of view of interpretation. Timpson has articulated this issue like this: "By assumption [of a reconstruction project], the world is such that the information-theoretic constraints are true, but this is too general and it says too little: it is consistent with a wide range of ways of understanding the quantum formalism." [29] (p. 177) In other words, once we have reconstructed the formalism, we should expect to be able to annex any interpretation to the result that we see fit. For this reason, the discussion here will instead take the following form:

- The goal of the discussion of the similarities between the quantum formalism and the classical example will be to provide a suggestive reason for why the above theorem can indeed be viewed as the counterpart of unitary evolution in quantum theory. This will rely on a result due to Lisi [25], who has proposed a heuristic derivation of the Schrödinger equations from similar assumptions as the ones that were required to prove the result of Section 3. Of course, and this is the important point here, such a discussion could only be suggestive (establishing too close a resemblance between the classical and the quantum case could only mean that we have made a mistake—the two cases are, after all, fundamentally different).
- The discussion of the *differences* between the two cases will be less suggestive. The argument will be structured around the thesis that *unlike* in the classical case, quantum theory is inconsistent with the assumption that each measurement has a determinate outcome. This argument will rely on a recent no-go theorem due to Frauchiger and Renner. [26], cf. [33] Since this no-go theorem is derived on the basis of the quantum formalism *itself*, this purports to show in what sense the formalism restricts the set of viable interpretations (this would not be visible if we stayed at the level of the reconstruction programs, for the reason articulated by Timpson).

Let me, then, begin the discussion of the similarities between the classical and the quantum case by briefly rehearsing the kinematical structure of the theory. Quantum mechanics is set in separable Hilbert spaces $\mathcal{H}$. *Observables* are defined to be the self-adjoint elements of $\mathcal{B}(\mathcal{H})$, the set of bounded linear operators on the Hilbert space; the space of these observables is a von Neumann algebra: an involutive Banach-algebra that's closed in the strong topology. *States* are defined as linear trace-class operators $\Im(\mathcal{H})$ on the Hilbert space (of trace 1). Notice that the state-space, too, is a vector-space: abstractly speaking, the state space is the *dual vector space* of the space of observables (playing such similar mathematical roles of being *structure-preserving functionals* makes for the analogy between number-state functions and quantum states). A special class of states, the so-called *pure states*, stand in one-to-one correspondence to the 1-dimensional subspaces of the Hilbert space (the so-called *rays*). Thus, we may (if we are dealing with pure states) use vectors in that subspace to represent the state (by convention, we take a vector of unit length although, strictly speaking, that still leaves a phase-ambiguity). To get probabilistic predictions from the theory, we rely on the Born rule, $P(A|\rho) = Tr(A\rho)$ (where A is an observable and ρ is a density operator). This specifies *probabilities* (if A is a projection operator) and *expectation values* (for general self-adjoint operators). If ψ is a pure state, the expectation value is given by a vector-space product of the form $(\psi, A\psi)$.

This mathematical structure, oversimplifying rather drastically, allows us to make probabilistic predictions for the outcomes of measurements, and hence it is sufficient to check quantum theory in experimental practice. But suppose that we aren't satisfied with making only such probabilistic predictions, and that we want some form of understanding for how the experimental results are brought about. Suppose, in other words, we want to be realists about the claim that "System S is in state ψ." How should we interpret this assertion?

On a first proposal, we might suggest that the state *represents* something in the world, in the sense that the world should be thought to be "isomorphic" ("structurally equivalent in the relevant ways") to ψ. While this may sound tempting, there is something strange going on here from a conceptual point of view. Ordinarily, we would suppose that *systems* exist but not that states exist—and what type of *thing* or *object* in the world could possibly be isomorphic to ψ? As, e.g., Halvorson [34] points out, if we confuse *states* and the *objects of which they are states*, we run into the problem that "if states are objects, then states themselves can be in states. But then, to be consistent, we should reify the states of those states, and these new states will have their own states, ad infinitum. In short, if you run roughshod over the grammatical rules governing the word "state", then you can expect some strange results." [34] (p. 6) Consider, as a response, a second realist strategy: the suggestion that states play the functional role within the theory of assigning to the system a rather specific property: *"being in state ψ."* This leads to what Halvorson calls a "state-to-property" link (he credits Wallace and Timpson [35] for this proposal). According to this, states no longer represent *objects*, but they represent a *property* of this object: *a system is said to be in state ψ, just in case it has the property of being in state ψ*. [34] (p. 24) However, it is not clear what exactly this implies with respect to ontology. As Halvorson remarks, "I suppose this claim is true. But I didn't need to learn any physics to draw that conclusion. This is nothing more than a disquotational theory of truth." [34] (p. 25) But the state-to-property link can be broadened: what certainly goes beyond the disquotational theory of truth, is the functional role states could play in assigning values to *other* observables. The set $\mathcal{B}(\mathcal{H})$ contains a whole fauna of self-adjoint operators, which represent physical properties. Relying on the eigenstate-eigenvalue link, a realist interpretation of the quantum state will, on this proposal, amount to the claim that states appropriately "track" the values (possibly unsharp!) that observables take for a given system. There is a host of technical complications connected to this proposal (most notably, the Kochen-Specker no-go theorem [36]). Here, however, I would only like to focus on the problem that has played center-stage in the discussion of realist interpretations of the quantum state: *the quantum measurement problem*.

To illustrate what the measurement problem consists in, we need to include the dynamical structure of quantum theory into our discussion. The first dynamical postulate of quantum theory is the *Schrödinger equation*, according to which states evolve unitarily. But Schrödinger's famous cat thought-experiment [37] illustrates that this cannot be: macroscopic systems may evolve unitarily into superpositions that aren't observed in experiment. In textbook presentations of the theory, one therefore typically encounters a further dynamical postulate: von Neumann's *collapse postulate*, which tells us that when the measurement happens, the state after this measurement is represented by the projection operator corresponding to the eigenspace in which the system was observed to be. But this also cannot be: even leaving aside quarrels about non-locality, if the state represents some objective feature of the world, then collapse is a real physical process, which should have been brought about for a certain reason. Since quantum theory itself doesn't provide such a reason, the collapse postulate, as Brown put it in the quote from the introduction, appears "mysterious." [23] (p. 17).

Anti-realist interpretations escape this dilemma by dropping the assumption of realism about the quantum state. This, of course, raises the question of what role the quantum state is then going to play? The purported insight of, in particular, *QBism* [2–7], *neo-Copenhagen views* [8–10], and the *relational interpretation* [11–13] is that the state does not represent an objective feature of the world, but should be regarded as a codification of observer-relative information. It is a mere *consequence* of this suggestion that different observers may have to update their opinions in different ways (i.e., subjected to different constraints). Hence, these approaches, at least *prima facie*, are less threatened by the dualism of dynamical evolutions—but only insofar as they can make sense of both types of evolution as forms of rational opinion updating.

It would be too big a task to focus on all these interpretations here. Instead, I would like to highlight how this purported insight—the democratization of state ascriptions across observers—plays out for the case of Rovelli's relational view, and how this insight bears on the measurement problem.

To this end, I will briefly outline the main commitments of the relational view. Now, on one way of reading Rovelli, his view is built around the following postulates:

- (*Equivalence of physical systems*) "All systems are equivalent: Nothing distinguishes a priori macroscopic systems from quantum systems." [11] (p. 4).
- (*Relative facts postulate*) Any system has a quantum state relative to other physical systems (which we, depending on context, consider as the "observing" or "reference" systems).

To avoid any potential confusion, it is important to note that the (*Relative facts postulate*) is only compatible with *some* of Rovelli's remarks, most prominently with his suggestion that "The core idea is to read the theory [of quantum mechanics] as a theoretical account of the way distinct physical systems *affect each other* when they interact (and not the way physical systems 'are')." [11] (*Relative facts postulate*) is inconsistent, however, with many other remarks of his, namely when he asserts what might be more aptly called the (*Empiricist facts postulate*):

- (*Empiricist facts postulate*) "A quantum description of the state of a system S exists only if some system O (considered as an observer) is actually "describing" S, or, more precisely, has interacted with S." [11] (p. 6).

Clearly, the (*Relative facts postulate*) and the (*Empiricist facts postulate*) express different propositions. I will further restrain attention to the version of the relational view that is based on the (*Relative facts postulate*) (although most of Rovelli's own ideas seem to rely on a commitment to the (*Empiricist facts postulate*)). There are several reasons for this (some of which will surface in Section 6). For the purposes of the current discussion, I would only like to note that the (*Relative facts postulate*) seems the more natural choice if we indeed take the classical example from Sections 2–4 as our intuition pump. The pertinent point is that the betting-states should be said to exist, relative to the different observers A and B, even if these observers don't know what these betting-states are (as would be suggested by the (*Empiricist facts postulate*)). B's best guess exists—in the same thin sense of "existence" in which any mathematical entity can be said to exist—even if B doesn't follow the calculation from Section 3. Similarly, the betting-state for A after the next game is either $\langle w+1,\ d,\ l\rangle$ or $\langle w,\ d+1,\ l\rangle$ or $\langle w,\ d,\ l+1\rangle$. Even though we don't know *which* one it will be, we are sure *that* one of these is correct. Hence, there should be no requirement that either A and B know what the correct betting-state is (relative to either of them), in order to say *that* there is such a state.

To illustrate the workings of the (*Relative facts postulate*), think about how mathematics is used in physics more broadly. Take, for instance, a classical particle that is dropped from a high altitude, while being subjected to air resistance. The particle's trajectory will, in certain models, be the solution to a (non-analytically solvable) differential equation of motion. Now, to say that this mathematical description is true because the interactions that occur (between the gravitational field, the particle, and the air) are nature's way of "solving this equation" is obviously nonsense (how could nature solve the equation, if it is not analytically solvable?). The ones solving the equations are certainly going to be the users of the theory. To say that this equation is "true" says something about the *solution* of the equation, but not anything about the *process* by which this solution is obtained—although the equation tells us something about the interaction, the interaction itself is irrelevant for what the equation says. The same applies to probability assignments on the relational approach that is based on the (*Relative facts postulate*). The *process* of calculating or acquiring information about relative states is as irrelevant, for the claim that there *are* such relative states, as the process of finding solutions to the equations of motion is for the claim that the equations of motion are true. Hence, the relational view that is based on the (*Relative facts postulate*) manages to incorporate the key conceptual point (that states are always relative to a reference system) in a way that is consistent with how mathematical models are applied in physics more broadly. Secondly, it is also worth emphasizing that assuming there to be facts about states relative to physical reference systems is not to say that unanimated objects use probability ascriptions to inform their betting behavior (just like nature doesn't solve differential equations). Of course, the state ascriptions are made, remembered, and exploited, by the users of the theory for their epistemic or decision-theoretic purposes (just as in the case of the particle being dropped, the equations are solved, and exploited, by agents for their epistemic or decision-theoretic purposes). The claim is, rather, that nature just ends up behaving in a way that is compatible with what would have been predicted if anyone had actually ascribed, recorded, or calculated these states (just like nature ends up behaving in a way that solves the equation of motion, independently of whether anyone actually *solves* the equation, or has any determinate record of the initial conditions). Only conscious agents rely on ascribing relative states, but consciousness itself is irrelevant for the claim that there are relative states. These remarks are intended to show that the relational view manages to deflate the subject/object dichotomy of QBism or neo-Copenhagen approaches (since the observers, on the relational approach, are thought of as purely physical systems). One requirement we certainly must place on the reference systems, however, is that they are the kinds of things that can *in principle* acquire information about a system, by virtue of *interacting* with it in the appropriate way (hence the (*Relative facts postulate*) requires the reference systems to be *physical* systems).

Departing now from these assumptions, let's return to the main thread that was left hanging above: how does the democratization of quantum state ascriptions, in the way that is prescribed by the (*Relative facts postulate*), bear on the quantum measurement problem? What is suggested by the remarks so far, is that the problem of understanding dynamical evolution in relational quantum mechanics will be solved by the modelling our answer on the question: *How should observers update their relative state ascriptions in the "best" way*? Let me reemphasize that it is a mere consequence of this view that we may be forced to admit that, just as in the classical case, different observers might have to update their opinions in different ways. If this would turn out to be the case, this would underscore the (*Relative facts postulate*).

Let's use the example of the Wigner's friend thought-experiment [28] to illustrate how this line of thought will be developed (the presentation here follows Rovelli in Ref. [11] (pp. 2–4)). Suppose that there are two physical reference systems—referred to as "observers"—O and P, as well as a further system S. So, according to the relational view, S has a state relative to both O and P. The experimental protocol is such that O is going to conduct a measurement on S, but P doesn't participate in the process. For concreteness, let S be a spin-half system, and let the measurement that O performs on S be a spin measurement along the z-axis. The relative states *after* the preparation procedure but *before* O's measurement are given by:

$$\psi(S, O) = \alpha\,|\text{up}\rangle \;+\; \beta\,|\text{down}\,\rangle$$

$$\psi(S+O,\,P) = \alpha\Big|\text{up} \otimes {''}\text{ready}''+\Big\rangle\; \beta\Big|\text{down} \otimes {''}\text{ready}''\Big\rangle$$

Here the obvious notational convention is that $\psi(S+O,\ P)$ denotes the state of the joint system S+O relative to P (the second term after the tensor product denotes the state of O). Note that because $\psi(S+O,\ P)$ is a product state O and P will initially agree on the relative state of the system S ("tracing out" the observer O from $\psi(S+O,\ P)$ yields the state $\psi(S,\ P)$ which is the same as $\psi(S,\ O)$).

Next, O carries out her measurement on S in the spin-z basis. After this measurement, the state of S relative to O will collapse, according to the *collapse postulate*, into one of the eigenstates "up" or "down." As was mentioned in the introduction, this can be justified by the proponent of the idea that the quantum state is observer-dependent by appealing to Lüder's rule [18]: oversimplifying slightly, the projected state agrees, in its Born probability assignments to observables, with the canonical generalization of the notion of conditional probability (conditioned on the observed outcome) to a quantum mechanical setting (in which the objects, to which probabilities are assigned, are elements of an ortho-modular lattice rather than a Boolean algebra). Therefore, *relative to O*, who has access to the measurement *outcome*, it is rational to use, as the basis for future predictions, the collapsed state [20] (pp. 171–175), [21], [22] (pp. 170–173).

If P, however, doesn't interact with S+O, then the standard quantum mechanical prediction for the state of S+O relative to P *after* O's measurement will be the unitarily evolved state:

$$\psi(S+O,\,P) = \alpha\Big|\text{up} \otimes {''}\text{up}'' + \beta\Big|\text{down} \otimes {''}\text{down}''$$

But now the question arises: can the proponents of observer-relative interpretations of the state justify unitary evolution as a form rational probability updating? By now it will be apparent that the suggestion is to view the theorem for optimal rational opinion change we have met in Section 3 as the classical counterpart of unitary evolution in quantum mechanics, in just the same way in which Bayes' theorem may be viewed as the classical counterpart of the projection postulate in quantum mechanics.

To substantiate this claim, I will use a result by Lisi [25], according to which P's question is presented in an identical form as the problem of optimized probability updating that was discussed in Section 3. Lisi, who explicitly puts his approach in the context of Rovelli's relational interpretation (cf. [38]), presents his argument in the language of the path-integral formalism developed by Feynman. [39] In this formalism, we consider all possible paths of a system between two fixed points, each of which is described by a configuration q(t). We then ascribe a probability p[q] to each path. Then, it can be shown that solving the following optimization problem is equivalent to the path integral formulation of quantum mechanics:

- *Optimization problem for the external observer in quantum mechanics.* We require that the entropy—$H = -\int Dq\, p[q] \log(p[q])$- is minimal, subject to the constraints that (1) the probabilities associated with each path sum up to 1: $\int Dq\, p[q] = 1$ (normalization), and (2) the expectation value of the action functional is fixed to be $S = \int Dq\, p[q]\, S[q]$.

Solving this problem (using the method of Lagrange multipliers), Lisi derives the form of the wavefunction ($\phi = e^{-iS/\hbar}$). From this, the Schrödinger equation, i.e., unitary time-evolution, follows in the standard manner. By virtue of requiring the expectation value of the action functional to be constant, Lisi refers to his "heuristic derivation" [25] (p. 1) as an attempt to reconstruct quantum theory from a universal action reservoir. [25] (p. 2) Despite acknowledging that his derivation is at best heuristic, he argues that "The main product of the work is the proposal of a new physical principle for the foundation and interpretation of quantum mechanics: a universal background action." [25] (p. 4).

Notice that this situation is completely analogous to the one we encountered in Sections 2 and 3! There are two observers (the *participating observers* A and O vs. the *non-participating observers* B and P), with access to different pieces of information (A and O know the *outcome* of the interaction, while B and P only know *that* the interaction has taken place). These differently situated observers are expected to update their information subject to different sets of conditions of optimality (relative to their respective epistemic situations). As before, the non-participating observer wishes to maximize entropy; and in both cases, quantum and classical, we end up with a dualism of dynamical evolutions.

From these assumptions and results, Rovelli's *main observation* follows:

> "**Main observation:** In quantum mechanics different observers may give different accounts of the same sequence of events." [11] (p. 4)

To summarize, the reason why they must do so is because: (1) of the presupposition that the relative states do not correspond to, or represent, real states of affairs, but should be understood as "encoding observer-dependent information about the system," and (2) different observers should update their information relative to different criteria of optimality and rationality.

The (*Relative facts postulate*) therefore provides a *natural* explanation for why there is more than one possibility to update one's opinion: different observers, depending on whether they know *the outcome* of an interaction (O's knowledge) or only *that* the interaction between O+S has taken place (P's knowledge), might have to update their beliefs in different ways in order to act the most rationally. In that sense, the suggestion that the quantum state is a codification of observer-relative information offers an elegant (dis-)solution to the measurement problem.

What we have seen so far, then, is that there is a suggestive and far-reaching analogy between our classical example and quantum mechanics. The proponents of the view that states are observer-relative, therefore, have the means to justify the view that quantum theory is a new form of (non-Boolean) probability theory. However, it is not immediately obvious what the democratization of quantum state ascriptions—as postulated by in the (*Relative facts postulate*)—implies from the point of view of interpretation. On the one hand, there clearly is no *logical* requirement that any such story of observer-dependent facts must be completed by some realist-type ontological story. But, of course, we might still *want* to find such a story. At least ideally, a stout non-representationalist should also provide an argument to the effect (not only that one *needn't* give a realist story but) that one *couldn't* give such a story. In Section 6, I will outline what I consider the strongest argument to establish precisely this conclusion.

6. Quantum vs. Classical: The Differences

As always, understanding an analogy, to a great extent, consists in the understanding of how it breaks down. While I have, so far, focused on the similarities between quantum and classical probability models, I will now discuss their differences.

The cue for understanding how quantum and classical probabilities differ, derives from the observation that in the *classical case,* it is natural to argue that A's betting-state is *privileged* over B's. After all, A has, unlike B, *actually* observed the outcome of the game, while B's state is only a "best guess." In the classical case, then, not all models are *epistemically equal,* in the sense that we might say that "A knows *more* than B." This subsequently allows us to assert that A's descriptions, i.e., the betting-states relative to A, *corresponds to something in reality,* while B's description has a merely epistemic function. For this reason, we may, in the classical case, envelop the probabilistic model in a straightforward ontological story of what the world is like.

So, we must investigate whether this situation also obtains in quantum mechanics: can either of the two perspectives (O's or P's) be privileged over the other? This, certainly, would undermine the (*Relative facts postulate*) from Section 5, which only has real bite if we regard both O's and P's relative state ascriptions as *equally valid* codifications of what is rational to believe from their respective perspectives. Section 6.1 argues why these privileging strategies are unsuccessful. Oversimplifying the matter slightly, the crux of the issue will be that *unlike* in the classical case, measurement outcomes, in quantum theory, can't (in general) be said to be *determinate.* This, after all, is the basis on which (in the classical case) epistemic privilege is handed out to the participating observer A. To establish that quantum mechanics differs from classical theories in precisely this respect, I will rely on a recent thought-experiment by Frauchiger and Renner [16] (many thanks to Richard Healey for bringing this to my attention).

But even so, there is still a third possibility that we will need to consider, namely that neither description is privileged, but that they are compatible with some underlying realism (a situation that obtains, e.g., for *velocity ascriptions* in special relativity).

I will try to illustrate how one might come to believe that *neither* of these three options—(1) O's state is privileged, (2) P's state is privileged, (3) O's and P's states are compatible with some underlying ontology—obtains. If the overall argument is indeed deemed successful, we will have a strong rationale for claiming that quantum theory can't, in general, be supplemented with a realist story (the argument here may be thought of as an elaboration and development of Rovelli's discussion of these issues in Ref. [11], in light of more recent technical results).

6.1. Privileging Strategies

Section 6.1.1 will discuss the suggestion that O's state is privileged. Section 6.1.2 will discuss the possibility that P's state is privileged. Both will be rejected.

6.1.1. Treating O's State as Privileged

The first stab we might take at privileging one observer would be to follow the classical analogy. According to this, we might argue that in our Wigner's friend example, only observer O's description (i.e., the collapsed state) is a true description of reality, while P's state has only an *epistemic,* but no directly *representational* function (i.e., that of quantifying P's degree of ignorance with respect to the true state, which is correctly represented by $\psi(S, O)$ after collapse has occurred). To motivate this proposal, one could argue, first, that there is an important asymmetry in the Wigner's friend example: O has, unlike P, *actually* received a measurement outcome, which mirrors our reasoning in the classical case (in which A's description was privileged). The problem with this proposal is that P can use her state to make predictions for future measurement outcomes, and for all we know, these predictions are empirically adequate. Therefore, the experimental facts undermine this strategy. [11] (p. 5) Since I don't know of any view that rejects this argument, I feel confident to move on to the next possibility. (The situation is more complex if either of the systems O or S is macroscopic. In this case, interference effects are seldom observed. However, rather than being forced to assume that the ascription of a superposition is *false*, P could rely on decoherence-theory to at least explain why no interference effects are observed.).

6.1.2. Treating P's State as Privileged

In a radical change of heart, we might abandon the idea that O's state is privileged and move towards the other extreme of proposing that P's state is privileged. On such a view, collapse never occurs: all there is, is the unitarily evolving state, as prescribed by the Schrödinger equation. Of course, this also requires an argument. In this section, I will consider—and reject—two possibilities.

6.1.2.1. Argument from Interference

The argument that's sometimes favored in the literature picks up the thread from Section 6.1.1. Using the unitarily evolved state, P can predict interference effects that can be confirmed experimentally, but that cannot be explained on the basis of the collapsed state. But while P's unitarily evolved state has an important role in guaranteeing the empirical adequacy of the theory, this may provide just enough reason not to "*under*"privilege P's state. It doesn't follow that we can "*over*"privilege this state.

One way to see this, is by noting that the argument from interference to reality is not logically valid. A counterargument can be constructed within the ontological models framework that was briefly introduced in Section 2.2.3. It can be shown that, e.g., Spekkens' toy-model, whose states must unambiguously be interpreted as epistemic, can reproduce interference effects, at least under certain constraints. [27] (pp. 3, 11) This provides the counterexample, which establishes that the inference from interference to reality is not logically valid (in Spekkens' words: "All this argument demonstrates ... is a lack of imagination concerning the interpretation of coherent superposition within an epistemic view" [27] (p. 11); cf. Leifer in Ref. [15] (pp. 78–79) for a more elaborate version of the argument).

6.1.2.2. Argument from Scientific Realism: Explanatory Virtues

One might respond to this by pointing out that the utility of the ontological models framework for foundational issues is controversial. Instead of cosmetically altering the formalism (for example, by adding an extra layer of ontological models) we should take the mathematical structures provided by the formalism itself seriously—i.e., *literally*. In that sense, the argument for privileging P's state could derive from the scientific realism debate.

This line of thought can be developed as follows. First, observe that insofar as the collapse postulate (as Brown [23] (p. 17) had put it) appears mysterious, we might reject it as an *ad hoc* modification of the theory, and this is unacceptable from the point of view of a realist interpretation of science. Furthermore, although the argument from interference to reality is not logically valid, it might still be the case that such an inference is *plausible*; after all, one might reason, if the collapse postulate is not "really" part of the quantum formalism, then the fact that P's state expresses some actual states of affairs is the best explanation for why quantum theory is so empirically successful. Hence, observer O's collapsed state should be underprivileged, and observer P's unitarily evolved state should be taken as the true description of the world.

This line of reasoning is endorsed by proponents of Everett's many-worlds interpretation [40,41], which has been argued to be the only strategy that reifies scientific realism and quantum theory [41] (pp. 35–39). In this section, I will outline how the case against this line of reasoning might unfold (thus, I use Everett's interpretation as a paradigmatic example of a realist interpretation, to illustrate what problems may arise in the context of such interpretations).

Let me, first, expand on the core idea behind the Everett interpretation. To fix the attention, let's return to the Wigner's friend example. According to Everettian views, the final state of the entire multiverse, before P's interaction with S+O but after O's interaction with S will be given by:

$$\psi(S+O+P) = \alpha\big|\text{up} \otimes \text{up} \otimes \text{ready}\big\rangle + \beta\big|\text{down} \otimes \text{down} \otimes \text{ready}\big\rangle$$

Proponents of the Everett interpretation then argue that after O's measurement, the world has split into two distinct branches, which are to be referred to as "worlds"—one in which O has received outcome *up* and one in which O's trans-worldly counterpart has received outcome *down*. According to Wallace's [41] influential construal of the Everett interpretation, the language of "worlds" is justified because two further conditions are satisfied:

- (*Non-interference*) The different terms in the superposition are "causally shielded" from one another. They are eigenvectors of some observable (thus they are necessarily orthogonal), and hence no interference effects can be observed if we measure in that basis. [41] (pp. 60–63).
- (*Functional instantiation of properties*) Each of the non-interfering terms of the superposition represent (= is functionally/structurally equivalent to) a world that consists of objects which instantiate (at least some) determinate properties (such as a particle having the determinate properties of being either "spin-up" or "spin down"). (*ibid.*)

The merits of this proposal are quite attractive. By equipping the formalism with a straightforwardly realist interpretation (in terms of real physical objects that instantiate at least some determinate properties) we gain the best possible *understanding* of what the world is *really like*, if quantum mechanics were *literally* true. Since these explanatory benefits are only accessible to the realist, the argument in favor of the reality of P's state derives from the scientific realism debate.

Of course, many-worlds interpretations are not free of problems. The two problems most widely recognized in the literature are the so-called *preferred basis problem* and the *probability problem*. Both have been discussed extensively in the literature and there is no need to repeat these arguments here. [41–45] Instead, I would like to very briefly mention one problem that can be generated due to the recent no-go result by Frauchiger and Renner. [26]

We begin by observing that the many-worlds approach still needs to make sense of the usage of the Born rule in experimental practice. One prominent strategy proceeds by giving the Born rule an *epistemic function* [41,43–45], namely in the following sense. A proponent of the many-worlds approach believes that only unitary evolution tracks real changes in the world. Thus, in our Wigner's friend example, there are (after O's measurement on S) *two* copies of P. The state of the multiverse $\psi(S+O+P)$ represents two worlds, each of which contains a still uncertain observer P in the "ready" state. In the case in which P happens to be a conscious agent, therefore, a question arises for P: which of the two versions of herself is she? This is referred to as "self-locating uncertainty" that arises in the many-worlds interpretation. The proposal then is that the Born rule provides the quantitative measure for P's degree of uncertainty: P can use this quantitative measure for her decision-theoretic purposes. [41,43–45].

Notice, however, that P's question only makes sense relative to the assumption that each version of O has received a definite outcome during her measurement on S. Here is why: there is a trivial sense in which it is silly to ask "Where am I?" In that trivial sense, the answer is always: "I am *here* ... in *my* world!" But there is sense in which this question is not trivial, namely if it is understood are referring to external circumstances: "Which of the different physically possible worlds is *my* world?" And this only makes sense if we have an account of what these different worlds are like. Only if there really are distinguishable worlds (in terms of their properties such as: there is one world in which O has received outcome "up", and one in which O has received "down") does it make sense for either version of P to ask "In which of the two worlds am I?" *Prima facie*, this doesn't sound like a problem; after all, this was precisely the reason to introduce (*Functional instantiation of properties*) in the first place (i.e., to justify talk about "worlds" by virtue of the idea that the different terms in the superposition could be said to represent objects that instantiate at least some determinate properties).

But now we phrase an attack against this as follows. Due to a recent no-go theorem by Frauchiger & Renner [26], who develop on Hardy's paradox [46,47], there are good reasons to believe that quantum mechanics is inconsistent with the assumption that, in general, each term in a superposition can be said to represent a world with determinate properties. The authors discuss an extended version of the Wigner's friend situation. In this extended version, there are *two* "friends" and two "external observers." They then set out to prove that no theory that is empirically equivalent to quantum mechanics can be both (1) self-consistent and (2) be committed to the claim that each measurement has a unique, determinate outcome. The details of their argument would distract from the current discussion (Appendix C gives sufficiently elaborate version that contains the relevant details; for criticisms of their argument, see [48–50]). The conceptual point is that this result provides a reason to believe that the Everettian commitment to (*Functional instantiation of properties*) might be inconsistent with quantum mechanics, at least in certain experimental set-ups. (In the preprint, Frauchiger and Renner suggested that their argument lends support to the Everettian view. In the published version, however, they have expressed doubts about whether the Everett interpretation is consistent with their no-go theorem, at least if branching is supposed to be objective. [26] (p. 10) The argument here (and in Appendix C) should strengthen these doubts. Cf. [33] (p. 7)).

From the point of view of interpretation, the consequences are twofold. For one thing, this shows that P cannot, in general, discriminate between the two worlds (because she cannot tell us what the different worlds are like). Therefore, P also cannot meaningfully ask questions about self-locating uncertainty (she can't question in which of the branches she has ended up in, if she can't tell us what these branches are like). Hence, we might ask the Everettian: if (*Functional instantiation of properties*) is indeed inconsistent with quantum mechanics, then, on a decision-theoretic interpretation of the Born rule, this rule aids users to make decisions... *about what?!* But clearly, this doesn't only challenge the decision-theoretic interpretation of the Born rule. It also undercuts the Everettian strategy in a deeper sense: *if* the motivation for the many-worlds interpretation was that it provides us with an *understanding*—in the peculiarly realist sense of providing an account of *what the world would have to be like if the theory were literally true*—that motivation has now been lost! And with it the argument we are currently considering—that P's state should be regarded as epistemically privileged *because* it provides us with such an understanding—disappears as well.

This concludes the discussion of the potential privileging strategies. If all these strategies are unsuccessful, the two descriptions—i.e., the states relative to O and P—should be seen as *equal*. This breaks the analogy to the classical case.

6.2. (Hidden) Commonalities?

There is one last case that we still need to consider: what if neither of the two states—the one relative to O and the one relative to P—are privileged, but that they are not actually *different* descriptions of what the world is like, in the sense that there is, really, some underlying commonality? In this most general way of putting the problem, this question is unanswerable. But we can identify two important aspects, which will be discussed in turn.

6.2.1. The Quantum State Is Not Epistemic!

The first brings us back to the ontological models framework that was introduced in Section 2.2.3: what if two different quantum states—i.e., those relative to O and P—are interpreted as assigning a non-zero probability to some underlying ontic state (from an ontic state-space that might be part of some hidden-variable framework)? Couldn't it be the case that two different quantum states each assign a non-zero probability to one and the same ontic state? On such a scenario, the two descriptions would certainly be compatible with a realist account, although neither description would be a "direct" representation of what the world is like. This possibility, however, is ruled out by the

PBR theorem [14]—in the ontological models framework, each ontic state is compatible with only one quantum state! Thus, different quantum state ascriptions, if both are valid, are incompatible with the assumption that the world is in some determinate state. At the same time, Section 2.2.3 already illustrated how one may consistently argue that a state is "ontic" (in this peculiar sense of being "non-epistemic") while shying away from attributing any representational function to such a state. (There is an important comment to be made here about the relationship between this argument and the distinction between different versions of relational quantum mechanics, which are based on what Section 5 referred to as the (*Relative facts postulate*) and the (*Empiricist facts postulate*). On the latter, though perhaps not on the former, the possibility of ascribing a state to oneself is ruled out [11] (p. 15). Thus, there is a trivial sense in which, on the empiricist version, O's and P's states are not compatible: both O's and P's states are about different systems. Once we move to the version of relational quantum mechanics based on the (*Relative facts postulate*), state ascriptions to oneself may become possible. Thus, it is *this* version of the relational view that potentially needs to exploit the PBR theorem to establish that relational quantum states are incompatible with the assumption of underlying realism.).

6.2.2. Measurement Outcomes Are Not Objective!

There is yet another sense in which the two state ascriptions, if neither is privileged, may be compatible with an ontology, namely an ontology of *measurement outcomes*. After all, even though we may have rejected a realist account of how measurement outcomes are brought about, we might still believe in the existence of these outcomes. Such a view is indeed part and parcel of Rovelli's own thinking about these matters. Although his view is built around the idea that relative state ascriptions vary with the reference systems, he nevertheless maintains that quantum theory is a theory *about* (in an ontological sense) measurement outcomes: "in [relational quantum mechanics], physical reality is taken to be formed by the individual quantum events (facts) through which interacting systems (objects) affect one another. Quantum events are therefore assumed to exist only in interactions and (this is the central point) the character of each quantum event is only relative to the system involved in the interaction." [51] (p. 2).

This claim is grounded in the alleged objectivity of measurement outcomes, which relates to Deutsch's influential discussion of the Wigner's friend thought experiment. [52] In the simple Wigner's friend example, one can show that although the *character* of the event is observer-relative (as O and P ascribe different relative states), the fact *that* they occurred is not [52], [51] (pp. 7–9). Hence, measurement outcomes can, in the Wigner's friend example, be thought of as objective, i.e., *observer-independent*. The problem with this, however, is that the technical aspects on which these ideas are founded are artefacts of certain specific cases, like the simple Wigner's friend example discussed above. In a more general context—like in the extended Wigner's friend experiment discussed by Frauchiger and Renner [26]—this conclusion is no longer true (again, the reader might wish to consult Appendix C). Therefore, even the thin ontology of measurement outcomes, can't consistently be postulated (at least not in general) as long as quantum theory remains our most successful empirical theory.

Finally, we have exhausted the space of possibilities: the above argument illustrates why the (classically expected) dualism of dynamical evolutions in quantum theory can't be ramified with our classical intuitions. *Unlike* in the classical case, in quantum mechanics the different descriptions, offered by different observers O and P, can't be endowed with a realist interpretation.

7. Conclusions: Representation Lost

The conclusion of this paper did not follow from any philosophical ideology. On the contrary, it was based on a conceptual analysis of technical results—both in formal epistemology, but also in quantum foundational research. This argument had three interdependent components: (1) Different observers will, because of the different pieces of information available to them, ascribe different quantum states to one and the same system. (2) In general, we cannot supplement the observer-relative quantum states with some kind of mechanical/substance-type story of what the world is like. (3) To make these radical-sounding claims more easily digestible, I presented a simple example, which also instantiated what I believe to be the most promising strategy to make sense of quantum theory. The toy-example, specifically, illustrates that operating within a framework of non-representational states is simply an artefact of certain modelling practices (and does not entail commitment to either skepticism or solipsism).

Of course, the usage of non-scientific examples or toy-models for foundational issues might be contested. Let me therefore conclude with some brief remarks about what I take to be the importance of such simple-minded examples as the one that was discussed above. The reason why it is hard to make sense of quantum theory is because the theory invites a battle of intuitions: between what science could not *possibly* be doing, but nevertheless *seems* to be doing ... In our strive for clarity, there is only so much we can do. We might, for one, inquire into the mathematical architecture of the theory (by arriving, for example, at such rightly celebrated results as the PBR theorem). But we can also aim to provide examples and counterexamples, which allows us to better separate what is truly *necessary* from what is merely *plausible*. The football example aims to do just this. Even if it isn't anything more than a crutch for our intuitions, it illustrates where these intuitions come from. And, at the very least, such examples can provide a proof of concept that a certain set of beliefs can consistently be upheld, even in light of strong results or intuitions that might suggest the opposite.

Acknowledgments: I would like to thank my supervisors, Markus Aspelmeyer and Martin Kusch, for their continuous support and (always constructive) feedback. Furthermore, I am grateful to Richard Healey for his insightful discussion of the Frauchiger & Renner no-go theorem (mistakes are mine). Finally, I would like to thank Carlo Rovelli for encouraging me to think about the ideas discussed in this paper.

Appendix A Proofs of the Theorems from Section 3

I will provide a sketch of the proofs for the three theorems that were mentioned in Section 3. For a complete discussion of the proofs, see Hughes and van Fraassen in Ref. [24], and van Fraassen in Ref. [20] (pp. 68–73).

In the following, let $\overline{x} = \langle x_1, \ldots, x_n \rangle$ denote a betting-state.

Theorem. *A transformation U is a symmetry if and only if there exist positive real numbers $u_1, \ldots, u_n$ such that $U(\overline{x})\langle = u_1 x_1, \ldots, u_n x_n \rangle$. [24] (p. 857), [20] (pp. 71–73).*

Proof. If there exist positive real numbers $u_1, \ldots, u_n$ such that for all $\overline{x}$, $U(\overline{x}) = \langle u_1 x_1, \ldots, u_n x_n \rangle$, then clearly U is a symmetry. To prove the converse, take the unit betting-state $\overline{1}$, whose components are all equal to 1. Let $U(\overline{1}) = \langle u_1, \ldots, u_n \rangle$. Since U is a symmetry: $U(\overline{x})/U(\overline{1}) = \overline{x}/\overline{1} = \overline{x}$. Let $U(\overline{x}) = \overline{x}\prime$. Then we may write: $\langle x_1'/u_1, \ldots, x_n'/u_n \rangle = \langle x_1, \ldots, x_n \rangle$. Since two betting-states are equal if and only if their components are equal, and since the components of betting-states must be positive real, the result follows. □

Theorem. *If two betting-states $\overline{x}(0)$ and $\overline{x}(t)$ are related via a uniform motion $\overline{x}(t) = U(t)(\overline{x}(0)) = \langle u_1(t)\, x_1(0), \ldots, u_n(t)\, x_n(0) \rangle$, then there exist positive real numbers $k_1, \ldots, k_n$ such that $u_i(t) = e^{k_i t}$ ([24] (p. 858), [20] (pp. 72–73)).*

Proof. From before, we know that $U(t)(\overline{x}(0)) = \langle u_1(t)\, x_1(0), \ldots, u_n(t)\, x_n(0) \rangle$, where the u_i's are positive real. Since U is a uniform motion, it follows that $u_i(t_1 + t_2) = u_i(t_1)u_i(t_2)$. Hence, we may define a function $f_i(t) = \ln(u_i(t))$, which is additive in t. From elementary calculus, we know that therefore $f_i(t) = k_i t + C$. Since $u_i(0) = 1$, it follows that $C = 0$, which proves the theorem. □

Theorem. *There exist constants v & w such that the $u_i(t)$'s are given by $u_i(t) = e^{vt} e^{ws_i t}$.*

Proof. We prove this by using the method of Lagrange multipliers. [24] (p. 861) The Lagrangian is given by: $L = \sum k_i x_i(t) - w(\sum s_i x_i(t) - r(t)) - v(\sum x_i(t) - n(t))$. Here, w and v are Lagrange multipliers. When we derive this with respect to component x_i and set the result equal to 0, we get: $k_i - ws_i - v = 0$. Hence the result follows. The two Lagrange multiplies are fixed by the conditions that (1) $(\sum e^{ws_i t} x_i(0))\, e^{vt} = n(t)$ and (2) $(\sum s_i\, e^{ws_i t} x_i(0))\, e^{vt} = r(t)$. □

Appendix B An Example for the Theorem in Section 3

In this appendix, I will try to illustrate why the theorem from Section 3 produces reasonable results. This investigation is spurred by the worry that since the components of the betting-states evolve exponentially, this implies (among other things) the rather counterintuitive result that the total number of games played n(t) evolves as a sum of exponential functions. To see why this result is less counterintuitive, and less threatening, than it might first appear to be, it will be best to just consider a concrete example.

So, let's assume the following boundary conditions: (1) At $t = 0$, the betting-state of a team relative to B is assumed to be $\langle 3, 3, 2\rangle$. Hence, the team has played 8 games. If we are assuming the standard choice of number-state function $s = \langle 3, 1, 0\rangle$ this means that the team has collected 12 points at the initial time. (2) Let the boundary condition for the final time $t = 1$ month be that (a) the team now has played 15 games in total, and (b) that it has picked up 22 points overall.

Using these boundary conditions, it is straightforward to calculate the values of the Lagrange multipliers to be: $v \approx 0.66167$ and $w \approx -0.02229$. Relying now on the theorem from Section 3, the final betting-state relative to B will therefore be given as $\langle 5.44, 5.69, 3.88\rangle$. Of course, it has to be admitted that this is not particularly useful—it is left open whether B should guess that the team is really in the betting-state $\langle 6, 4, 5\rangle$ or in the betting-state $\langle 5, 7, 3\rangle$ (which were the only possible betting-states to begin with; the remaining two betting-states that the final number-state ascription declares possible—$\langle 7, 1, 7\rangle$ and $\langle 4, 10, 1\rangle$—are ruled out by the initial conditions, since the team cannot have fewer draws or losses than it had started out with).

However, to probe whether result is at least reasonable, we can use it to illustrate what the overall evolution of n(t)—the total number of games as a function of time—looks like. This function is just the sum of the components of the betting-state, i.e., it is a sum of exponentially evolving functions in time. n(t) is plotted in Figure A1 for $0 < t < 1$:

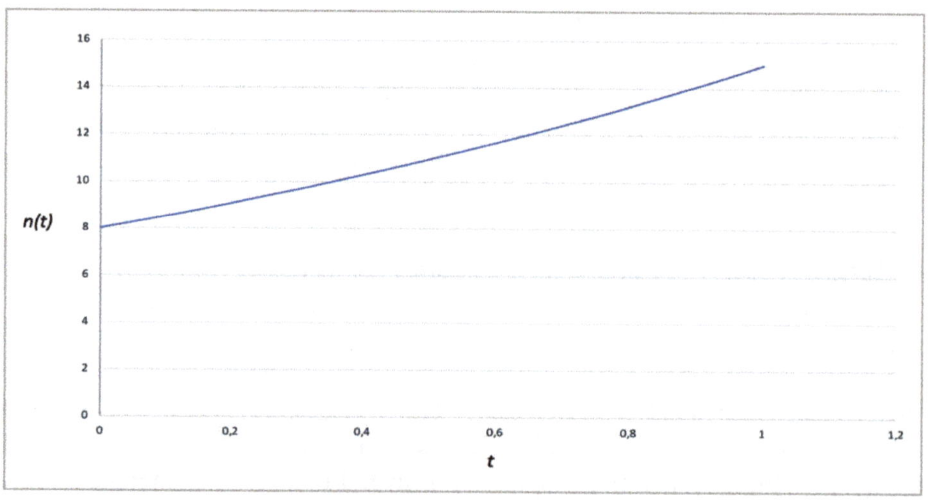

Figure A1. Total number of games as a function of time.

Hence, the evolution of the total number of games, for the time-frame in question, is very close to being linear. This therefore illustrates that the counterintuitive result—that $n(t)$ is a sum of exponential functions—will not be threatening in concrete examples. Therefore, the calculations from Section 3 produce reasonable results.

Appendix C The Frauchiger and Renner 2016 Thought Experiment

The thought-experiment constructed by Frauchiger and Renner [26] consists in an extended version of Wigner's friend example. There are two friends (F1 & F2), Wigner (W) and his assistant (A). F1 prepares a *quantum coin* at t_0 and depending on the outcome of her measurement (*heads* or *tails*) at t_1, she sends a spin-1/2 particle to the second friend F2. This spin particle is going to be prepared in either the z+ or the x+ direction, depending on whether F1 has observed outcome *heads* or *tails* respectively. F2 then conducts a measurement in the z-basis at t_2 on the particle she has received from F1. At t_3 the assistant A conducts a measurement on F1 in the basis *ok/fail* (which is an equally weighted superposition of the *heads/tails* basis—with a minus and a plus sign respectively—on the Hilbert space of F1). At t_4 Wigner conducts a measurement in the *ok/fail* basis on F2. The experiment is repeated many times, until Wigner and A have both received the outcome *ok* in their respective measurements on F2 and F1.

We then consider the final state of the entire system, which can be written in five equivalent ways, given the initial state of the quantum coin (unnecessary normalizing factors and factors in the tensor product have been omitted):

$$(A) : \psi \sim (\text{tails}_{F1} \bigotimes \text{up}_{F2} + \text{tails}_{F1} \bigotimes \text{down}_{F2} + \text{heads}_{F1} \bigotimes \text{down}_{F2})$$

$$(B) : \psi \sim (\sqrt{2}\,\text{fail}_{F1} \bigotimes \text{down}_{F2} + \text{tails}_{F1} \bigotimes \text{up}_{F2})$$

$$(C) : \psi \sim (\text{tails}_{F1} \bigotimes \text{fail}_{A} + \frac{1}{\sqrt{2}}\text{heads}_{F1} \bigotimes (\text{ok}_{A} + \text{fail}_{A}))$$

$$(D) : \psi \sim (\left(\frac{1}{\sqrt{2}} - 1\right)\text{ok}_{F1} \bigotimes \text{ok}_{A} + \left(\frac{1}{\sqrt{2}} + 1\right)\text{fail}_{F1} \bigotimes \text{fail}_{A})$$

$$(E) : \psi \sim (3\,\text{fail}_{W} \bigotimes \text{fail}_{A} + \text{fail}_{W} \bigotimes \text{ok}_{A} + \text{ok}_{W} \bigotimes \text{fail}_{A} + \text{ok}_{W} \bigotimes \text{ok}_{A})$$

From these results, we will now prove the claim from Section 6.1.2.2 that (*Functional instantiation of properties*) is inconsistent with quantum mechanics. From an Everettian standpoint, we assume that each term in the superpositions constitutes a world in which the objects instantiate the properties corresponding to the eigenvalue of the observable of which they are eigenvectors. With this assumption, we can generate the contradiction:

- By (B) and (*Functional instantiation of properties*) W concludes that, in one world, there is a version of F2 whose system instantiates *spin-up*. In another world, the particle that was measured by F2 instantiates *spin-down*. If W now wonders which version of F2 he shares a world with (what Section 6.1.2 called *self-locating uncertainty*), he will reason as follows.
- (*Case 1*) If the version of W (who is uncertain) is in the world in which F2's system instantiates *spin-up*, then by (A) this is also the same world in which F1's coin instantiates *tails*. But then, by (C), this is also a world in which A instantiates property *fail*.
- (*Case 2*) If the version of W (who is uncertain) is in the world in which F2's system instantiates *spin-down*, then by (B) this is also a world in which F1 instantiates outcome *fail*. But then by (D) this is also a world in which A instantiates property *fail*.

Since, in all possible worlds, W concludes that his assistant A instantiates property *fail*, the conjunction of quantum mechanics and (*Functional instantiation of properties*) predicts that whatever world W is uncertain to have ended up in, neither of those could be a world in which the experiment has stopped. But by (E), $\text{ok}_W \otimes \text{ok}_A$ has a non-zero probability. Therefore, quantum mechanics implies that there is a non-zero probability that the experiment will stop. Therefore, (*Functional instantiation of properties*) is inconsistent with quantum mechanics.

References

1. Petersen, A. The Philosophy of Niels Bohr. *Bull. At. Sci.* **1963**, *19*, 8–14. [CrossRef]
2. Caves, C.M.; Fuchs, C.A.; Schack, R. Quantum probabilities as Bayesian probabilities. *Phys. Rev. A* **2002**, *65*, 022305. [CrossRef]
3. Caves, C.M.; Fuchs, C.A.; Schack, R. Subjective probability and quantum certainty. *Stud. Hist. Philos. Mod. Phys.* **2007**, *38*, 893–897. [CrossRef]
4. Fuchs, C.A. Quantum mechanics as quantum information (and only a little more). In *Quantum theory: Reconsiderations of Foundations*; Khrenikov, A., Ed.; Växjö University Press: Växjö, Sweden, 2002; Available online: https://arxiv.org/abs/quant-ph/0205039 (accessed on 10 December 2018).
5. Fuchs, C.A. Quantum States: What the Hell Are They? (The Post-Växjö Phase-Transition). 2002. Available online: http://perimeterinstitute.ca/personal/cfuchs/PhaseTransition.pdf (accessed on 10 December 2018).
6. Fuchs, C.A. Notwithstanding Bohr, the Reasons for QBism. 2017. Available online: https://arxiv.org/abs/1705.03483 (accessed on 10 December 2018).
7. Fuchs, C.A.; Mermin, N.D.; Schack, R. An introduction to QBism with an application to the locality of quantum mechanics. *Am. J. Phys.* **2014**, *82*, 749–754. [CrossRef]
8. Zeilinger, A. A foundational principle for quantum mechanics. *Found. Phys.* **1999**, *29*, 631–643. [CrossRef]
9. Brukner, Č.; Zeilinger, A. Information and the fundamental elements of the structure of quantum theory. In *Time, Quantum, Information*; Castell, L., Ischbeck, O., Eds.; Springer: Berlin/Heidelberg, Germany, 2003.
10. Brukner, Č. On the quantum measurement problem. *arXiv* **2015**, arXiv:1507.05255.
11. Rovelli, C. Relational Quantum Mechanics. *Int. J. Theor. Phys.* **1996**, *35*, 1637–1678. [CrossRef]
12. Laudisa, F.; Rovelli, C. *Relational Quantum Mechanics*, Summer 2013 Edition; The Stanford Encyclopedia of Philosophy; Zalta, E.N., Ed.; Available online: https://plato.stanford.edu/archives/sum2013/entries/qm-relational/ (accessed on 10 December 2018).
13. Rovelli, C. Space is Blue and Birds Fly Through It. *Philos. Trans. R. Soc. A* **2018**, *376*, 20170312. [CrossRef]
14. Pusey, M.F.; Barrett, J.; Rudolph, T. The quantum state cannot be interpreted statistically. *arXiv* **2011**, arXiv:1111.3328.
15. Leifer, M. Is the quantum state real? An extended review of ψ-ontology theorems. *Quanta* **2014**, *3*, 67–155. [CrossRef]
16. Dennet, D. *Consciousness Explained*; Penguin Press: London, UK, 1991.
17. Von Neumann, J. *Mathematical Foundations of Quantum Mechanics*; English Translation: 1955; Princeton University Press: Princeton, NJ, USA, 1932.
18. Lüders, G. Über die Zustandsänderung durch den Messprozess. *Annalen der Physik* **1951**, *8*, 322–328. [CrossRef]
19. Bub, J. Von Neumann's projection postulate as a probability conditionalization rule in quantum mechanics. *J. Philos. Log.* **1977**, *6*, 381–390. [CrossRef]
20. Van Fraassen, B. *Quantum Mechanics: An Empiricist View*; Oxford University Press: Oxford, UK, 1991.
21. Ruetsche, L.; Earman, J. *Interpreting Probabilities in Quantum Field Theory and Quantum Statistical Mechanics*; Breisbart, C., Hartmann, S., Eds.; Probabilities in Physics; Oxford University Press: Oxford, UK, 2011.
22. Ruetsche, L. *Interpreting Quantum Theories*; Oxford University Press: Oxford, UK, 2011.
23. Brown, H. The Reality of the Wavefunction: Old Arguments and New. 2017. Available online: http://philsci-archive.pitt.edu/12978/ (accessed on 10 December 2018).
24. Hughes, R.I.G.; van Fraassen, B. Symmetry Arguments in Probability Kinematics. *PSA Proc. Bienn. Meet. Philos. Sci. Assoc.* **1984**, 851–869. [CrossRef]
25. Lisi, G. Quantum Mechanics from a Universal Action Reservoir. 2006. Available online: https://arxiv.org/abs/physics/0605068 (accessed on 10 December 2018).
26. Frauchiger, D.; Renner, R. Quantum theory cannot consistently describe the use of itself. *Nat. Commun.* **2018**, *9*, 3711. [CrossRef] [PubMed]
27. Spekkens, R.W. Evidence for the epistemic view of quantum states: A toy theory. *Phys. Rev. A* **2004**, *75*, 032110. [CrossRef]

28. Wigner, E.P. Remarks on the mind-body question. In *The Scientist Speculates*; Good, I.J., Ed.; 1961; Reprinted in Wheeler and Zurek 1983 (pp. 168–181); In *Quantum Theory and Measurement*; Princeton University Press: Princeton, NJ, USA; pp. 284–302.
29. Timpson, C.G. *Quantum Information Theory & the Foundations of Quantum Mechanics*; Oxford University Press: Oxford, UK, 2013.
30. Hardy, L. Quantum theory from five reasonable axioms. *arXiv* **2001**, arXiv:quant-ph/0101012.
31. Clifton, R.; Bub, J.; Halvorson, H. Characterizing quantum theory in terms of information theoretic constraints. *Phys. Lett. A* **2003**, *292*, 1–11. [CrossRef]
32. Höhn, P.A. Toolbox for reconstructing quantum theory from rules on information acquisition. *Quanta* **2017**, *1*, 38. [CrossRef]
33. Brukner, Č. A no-go theorem for observer-independent facts. *Entropy* **2018**, *20*, 350. [CrossRef]
34. Halvorson, H. To Be a Realist about Quantum Theory. 2018. Available online: http://philsci-archive.pitt.edu/14310/ (accessed on 10 December 2018).
35. Wallace, D.; Timpson, C.G. Quantum Mechanics on Spacetime I: Spacetime State Realism. *Br. J. Philos. Sci.* **2010**, *61*, 697–727. [CrossRef]
36. Kochen, S.; Specker, E.P. The Problem of Hidden Variables in Quantum Mechanics. In *The Logico-Algebraic Approach to Quantum Mechanics*; Springer: Dodrecht, The Netherlands, 1975; Volume 1.
37. Schrödinger, E. Die gegenwärtige Situation in der Quantenmechanik (The present situation in quantum mechanics). *Naturwissenschaften* **1935**, *23*, 807–812. [CrossRef]
38. Munkhammar, J. Canonical Relational Quantum Mechanics from Information Theory. *arXiv* **2011**, arXiv:1101.1417.
39. Feynman, R.P.; Hibbs, A.R. *Quantum Mechanics and Path Integrals*; McGraw-Hill: New York, NY, USA, 1965.
40. Everett, H. 'Relative State' Formulation of Quantum Mechanics. *Rev. Mod. Phys.* **1957**, *29*, 454–462. [CrossRef]
41. Wallace, D. *The Emergent Multiverse*; Oxford University Press: Oxford, UK, 2012.
42. Kent, A. Against many-worlds interpretations. *Int. J. Theor. Phys.* **1990**, *A5*, 1745–1762. [CrossRef]
43. Deutsch, D. Quantum theory of probability and decisions. *Proc. R. Soc. Lond.* **1999**, *A455*, 3129–3137. [CrossRef]
44. Carroll, S.M.; Sebens, C.T. Many Worlds, the Born Rule, and Self Locating Uncertainty. In *Quantum Theory: A Two-Time Success Story, Yakir Aharonov Festschrift*; Struppa, D.C., Tollaksen, J.M., Eds.; Springer: Berlin/Heidelberg, Germany, 2013; p. 157.
45. Wallace, D. Everettian rationality: Defending Deutsch's approach to probability in the Everett interpretation. *Stud. Hist. Philos. Mod. Phys.* **2003**, *34*, 415–439. [CrossRef]
46. Hardy, L. Quantum mechanics, local realistic theories, and Lorentz-invariant realistic theories. *Phys. Rev. Lett.* **1992**, *68*, 2981–2984. [CrossRef]
47. Hardy, L. Nonlocality for two particles without inequalities for almost all entangled states. *Phys. Rev. Lett.* **1993**, *71*, 1665–1668. [CrossRef]
48. Bub, J. In Defense of a "Single-World" Interpretation of Quantum Mechanics. *Stud. Hist. Philos. Sci. Part B Stud. Hist. Philos. Mod. Phys.* **2018**. [CrossRef]
49. Sudbury, A. Single-World Theory of the Extended Wigner's Friend Experiment. *Found. Phys.* **2017**, *47*, 658–669. [CrossRef]
50. Baumann, V.; Hansen, A.; Wolf, S. The measurement problem is the measurement problem is the measurement problem. *arXiv* **2016**, arXiv:1611.01111.
51. Smerlak, M.; Rovelli, C. Relational EPR. 2006. Available online: https://arxiv.org/abs/quant-ph/0604064 (accessed on 10 December 2018).
52. Deutsch, D. Quantum theory as a universal physical theory. *Int. J. Theor. Phys.* **1985**, *24*, 1–41. [CrossRef]

Permissions

All chapters in this book were first published in MDPI; hereby published with permission under the Creative Commons Attribution License or equivalent. Every chapter published in this book has been scrutinized by our experts. Their significance has been extensively debated. The topics covered herein carry significant findings which will fuel the growth of the discipline. They may even be implemented as practical applications or may be referred to as a beginning point for another development.

The contributors of this book come from diverse backgrounds, making this book a truly international effort. This book will bring forth new frontiers with its revolutionizing research information and detailed analysis of the nascent developments around the world.

We would like to thank all the contributing authors for lending their expertise to make the book truly unique. They have played a crucial role in the development of this book. Without their invaluable contributions this book wouldn't have been possible. They have made vital efforts to compile up to date information on the varied aspects of this subject to make this book a valuable addition to the collection of many professionals and students.

This book was conceptualized with the vision of imparting up-to-date information and advanced data in this field. To ensure the same, a matchless editorial board was set up. Every individual on the board went through rigorous rounds of assessment to prove their worth. After which they invested a large part of their time researching and compiling the most relevant data for our readers.

The editorial board has been involved in producing this book since its inception. They have spent rigorous hours researching and exploring the diverse topics which have resulted in the successful publishing of this book. They have passed on their knowledge of decades through this book. To expedite this challenging task, the publisher supported the team at every step. A small team of assistant editors was also appointed to further simplify the editing procedure and attain best results for the readers.

Apart from the editorial board, the designing team has also invested a significant amount of their time in understanding the subject and creating the most relevant covers. They scrutinized every image to scout for the most suitable representation of the subject and create an appropriate cover for the book.

The publishing team has been an ardent support to the editorial, designing and production team. Their endless efforts to recruit the best for this project, has resulted in the accomplishment of this book. They are a veteran in the field of academics and their pool of knowledge is as vast as their experience in printing. Their expertise and guidance has proved useful at every step. Their uncompromising quality standards have made this book an exceptional effort. Their encouragement from time to time has been an inspiration for everyone.

The publisher and the editorial board hope that this book will prove to be a valuable piece of knowledge for researchers, students, practitioners and scholars across the globe.

List of Contributors

Andrei Khrennikov
International Center for Mathematical Modeling in Physics and Cognitive Sciences, Linnaeus University,
SE-351 95 Växjö, Sweden

Marek Danielewski
Faculty of Materials Science & Ceramics, AGH UST, Mickiewicza 30, 30-059 Kraków, Poland

Lucjan Sapa
Faculty of Applied Mathematics, AGH UST, Mickiewicza 30, 30-059 Kraków, Poland

Zhan-Yun Wang
School of Electronic Engineering, Xi'an University of Posts and Telecommunications, Xi'an 710121, China

Yi-Tao Gou
School of Physics, Northwest University, Xi'an 710127, China

Jin-Xing Hou
Institute of Modern Physics, Northwest University, Xi'an 710127, China
Shaanxi Key Laboratory for Theoretical Physics Frontiers, Xi'an 710127, China

Li-Ke Cao and Xiao-Hui Wang
School of Physics, Northwest University, Xi'an 710127, China
Shaanxi Key Laboratory for Theoretical Physics Frontiers, Xi'an 710127, China

Chris Fields
Independent Researcher, 23 rue des Lavandières, 11160 Caunes Minervois, France

Avishy Carmi and Daniel Moskovich
Center for Quantum Information Science and Technology, Ben-Gurion University of the Negev, Beersheba 8410501, Israel
Faculty of Engineering Sciences, Ben-Gurion University of the Negev, Beersheba 8410501, Israel

Xiangluo Wang
School of Information Technology, Luoyang Normal University, Luoyang 471934, China

Chunlei Yang, Guo-Sen Xie and Zhonghua Liu
School of Information Engineering, Henan University of Science and Technology, Luoyang 471023, China

Ariel Caticha
Physics Department, University at Albany-SUNY, Albany, NY 12222, USA

Omar Jiménez
Centro de Óptica e Información Cuántica, Facultad de Ciencias, Universidad Mayor, Camino La Pirámide N°5750, Huechuraba, Santiago, Chile

Miguel Angel Solís-Prosser and Aldo Delgado
Departamento de Física, Universidad de Concepción, 160-C Concepción, Chile

Leonardo Neves
Departamento de Física, Universidade Federal de Minas Gerais, Belo Horizonte-MG 31270-901, Brazil

Pu Wang
State Key Laboratory of Quantum Optics and Quantum Optics Devices, Institute of Opto-Electronics, Shanxi University, Taiyuan 030006, China

Xuyang Wang and Yongmin Li
State Key Laboratory of Quantum Optics and Quantum Optics Devices, Institute of Opto-Electronics, Shanxi University, Taiyuan 030006, China
Collaborative Innovation Center of Extreme Optics, Shanxi University, Taiyuan 030006, China

Julio A. López-Saldívar
Instituto de Ciencias Nucleares, Universidad Nacional Autónoma de México, Apdo. Postal 70-543, Ciudad de México 04510, Mexico
Moscow Institute of Physics and Technology, State University, Institutskii Per. 9, Dolgoprudnyi, Moscow 141700, Russia

Octavio Castaños
Instituto de Ciencias Nucleares, Universidad Nacional Autónoma de México, Apdo. Postal 70-543, Ciudad de México 04510, Mexico

Margarita A. Man'ko
Lebedev Physical Institute, Leninskii Prospect 53, Moscow 119991, Russia

Vladimir I. Man'ko
Moscow Institute of Physics and Technology, State University, Institutskii Per. 9, Dolgoprudnyi, Moscow 141700, Russia
Lebedev Physical Institute, Leninskii Prospect 53, Moscow 119991, Russia
Department of Physics, Tomsk State University, Lenin Avenue 36, Tomsk 634050, Russia

Bahaaudin Mohammadnoor Raffah
Department of Physics, Faculty of Sciences, King Abdulaziz University, Jeddah 21589, Saudi Arabia

Kamal Berrada
Department of Physics, College of Science, Al Imam Mohammad Ibn Saud Islamic University (IMSIU), Riyadh 11623, Saudi Arabia

Mladen Pavičić
Center of Excellence for Advanced Materials and Sensors, Research Unit Photonics and Quantum Optics, Institute Ruder Bošković, Zagreb 10000, Croatia

Jihwan Kim, Donghoon Ha and Younghun Kwon
Department of Applied Physics, Hanyang University, Ansan, Kyunggi-Do 425-791, Korea

Raffael Krismer
Department of Philosophy, University of Vienna, 1090 Vienna, Austria

Index

Printed in the USA
CPSIA information can be obtained
at www.ICGtesting.com
LVHW080044211023
761596LV00010BA/1907

9 798888 360750